# Invariant Imbedding T-matrix Method for Light Scattering by Nonspherical and Inhomogeneous Particles

# Invariant Imbedding T-matrix Method for Light Scattering by Nonspherical and Inhomogeneous Particles


*Bingqiang Sun*

Assistant Research Scientist, Department of Atmospheric Sciences, Texas A&M University, College Station, Texas, United States

*Lei Bi*

Assistant Research Scientist, Department of Atmospheric Sciences, Texas A&M University, College Station, Texas, United States

*Ping Yang*

Professor, Department of Atmospheric Sciences Texas A&M University College Station, Texas, United States

*Michael Kahnert*

Adjunct Professor, Department of Space, Earth and Environment Chalmers University of Technology, Gothenburg, Sweden; Research Department, Swedish Meteorological and Hydrological Institute, Norrköping, Sweden

*George Kattawar*

Professor Emeritus, Department of Physics and Astronomy & Institute for Quantum Science & Engineering, Texas A&M University, College Station, Texas, United States



ELSEVIER

Elsevier
Radarweg 29, PO Box 211, 1000 AE Amsterdam, Netherlands
The Boulevard, Langford Lane, Kidlington, Oxford OX5 1GB, United Kingdom
50 Hampshire Street, 5th Floor, Cambridge, MA 02139, United States



**Library of Congress Cataloging-in-Publication Data**
A catalog record for this book is available from the Library of Congress

**British Library Cataloguing-in-Publication Data**
A catalogue record for this book is available from the British Library

ISBN: 978-0-12-818090-7

For information on all Elsevier publications visit
our website at https://www.elsevier.com/books-and-journals

*Publisher:* Candice Janco
*Acquisition Editor:* Laura Kelleher
*Editorial Project Manager:* Devlin Person
*Production Project Manager:* Maria Bernard
*Designer:* Mark Rogers

Typeset by SPi Global, India

# Contents

# Preface

The theories and numerical simulations of electromagnetic wave scattering by nonspherical and inhomogeneous particles have found diverse applications in atmospheric sciences, astronomy, engineering, chemistry, and biophysics. The subject of electromagnetic wave scattering has evolved from early studies of simple nonspherical geometries such as an infinite circular cylinder or a spheroid to arbitrarily shaped nonspherical and inhomogeneous particles. However, it is well known that obtaining the analytical solution of Maxwell's equations for an arbitrary particle is unlikely possible, although the mathematical formulation for light scattering by a homogeneous sphere was clearly and well established with the development of the Lorenz-Mie theory. From an application perspective, two associated aspects are critical to the computational capabilities of an electromagnetic-scattering solver: the stability of the algorithm and an advanced numerical implementation. The intent of this book is to present a new powerful computational tool, namely, the invariant imbedding T-matrix method (IITM), for computing the scattering and absorption properties of arbitrarily shaped nonspherical and inhomogeneous particles.

The motivation of writing this book is twofold. First, the T-matrix method may be the most accurate and efficient method for solving the scattering of an electromagnetic wave by a nonspherical particle. Second, for about three decades, the T-matrix method was considered to be practically applicable to axially symmetric and homogeneous particles, while significant efforts were devoted to applying the T-matrix method to nonsymmetric and inhomogeneous particles. At present, an accurate and versatile T-matrix implementation based on the extended boundary condition method (EBCM) (Waterman, 1971) for generally nonspherical and inhomogeneous particles is not available, particularly in the case of large size parameters. Our research efforts to apply the invariant imbedding T-matrix method started in 2013, although the invariant imbedding principle was introduced to obtain the T-matrix recurrence relation by Johnson (1988) in the framework of an electromagnetic volume integral equation. By combining several advances of the T-matrix method and solving relevant instability issues, the power of this technique allows for computing the optical properties of randomly oriented nonspherical and inhomogeneous particles with sizes much larger than the incident wavelength. This book is expected to be useful for active researchers and graduate students who have interests in light scattering and its applications in many disciplines.

The invariant imbedding technique was developed to handle the diffuse reflection of scattering and absorbing semiinfinite media by Ambarzumian (1943). Chandrasekhar (1960) generalized the technique to handle both the semi-infinite and finite medium cases and thus coined the name "principles of invariance" in Chapter IV of his classic book *Radiative Transfer*. Bellman and Wing (1975) then used "invariant imbedding" to coin the name of the technique and applied it to neutron transport theory. In particular, Bellman and Wing (1975) systematically described the invariant imbedding technique in their book *An Introduction to Invariant Imbedding*. In addition to radiative transfer, the invariant imbedding technique was first applied to electromagnetic wave scattering by Johnson (1988). With an advanced numerical implementation (Bi and Yang, 2014), the IITM was shown to be a powerful tool with far-reaching capabilities in numerical simulations of electromagnetic scattering by arbitrarily shaped particles and recently attracted a lot of attention in the electromagnetic scattering and atmospheric radiation communities. To better share this technique with these communities, we organize the principles and present applications of the IITM in this book.

Basically, the IITM is one of the most efficient methods to compute the T-matrix of a dielectric particle. Consequently, all properties with respect to the T-matrix, such as its symmetry and the analytical scattering phase matrix based on the T-matrix, are available to the IITM. Thus, the T-matrix method is a semianalytical method to solve light scattering by a scattering particle. This book gives a systematic introduction of the IITM in five chapters: terminology highlights, light scattering representation, the T-matrix concepts, the IITM algorithm, and IITM applications.

To better demonstrate the potential advantages of the IITM in later chapters, Chapter 1 highlights four topics: particle shape and inhomogeneity, size parameter, random orientation, and the invariant imbedding principle. Chapter 2 gives a systematic introduction to the representation of an electromagnetic wave and quantifying electromagnetic scattering by a dielectric particle. Maxwell's equations and their boundary conditions are reviewed in Section 2.1.1. Sections 2.1.2 and 2.1.3 present the energy density and the Poynting vector of a propagating electromagnetic wave. The polarization representation of an electromagnetic wave is given in Section 2.1.4. Section 2.2 systematically describes polarization effects caused by a dielectric particle. The topics include the amplitude scattering matrix and the scattering phase matrix and their symmetry properties and the extinction and scattering cross sections associated with single scattering and bulk scattering properties. A brief review of rigorous scattering computational methods is given in Section 2.3.

Chapter 3 gives a full introduction to the T-matrix method. A brief history and development of the T-matrix method is given in Section 3.1. Sections 3.2.1 and 3.2.2 describe how, in the T-matrix method, the electromagnetic field can be expanded by using vector spherical wave functions. The corresponding definition of the T-matrix is shown in Section 3.2.3. Section 3.2.4 implements rotations and translations of the T-matrix in terms of the vector spherical wave functions. The EBCM is briefly reviewed in Section 3.3. The symmetric relations associated with reciprocity and morphology are given in Sections 3.4.1 and 3.4.2, respectively. Light scattering by a dielectric particle in completely and partially random orientations is derived in detail in Sections 3.4.3–3.4.5.

Chapter 4 presents a comprehensive development of the IITM. The IITM is a volume integral method, while the EBCM is a surface integral method. Consequently, the volume integral equation is introduced in Section 4.1.1, and Section 4.1.2 proves the equivalence between the volume and surface integral equations. The dyadic

Green's function expansion using the vector spherical wave functions is derived in Section 4.1.3. The expansion using the vector spherical wave functions is rearranged for convenience, and further derivation is in Section 4.1.4. The differential and difference forms of the T-matrix using the invariant imbedding technique are derived in Sections 4.2.1–4.2.3. In these sections, the equivalence between the two forms is proved by allowing the radial difference to approach zero. The verification of the differential form in a spherical situation is given in Section 4.2.4. Section 4.3 discusses issues related to the IITM, such as numerical stability and memory requirements; how to determine the starting point, the processing points, and the ending point; and factors affecting the truncation and imbedding steps.

Chapter 5 describes applications of the IITM to different morphologies. Section 5.1 discusses the effect of different quadrature rules and step sizes on spherical particles. Spheroids and cylinders are used as examples of axially symmetric particles to show the accuracy and the efficiency of the IITM in Section 5.2. Section 5.3 shows IITM simulations of finite-fold rotationally symmetric particles by focusing on hexagonal ice crystals. Section 5.4 summarizes IITM applications to asymmetric particles by focusing on aggregates and irregular hexahedra. Similarly, Section 5.5 focuses on inhomogeneous particles such as nested hexahedra.

In summary, not only are the chapters arranged in progressive order, but also each chapter is relatively independent. Chapter 2 gives the basic knowledge of light scattering. Chapter 3 is a thorough introduction to the T-matrix method. The systematic description of the IITM is in Chapter 4. Application examples of the IITM are presented in Chapter 5. In organizing the theoretical formulation of light scattering process in Chapter 2 and the T-matrix concept in Chapter 3, we have referred to several classical light scattering books, such as *Light Scattering by Small Particles* by van de Hulst (1957); *Absorption and Scattering of Light by Small Particles* by Bohren and Huffman (1983); *Scattering of Electromagnetic Waves: Theories and Applications* by Tsang et al. (2000); *Scattering, Absorption, and Emission of Light by Small Particles* by Mishchenko et al. (2002); and *Electromagnetic Scattering by Particles and Particle Groups* by Mishchenko (2014), and acknowledge these books here together.

We are very grateful to several individuals whose help directly impacted the final production of this book: Dr. Steven Schroeder carefully edited the manuscript and offered a number of insightful suggestions to improve the book; Dr. Jiachen Ding made significant contributions to Chapter 5; Adam Bell proofread the manuscript; and Ms. Devlin Person, an editorial project manager at Elsevier, patiently worked with the authors to ensure that the book manuscript was delivered within the originally proposed time frame.

Last but not least, we would like to take this opportunity to thank our families for supporting our effort in writing this book, which required a significant amount of quality time after our normal work hours and during weekends.

# Introduction

# 1

Since van de Hulst published his classical book entitled *Light Scattering by Small Particles* in 1957, a number of monographs have been published that summarize theoretical and computational developments in light scattering research from their own unique perspectives (e.g., *Absorption and Scattering of Light by Small Particles* by Bohren and Huffman, 1983; *Scattering of Electromagnetic Waves: Theories and Applications* by Tsang et al., 2000; *Scattering, Absorption, and Emission of Light by Small Particles* by Mishchenko et al., 2002; *Light Scattering by Systems of Particles* by Doicu et al., 2006; and *Electromagnetic Scattering by Particles and Particle Groups: An Introduction* by Mishchenko, 2014). Such tremendous efforts were motivated by the urgent need to apply the theory of light scattering to multiple scientific disciplines, including particle characterization, biomedical sciences, atmospheric remote sensing, the atmospheric radiant energy budget in climate science, ocean optics, astronomy, and optical engineering. However, the computational problems of the scattering and absorption of electromagnetic waves by nonspherical and inhomogeneous particles have not yet been satisfactorily solved when the particle size becomes large relative to the wavelength of incident radiation and when the particle is randomly oriented. Due to the complexity involved in mathematical physics, van de Hulst (1957) summarized slow progress of research efforts in developing the Lorenz-Mie theory by stating that "it is a long way from the formula containing the solution to reliable numbers and curves." Similarly, significant efforts in the light scattering research community have been continually devoted to expanding computational capabilities that solve Maxwell's equations for the solution of light scattering by nonspherical and inhomogeneous particles. As a complementary contribution to previous accomplishments, this book is a treatise on this subject, in which we document recent advances achieved by exploring the concepts and application of the invariant-imbedding T-matrix (IITM) method ( Johnson, 1988; Bi et al., 2013; Bi and Yang, 2014; Doicu and Wriedt, 2018).

The IITM is a rigorous computational technique, which has been demonstrated to be most suitable for computing the optical properties of randomly oriented, arbitrarily shaped, and inhomogeneous particles. Although our work is primarily motivated by applications of light scattering research to remote sensing of cirrus clouds and atmospheric aerosols, the concept and developed technique are essentially applicable to electromagnetic wave problems in other research areas including biomedical optics and marine sciences. The main purpose of this book is to provide a self-consistent summary of the IITM and to highlight some canonic IITM simulations. Similar to the subject limitation in most previous research, this book restricts its focus to dielectric particles in a nonabsorbing medium under the illumination of a polarized plane wave. The solution of light scattering by dielectric particles is determined by the

**Invariant Imbedding T-matrix Method for Light Scattering by Nonspherical and Inhomogeneous Particles**
**https://doi.org/10.1016/B978-0-12-818090-7.00001-2**

value(s) of refractive index related to permittivity and its spatial distribution characterized by shape and the size parameter, which is defined in Section 1.2.

To obtain an overall understanding of the features of the IITM, in this chapter, we briefly recapture the basic definitions related to the modeling capabilities of computational techniques.

## 1.1 Particle shape and inhomogeneity

In the literature, "nonspherical" and "inhomogeneous" are frequently used to describe the particle shape and refractive index distribution involved in the light scattering computation. In the context of Maxwell's equations, the shape of the particle is essentially the spatial distribution of refractive indices. Many techniques developed for analyzing light scattering problems have varying ranges of applicability to particles with differing shapes and inhomogeneity. In general, the complexity of analytical treatment involved in the computation is inversely proportional to the suitable range of applicability to particle shapes and inhomogeneity. For example, the method of separation of variables is only applicable to spheres, spheroids, and infinite cylinders with smooth regular cross-sectional shapes. Classical numerical methods, such as the finite-difference time-domain (FDTD) (Yee, 1966) and the discrete-dipole approximation (DDA) (Purcell and Pennypacker, 1973) methods, are sufficiently flexible to handle a wide range of particle shapes and inhomogeneity. Semianalytical methods, such as the extended boundary condition method (EBCM) (Waterman, 1965, 1971; Mishchenko et al. 2002) and the superposition method (Mackowski and Mishchenko, 1996, 2011; Mackowski, 2014), have a domain of applicability between the separation of variables and purely numerical methods.

The key feature of the IITM is that it treats an arbitrary homogeneous nonspherical particle as an inhomogeneous spherical particle, namely, a spherical volume enclosing the nonspherical particle embedded in the surrounding medium such as air. The refractive index of the nonspherical particle is different from that of the remaining portion in the sphere. "Inhomogeneous" is a more generalized term than overall shape to describe the spatial distribution pattern of refractive indices. Therefore, the light scattering problem of an arbitrary nonspherical particle is a special scenario of the light scattering of an arbitrarily inhomogeneous sphere. Understanding the scattering of light by a particular nonspherical particle, such as an inhomogeneous sphere, might not lead to new insights into general computational techniques. However, it will be shown that this feature makes the IITM a universal tool for computing light scattering by arbitrarily shaped nonspherical particles. Such universality in handling particle geometry is similar to that of the DDA and FDTD methods, but specific advantages of IITM are further summarized in the sections later.

## 1.2 Size parameter

The size parameter ($kR$; $k=2\pi/\lambda$) of a particle is defined as $2\pi$ times the ratio of characteristic particle size ($R$) to the wavelength ($\lambda$) of the incident radiation in the medium.

In particular, here, we understand $R$ as the minimum radius of the circumscribed sphere of a nonspherical particle. On the one hand, this parameter is closely related to the scattering mechanism; on the other hand, the modeling capabilities of exact computational techniques highly depend on this parameter. First, as the size parameter approaches practical infinity (the size of the particle is much larger than the incident wavelength), scattering characteristics associated with geometric optics become more evident, as can be predicted from a ray tracing process, although the validity of geometric optics is vague under most scenarios. Second, if we expand the incident plane wave in terms of vector wave harmonics, the infinite series must be truncated to include a sufficient number of terms to guarantee the convergence of the plane waves in the spatial region within the radius $R$. This number is proportional to $kR+(kR)^{1/3}$, which can be understood from the localization principle of geometric waves (van de Hulst, 1957). From the computational perspective, the size parameter also determines the number of volume or surface elements needed to discretize the particle volume or surface. The number of unknowns has an impact on the computational demand on computer memory and central processing unit (CPU) time and also affects the numerical stability of the algorithm. For methods based on the expansion of the incident and scattered waves, the size parameter determines the truncation number.

Based on current modeling capabilities, the solution of light scattering by an arbitrarily shaped nonspherical particle for small size parameters (e.g., $<\sim 20$) can be easily obtained by using several different computational methods. Due to the increasing power of supercomputers, it is now feasible to obtain reliable solutions for irregular particles with sizes much larger than the incident wavelength. For example, by using the Amsterdam DDA (ADDA) software package, a solution for a sphere with size parameter 320 (roughly 100-wavelength diameter) and refractive index 1.05 was reported (Yurkin and Hoekstra, 2011). As another example, the light scattering by an asymmetric Chebyshev particle has been solved using a multilevel fast multipole algorithm with a size parameter up to 600 (Ergul and Gurel, 2014). The pseudospectral time-domain (Liu, 1997; Liu et al., 2012) and discontinuous Galerkin time-domain methods have been successfully applied to nonspherical particles with size parameters up to 200 (Cockburn et al., 2000; Panetta et al., 2013). At present, because of tremendous computational resources required (number of CPUs, total amount of memory, and computational time), these methods are applied to large size parameters for case studies rather than extensive simulations in practical applications. In contrast to the previous methods, the maximum size parameter for which the invariant-imbedding method is numerically feasible depends on the symmetry of the particle. For particles without any symmetry, the applicable maximum size parameter is smaller. However, the main advantage of the IITM method is that it can more efficiently handle particle's random orientations than the other numerical methods introduced earlier.

## 1.3 Random orientations

The concept of random orientations has been widely used in light scattering and radiative transfer calculations to describe a hypothetical but quite realistic ensemble of particles. A more in-depth discussion of this concept is given by Mishchenko and Yurkin (2017). In the atmosphere, particles such as ice crystals and aerosols more often have random orientations than preferred orientations. There has been a pressing need to develop an efficient and accurate algorithm to compute light scattering for randomly oriented nonspherical and inhomogeneous particles. Moreover, the number of orientations that must be taken into consideration increases dramatically as the size parameter increases. The random orientation condition of large particles poses a significant challenge for most computational methods, which calculate the optical properties of particles by averaging over many distinct orientations. This approach usually leads to a conundrum in estimating the number of orientations needed to achieve convergence in some optical properties. In this framework, many computational methods are not practical for large size parameters because they require a large number of orientations to guarantee numerical convergence. Such convergence is further affected by refractive index or particle inhomogeneity, although this factor has not been systematically investigated. Unlike other computational methods, the T-matrix method obtains the optical properties of randomly oriented particles through an analytical orientation average algorithm, which was published in a seminal paper by Mishchenko (1991). However, the stability of the traditional approach to calculating the T-matrix strongly depends on the particle shape. Thus, it is a natural motive to develop a solver that can obtain the optical properties of randomly oriented arbitrary inhomogeneous particles. This power of the IITM results from its incorporation of advancements related to the analytical features of the T-matrix methods. First, the analytical orientation average algorithm can account for either finite-fold symmetry or no symmetry. Second, the IITM can be combined with the Lorenz-Mie theory or the EBCM to accelerate the speed of relevant computations.

## 1.4 Invariant-imbedding principle

The invariant-imbedding principle is an outgrowth of the principle of invariance, which had already gained popularity in radiative transfer and computational mathematics (e.g., Bellman and Wing, 1975). For electromagnetic wave scattering, unlike solving the Maxwell equations in conjunction with prescribed boundary conditions, the invariant-imbedding principle transforms the boundary value problem into an initial value problem in the framework of an electric-field volume integral equation. Although this formalism to compute the T-matrix was pioneered by Johnson (1988), that paper drew scant attention, most likely due to the lack of a stable and efficient numerical implementation for arbitrarily shaped nonspherical particles. We have explored the power of this concept in light scattering by nonspherical particles in recent years (Bi et al., 2013, 2018; Bi and Yang, 2013, 2014; Liu et al., 2015; Yang et al. 2019). The numerical accuracy and efficiency of this method make it highly suitable to compute the orientation-averaged optical properties of arbitrarily shaped and inhomogeneous particles. As stated in Section 1.1, any particle in the framework of the IITM can be treated as an inhomogeneous sphere. By discretizing the inhomogeneous sphere into a number of layers, the scattering of light by the nonspherical particle can be transformed into a family of scattering problems with multiple layers. Each light scattering problem becomes a stepping stone to recursively compute the solution for the whole nonspherical particle of interest. Of course, the computational burden of IITM is much larger than that of the surface integral-based T-matrix technique. However, as a benefit, a more stable algorithm and a more flexible scheme are provided by IITM to treat inhomogeneity. On the other hand, a large computational burden can be reduced by using modern computers that support parallel computing, which will also be specifically discussed.

In Chapter 2, we will start with the basics of electromagnetic waves, relevant physical quantities of light scattering, and a brief summary of the rigorous methods that have been developed to solve light scattering by nonspherical and inhomogeneous particles. We realize that these basics can be found in available monographs. However, we wish to present a self-contained light scattering formulation. In Chapter 3, we summarize the concept and mathematical properties of the T-matrix. Special attention is given to the T-matrix formulation for asymmetric particles. Chapter 4 contains complete documentation of the IITM and is the main part of this book. Chapter 5 discusses numerical examples. Readers not interested in the technical part of the computational methods may turn to this chapter to explore the practical capabilities of the IITM.

# Fundamentals

# 2

Radiative properties of atmospheric particles, such as water droplets, ice crystals and aerosols, and oceanic particles such as phytoplankton and zooplankton, are very important in the study of remote sensing and climate change. Scattering from such an ensemble of particles can usually be broken down into two main categories, namely, active sources that include laser sources or passive sources that include solar radiation and terrestrial thermal emission. One must be extremely careful when describing the source type used in the analysis because erroneous conclusions can be reached if one tries to interchange the two. For example, laser radiation has both large lateral and longitudinal coherence lengths, while on the other hand, sunlight has both short lateral ($\sim$61 μm) and longitudinal (dependent on the bandwidth of the source) coherent lengths (Wolf, 2007). To perform a first-principles simulation of a source interacting with a large ensemble of particles, one must solve Maxwell's equations for the entire system, which would be a formidable and practically impossible task. However, by judicious use of some fidelity reducing assumptions involving far field and incoherence, the description of multiple scattering of atmospheric and oceanic particles can be reduced to the classic radiative transfer equation (Tsang et al., 1985; Mishchenko et al., 2002). The radiative transfer equation has been extensively discussed in many classical textbooks (e.g., Chandrasekhar, 1960; Preisendorfer, 1965; Sobolev, 1975; van de Hulst, 1980; Liou, 1980; Goody and Yung, 1989; Mishchenko et al., 2006). Multiple scattering by particles is determined by the single-scattering properties such as the scattering matrix and extinction and scattering cross sections. Single scattering was systematically described first by van de Hulst (1957) and was further explored and summarized by others (e.g., Bohren and Huffman, 1983; Tsang et al., 2000; Mishchenko et al., 2002).

This chapter summarizes principles of single scattering. We introduce electromagnetic waves in Section 2.1. Section 2.2 gives the physical quantities quantifying light scattering. Section 2.3 gives a brief summary of numerically rigorous methods for solving light scattering by nonspherical particles.

## 2.1 Electromagnetic waves

### 2.1.1 Maxwell's equations

For macroscopic media, the governing principles for all electromagnetic phenomena are covered in the four partial differential or integral Maxwell's equations, constructed by James Clerk Maxwell (1861). Maxwell's equations are a remarkable unification of electricity and magnetism, called electromagnetism. Moreover, Maxwell's equations are completely compatible with special relativity, developed by Albert Einstein in 1905 ( Jackson, 1998). Electromagnetism is comprehensively

**Invariant Imbedding T-matrix Method for Light Scattering by Nonspherical and Inhomogeneous Particles**
**https://doi.org/10.1016/B978-0-12-818090-7.00002-4**

described in the legendary book *Classical Electrodynamics* by Jackson (1998). However, we will focus on the theory of light scattering by atmospheric and oceanic particles in this book.

The differential forms of the macroscopic Maxwell equations under SI units in a medium are ( Jackson, 1998)

$$\nabla \cdot \mathbf{D} = \rho, \tag{2.1.1}$$

$$\nabla \times \mathbf{E} = -\frac{\partial \mathbf{B}}{\partial t}, \tag{2.1.2}$$

$$\nabla \cdot \mathbf{B} = 0, \tag{2.1.3}$$

$$\nabla \times \mathbf{H} = \mathbf{J} + \frac{\partial \mathbf{D}}{\partial t}, \tag{2.1.4}$$

with the constitutive relations

$$\mathbf{D} = \varepsilon_0 \mathbf{E} + \mathbf{P}, \tag{2.1.5}$$

$$\mathbf{H} = \frac{\mathbf{B}}{\mu_0} - \mathbf{M}, \tag{2.1.6}$$

and if the medium is a conductor, one can have another relation, often referred to as Ohm's law:

$$\mathbf{J} = \sigma \mathbf{E}. \tag{2.1.7}$$

The quantities in the earlier equations are defined in Table 2.1.

Note that the space and time arguments $\left(\vec{r}, t\right)$ in all physical variables are suppressed in the earlier equations. The electric field **E** and magnetic induction **B**,

**Table 2.1** Basic electromagnetic quantities.

| | |
|---|---|
| **E** | Electric field |
| **D** | Electric displacement |
| **B** | Magnetic induction |
| **H** | Magnetic field |
| **P** | Induced electric polarization |
| **M** | Induced magnetization |
| **J** | Free electric current density |
| $\rho$ | Free electric charge density |
| $\varepsilon_0$ | Permittivity in vacuum |
| $\mu_0$ | Permeability in vacuum |
| $\sigma$ | Conductivity |

rather than **D** and **H**, are the fundamental quantities. Eq. (2.1.1) is known as Gauss's law for the electric displacement. It states that the source of the electric displacement is the electric charge density. The corresponding magnetic Gauss's law, Eq. (2.1.3), states that there are no free magnetic monopoles for magnetic induction. The field lines of magnetic induction either extend infinitely, or they form closed loops. These two basic laws of electromagnetism hold in all Lorentz frames. In fact, if we make use of the transformation properties of the electromagnetic field under Lorentz transformations, and we demand that Eqs. (2.1.1), (2.1.3) be form invariant under Lorentz transformations, then we obtain Eqs. (2.1.2), (2.1.4). Thus, (2.1.2) and (2.1.4) follow from (2.1.1) and (2.1.3) in conjunction with a symmetry principle, namely, the Lorentz invariance of (2.1.1) and (2.1.3). A detailed discussion can be found in the standard literature on relativity theory (e.g., Misner et al., 1973).

Eq. (2.1.2) is known as Faraday's induction law. Eq. (2.1.4) is an extended version of Ampère's law, which is given by

$$\nabla \times \mathbf{H} = \mathbf{J}. \tag{2.1.8}$$

However, Ampère's law is only valid in a stationary current with $\nabla \cdot \mathbf{J} = 0$. For a non-stationary current, the charge and current must satisfy the following continuity relation:

$$\nabla \cdot \mathbf{J} + \frac{\partial \rho}{\partial t} = 0. \tag{2.1.9}$$

Using Eq. (2.1.1), the divergence free relation similar to the stationary current relation can be given as ( Jackson, 1998, section 6.1)

$$\nabla \cdot \left( \mathbf{J} + \frac{\partial \mathbf{D}}{\partial t} \right) = 0. \tag{2.1.10}$$

Maxwell referred to $\frac{\partial \mathbf{D}}{\partial t}$ as the "displacement current" and generalized Ampère's law from the stationary current form of Eq. (2.1.8) to the time-dependent field of Eq. (2.1.4). The displacement current term is essential to Maxwell's equations because Eqs. (2.1.2), (2.1.4) construct the mutual induction for rapidly fluctuating electric and magnetic fields. Consequently, Maxwell formalized the unification of electric and magnetic fields and predicted the existence of an electromagnetic wave. The experimental proof for the existence of electromagnetic waves was convincingly given by Hertz (1887).

The differential operators in Maxwell's equations are only defined at those locations at which the fields, sources, and constitutive parameters are continuously differentiable functions of space and time. This is generally not the case at the interface between different media. Thus, we need additional boundary conditions to determine how the fields change across the adjoining surface. Such boundary conditions are derived in the literature from the volume or surface integral form of Maxwell's equations by modeling the surface as an infinitesimal layer as shown in Fig. 2.1, where

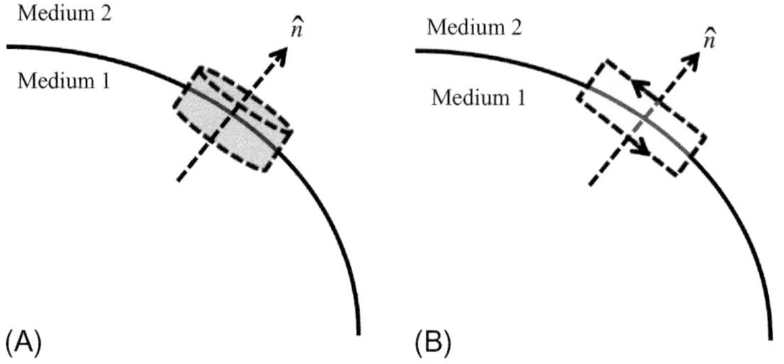


**Fig. 2.1** The integral volume in panel (A) over the divergence equations and the integral surface in panel (B) over the curl equations.

the volume and the surface integrals with respect to the normal directions $\hat{n}$ (from medium 1 to medium 2) are infinitesimal, shown in panels (A) and (B), respectively. However, such a treatment can only be justified if the fields remain continuous at the infinitesimal layer. On the other hand, if we model the boundary as a surface at which the fields undergo a discontinuous jump, then it is necessary to introduce the boundary conditions as a postulate based on our experimental knowledge. A detailed discussion can be found in Rothwell and Cloud (2001). Here, the boundary conditions are directly given as additional laws supplementing Maxwell's equations. The unit vector $\hat{n}$, normal to the interface, points from medium 1 to medium 2 as shown in Fig. 2.1. The normal components of the fields at the interface satisfy the conditions:

$$\hat{n}\cdot(\mathbf{D}_2-\mathbf{D}_1)=\rho_s, \tag{2.1.11}$$

$$\hat{n}\cdot(\mathbf{B}_2-\mathbf{B}_1)=0, \tag{2.1.12}$$

where $\rho_s$ is the surface charge density. The tangential components of the fields at the interface can be determined as

$$\hat{n}\times(\mathbf{E}_2-\mathbf{E}_1)=\mathbf{0}, \tag{2.1.13}$$

$$\hat{n}\times(\mathbf{H}_2-\mathbf{H}_1)=\mathbf{J}_s, \tag{2.1.14}$$

where $\mathbf{J}_s$ is the surface current density.

Eqs. (2.1.1)–(2.1.4) are Maxwell's equations in a medium, Eqs. (2.1.5), (2.1.6) are the corresponding constitutive relations, and Eqs. (2.1.11)–(2.1.14) specify the boundary conditions when there is an interface separating two media. However, the constitutive relations are not complete unless the relations between **P** and **E** or **D** and between **M** and **B** or **H** are given. The polarization **P** and magnetization **M** are determined by the characteristics of the medium.

In a vacuum, there does not exist a polarization or magnetization process, and there is no free charge or free current density. The derived quantities **D** and **H** are reduced to simple constitutive relations:

$$\mathbf{D} = \varepsilon_0 \mathbf{E}, \tag{2.1.15}$$

$$\mathbf{H} = \frac{\mathbf{B}}{\mu_0}. \tag{2.1.16}$$

Maxwell's equations in a vacuum are as follows:

$$\nabla \cdot \mathbf{E} = 0, \tag{2.1.17}$$

$$\nabla \times \mathbf{E} = -\frac{\partial \mathbf{B}}{\partial t}, \tag{2.1.18}$$

$$\nabla \cdot \mathbf{B} = 0, \tag{2.1.19}$$

$$\nabla \times \mathbf{B} = \frac{1}{c^2}\frac{\partial \mathbf{E}}{\partial t}, \tag{2.1.20}$$

where $c = 1/\sqrt{\varepsilon_0 \mu_0}$ and $c$ is the velocity of light in a vacuum. If the response of a medium to an applied field is assumed to be linear and the medium is isotropic and nondispersive, then the polarization and magnetization are experimentally shown to have a simple linear relationship with the applied field:

$$\mathbf{P} = \varepsilon_0 \chi_e \mathbf{E}, \tag{2.1.21}$$

$$\mathbf{M} = \chi_m \mathbf{H}, \tag{2.1.22}$$

where $\chi_e$ and $\chi_m$ are the electric susceptibility and magnetic susceptibility, respectively. For a linear, isotropic, and nondispersive medium, the complete constitutive relations are

$$\mathbf{D} = \varepsilon_0 (1 + \chi_e)\mathbf{E} = \varepsilon \mathbf{E}, \tag{2.1.23}$$

$$\mathbf{B} = \mu_0 (1 + \chi_m)\mathbf{H} = \mu \mathbf{H}, \tag{2.1.24}$$

where $\varepsilon$ and $\mu$ are the permittivity and permeability in the medium. If the medium is anisotropic, the permittivity should be a second-order tensor $\overleftrightarrow{\varepsilon}$. If the response is nonlinear, the relations between **D** and **E** become complicated and are not discussed here. If the medium is ferromagnetic, the relations between **B** and **H** are nonlinear and depend on the process of magnetization. These complex cases are not discussed in the book. However, the assumption we made in (2.1.23) and (2.1.24) that the medium be nondispersive is very restrictive. Thus, Eqs. (2.1.23), (2.1.24) are of limited practical use. We will discuss the generalization of the constitutive relations to dispersive media when we consider the formulation of Maxwell's equations in the frequency domain.

### 2.1.2 Energy and energy flow in an electromagnetic field

Using a Fourier transform, an arbitrary field can be decomposed into different time-harmonic components. We express the electromagnetic field in time harmonics with the term $\exp(-i\omega t)$, where $\omega = 2\pi/T$ is the angular frequency and $T$ is the period of the wave. Since Maxwell's equations are linear, we can always take the real part of a complex function to obtain the physical quantity. For instance, the electric field can be physically stated as ( Jackson, 1998):

$$\begin{aligned}\mathbf{E}\left(\vec{r},t\right) &= Re\left[\mathbf{E}\left(\vec{r}\right)\exp\left(-i\omega t\right)\right]\\ &= \frac{1}{2}\left[\mathbf{E}\left(\vec{r}\right)\exp\left(-i\omega t\right) + \mathbf{E}^*\left(\vec{r}\right)\exp\left(i\omega t\right)\right],\end{aligned} \tag{2.1.25}$$

where "*" denotes the complex conjugate. For a quadratic operation involving $\mathbf{A}\left(\vec{r},t\right)$ and $\mathbf{B}\left(\vec{r},t\right)$ in electromagnetism, one can have the following expressions:

$$\mathbf{A}\left(\vec{r},t\right)\bigcirc\mathbf{B}\left(\vec{r},t\right) = \frac{1}{2}Re\left[\mathbf{A}^*\left(\vec{r}\right)\bigcirc\mathbf{B}\left(\vec{r}\right) + \mathbf{A}\left(\vec{r}\right)\bigcirc\mathbf{B}\left(\vec{r}\right)\exp\left(-2i\omega t\right)\right], \tag{2.1.26}$$

where "$\bigcirc$" denotes an operation, such as a dot product, cross product, or a dyadic operation. For a physical quantity $\mathbf{C}\left(\vec{r},t\right)$, the time average is defined as

$$\left\langle \mathbf{C}\left(\vec{r},t\right)\right\rangle = \frac{1}{T}\int_0^T \mathbf{C}\left(\vec{r},t\right)dt, \tag{2.1.27}$$

where $T$ is the period of a wave. Consequently, the time average of a quantity after a quadratic operation can be determined as ( Jackson, 1998; p. 264)

$$\left\langle \mathbf{A}\left(\vec{r},t\right)\bigcirc\mathbf{B}\left(\vec{r},t\right)\right\rangle = \frac{1}{2}Re\left[\mathbf{A}^*\left(\vec{r}\right)\bigcirc\mathbf{B}\left(\vec{r}\right)\right]. \tag{2.1.28}$$

The time average of a quadratic operation on time-harmonic fields can be obtained by taking half of the real part of the same operation of one complex quantity (**A** or **B**) with the complex conjugate of the other quantity (**B** or **A**).

Using Eqs. (2.1.2), (2.1.4), we can directly obtain

$$\mathbf{E}\cdot\frac{\partial\mathbf{D}}{\partial t} + \mathbf{H}\cdot\frac{\partial\mathbf{B}}{\partial t} = \mathbf{E}\cdot(\nabla\times\mathbf{H} - \mathbf{J}) - \mathbf{H}\cdot(\nabla\times\mathbf{E}). \tag{2.1.29}$$

After some algebra and reorganization, we get

$$-\mathbf{J}\cdot\mathbf{E} = \mathbf{E}\cdot\frac{\partial\mathbf{D}}{\partial t} + \mathbf{H}\cdot\frac{\partial\mathbf{B}}{\partial t} + \nabla\cdot(\mathbf{E}\times\mathbf{H}). \tag{2.1.30}$$

Eq. (2.1.30) is actually the differential form of energy conservation of an electromagnetic field. One can define the energy flow density vector **S** (which is called the Poynting vector) and the energy density $W$ as follows:

$$\mathbf{S}=\mathbf{E}\times\mathbf{H}, \tag{2.1.31}$$

$$\frac{\partial W}{\partial t}=\mathbf{E}\cdot\frac{\partial \mathbf{D}}{\partial t}+\mathbf{H}\cdot\frac{\partial \mathbf{B}}{\partial t}. \tag{2.1.32}$$

The differential and integral forms of the conservation can then be succinctly given:

$$-\mathbf{J}\cdot\mathbf{E}=\frac{\partial W}{\partial t}+\nabla\cdot\mathbf{S}, \tag{2.1.33}$$

$$-\int_{\mathrm{V}}\mathbf{J}\cdot\mathbf{E}d\mathrm{V}=\frac{d}{dt}\int_{\mathrm{V}}Wd\mathrm{V}+\oint_{\sigma}\mathbf{S}\cdot d\vec{\sigma}, \tag{2.1.34}$$

where $V$ is a finite volume and $\sigma$ is the surface enclosing volume $V$. The rate of doing work by the electromagnetic field over a charge $q$ can be given as

$$\mathbf{F}\cdot\mathbf{v}=q(\mathbf{E}+\mathbf{v}\times\mathbf{B})\cdot\mathbf{v}=q\mathbf{v}\cdot\mathbf{E}, \tag{2.1.35}$$

where **F** is the Lorentz force and **v** is the charge velocity. Consequently, the magnetic field does not do work due to the nature of the magnetic field. For a continuous charge distribution in volume $V$, the rate of doing work by the electric field can be given as

$$q\mathbf{v}\cdot\mathbf{E}\rightarrow\int_{\mathrm{V}}\rho\mathbf{v}\cdot\mathbf{E}d\mathrm{V}=\int_{\mathrm{V}}\mathbf{J}\cdot\mathbf{E}d\mathrm{V}. \tag{2.1.36}$$

If the electric field does work on the charges or the current, the energy of the field decreases. The physical meanings of the conservation forms are evident. If the finite volume goes to infinity, there is no energy flow and one can have

$$-\int_{\mathrm{V}}\mathbf{J}\cdot\mathbf{E}d\mathrm{V}=\frac{d}{dt}\int_{\mathrm{V}}Wd\mathrm{V}. \tag{2.1.37}$$

For a linear, nondispersive medium as shown in Eqs. (2.1.23), (2.1.24), the energy density can be explicitly given by

$$W=\frac{1}{2}\left(\varepsilon\mathbf{E}^2+\frac{\mathbf{H}^2}{\mu}\right). \tag{2.1.38}$$

It follows that the time-average energy density and Poynting vector in complex notation can be given as

$$\langle W\rangle = \frac{1}{4}Re\left[\varepsilon\mathbf{E}^*\cdot\mathbf{E} + \frac{\mathbf{H}^*\cdot\mathbf{H}}{\mu}\right], \tag{2.1.39}$$

$$\langle\mathbf{S}\rangle = \frac{1}{2}Re[\mathbf{E}^*\times\mathbf{H}]. \tag{2.1.40}$$

## 2.1.3 Propagation of electromagnetic waves

An electromagnetic wave can be fully described by Maxwell's equations. We assume that the medium is linear, nondispersive, and nonconductive and has constant permittivity and permeability. Moreover, the medium is in free space so it does not have any charge or current distribution. In this case, Maxwell's equations in an infinite medium can be given as

$$\nabla\cdot\mathbf{E} = 0, \quad \nabla\times\mathbf{E} = -\mu\frac{\partial\mathbf{H}}{\partial t}, \tag{2.1.41}$$

$$\nabla\cdot\mathbf{H} = 0, \quad \nabla\times\mathbf{H} = \varepsilon\frac{\partial\mathbf{E}}{\partial t}. \tag{2.1.42}$$

Through Eqs. (2.1.41), (2.1.42), the wave equations for the electric and magnetic fields in a nondispersive medium are

$$\nabla\cdot\mathbf{E} = 0, \ \nabla^2\mathbf{E} - \mu\varepsilon\frac{\partial^2\mathbf{E}}{\partial t^2} = 0, \tag{2.1.43}$$

$$\nabla\cdot\mathbf{H} = 0, \ \nabla^2\mathbf{H} - \mu\varepsilon\frac{\partial^2\mathbf{H}}{\partial t^2} = 0. \tag{2.1.44}$$

where $\upsilon = 1/\sqrt{\mu\varepsilon}$ is the propagation velocity of the electromagnetic wave in the medium and also $\upsilon = c/m$, where $m$ is the refractive index of the medium.

In a dispersive medium, the constitutive relations are given by convolution integrals as

$$\mathbf{D}\left(\vec{r},t\right) = \int_{-\infty}^{t} d\tau\varepsilon\left(\vec{r},t-\tau\right)\mathbf{E}\left(\vec{r},\tau\right), \tag{2.1.45}$$

$$\mathbf{B}\left(\vec{r},t\right) = \int_{-\infty}^{t} d\tau\mu\left(\vec{r},t-\tau\right)\mathbf{H}\left(\vec{r},\tau\right), \tag{2.1.46}$$

$$\mathbf{J}\left(\vec{r},t\right) = \int_{-\infty}^{t} d\tau\sigma\left(\vec{r},t-\tau\right)\mathbf{E}\left(\vec{r},\tau\right). \tag{2.1.47}$$

The simple Eqs. (2.1.23), (2.1.24), (2.1.7) for nondispersive media follow from the earlier equations when we assume that $\varepsilon\left(\vec{r},t\right)=\varepsilon\left(\vec{r}\right)\delta(t)$ and, similarly for $\mu\left(\vec{r},t\right)$ and $\sigma\left(\vec{r},t\right)$, where $\delta(t)$ denotes the Dirac delta-function.

The treatment of dispersive media is significantly easier in the frequency domain than in the time domain. We assume that the fields and sources are time harmonic, that is, in the frequency domain as

$$\mathbf{E}\left(\vec{r},\omega\right)=Re\left[\mathbf{E}\left(\vec{r}\right)\exp(-i\omega t)\right], \tag{2.1.48}$$

and similarly for **D**, **B**, **H**, **J**, and $\rho$. We introduce the complex Fourier transforms of the permittivity as

$$\varepsilon\left(\vec{r},\omega\right)=\int_0^\infty dt\varepsilon\left(\vec{r},t\right)\exp(i\omega t), \tag{2.1.49}$$

and similarly for $\mu\left(\vec{r},\omega\right)$ and $\sigma\left(\vec{r},\omega\right)$. Accordingly, the convolution integrals in Eqs. (2.1.45)–(2.1.47) for the time-domain constitutive relations become simple products in the frequency domain as

$$\mathbf{D}\left(\vec{r},\omega\right)=\varepsilon\left(\vec{r},\omega\right)\mathbf{E}\left(\vec{r},\omega\right), \tag{2.1.50}$$

$$\mathbf{B}\left(\vec{r},\omega\right)=\mu\left(\vec{r},\omega\right)\mathbf{H}\left(\vec{r},\omega\right), \tag{2.1.51}$$

$$\mathbf{J}\left(\vec{r},\omega\right)=\sigma\left(\vec{r},\omega\right)\mathbf{E}\left(\vec{r},\omega\right). \tag{2.1.52}$$

If the medium is assumed to be nondispersive in the time domain, that is, $\varepsilon\left(\vec{r},t\right)=\varepsilon\left(\vec{r}\right)\delta(t)$, then in the frequency domain the Fourier transform of the permittivity becomes independent of frequency, $\varepsilon\left(\vec{r},\omega\right)=\varepsilon\left(\vec{r}\right)$. Thus, we account for the fact that the dielectric properties of macroscopic media are generally frequency dependent. Moreover, Maxwell's equations in Eqs. (2.1.1)–(2.1.4) for time-harmonic fields become

$$\nabla\cdot\mathbf{D}\left(\vec{r},\omega\right)=\rho\left(\vec{r},\omega\right), \tag{2.1.53}$$

$$\nabla\times\mathbf{E}\left(\vec{r},\omega\right)=i\omega\mathbf{B}\left(\vec{r},\omega\right), \tag{2.1.54}$$

$$\nabla\cdot\mathbf{B}\left(\vec{r},\omega\right)=0, \tag{2.1.55}$$

$$\nabla\times\mathbf{H}\left(\vec{r},\omega\right)=\mathbf{J}\left(\vec{r},\omega\right)-i\omega\mathbf{D}\left(\vec{r},\omega\right). \tag{2.1.56}$$

With the constitutive relations, the last equation becomes

$$\begin{aligned}\nabla\times\mathbf{H}\left(\vec{r},\omega\right)&=-i\omega\left[\varepsilon\left(\vec{r},\omega\right)+i\sigma\left(\vec{r},\omega\right)/\omega\right]\mathbf{E}\left(\vec{r},\omega\right)\\&=-i\omega\varepsilon'\left(\vec{r},\omega\right)\mathbf{E}\left(\vec{r},\omega\right).\end{aligned}\tag{2.1.57}$$

In nonconducting media where $\sigma=0$, $\varepsilon'\left(\vec{r},\omega\right)=\varepsilon\left(\vec{r},\omega\right)$. Further in the absence of free charges, $\rho=0$, and Maxwell's equations for time-harmonic fields become

$$\nabla\cdot\mathbf{E}=0,\ \ \nabla\times\mathbf{E}=i\omega\mu\mathbf{H},\tag{2.1.58}$$

$$\nabla\cdot\mathbf{H}=0,\ \ \nabla\times\mathbf{H}=-i\omega\varepsilon\mathbf{E},\tag{2.1.59}$$

where the arguments $\left(\vec{r},\omega\right)$ for the fields, the permeability, and the permittivity are suppressed and the same notations as in the time domain are used without causing confusion. If we assume that the medium is homogeneous, or, at least, piecewise homogeneous, then $\varepsilon$ and $\mu$ are independent, or piecewise independent of position. Then, the wave equations of the electromagnetic field are reduced to the vector Helmholtz equations:

$$\nabla\cdot\mathbf{E}=0,\ \ \nabla^2\mathbf{E}+\omega^2\mu\varepsilon\mathbf{E}=0,\tag{2.1.60}$$

$$\nabla\cdot\mathbf{H}=0,\ \ \nabla^2\mathbf{H}+\omega^2\mu\varepsilon\mathbf{H}=0.\tag{2.1.61}$$

where Eqs. (2.1.60), (2.1.61) are not independent. The electric field can be completely given in terms of Eq. (2.1.60) and the magnetic field can be given in terms of the electric field as

$$\mathbf{H}=-\frac{i}{\omega\mu}\nabla\times\mathbf{E},\tag{2.1.62}$$

or Eq. (2.1.61) determines the magnetic field and then the electric field is

$$\mathbf{E}=\frac{i}{\omega\varepsilon}\nabla\times\mathbf{H}.\tag{2.1.63}$$

Using complex notation, the simplest solution satisfying Maxwell's equations or the Helmholtz equations is a plane wave

$$\mathbf{E}\left(\vec{r},t\right)=\mathbf{E}_0\exp\left[i\left(\vec{k}\cdot\vec{r}-\omega t\right)\right],\ \ \vec{k}\cdot\mathbf{E}_0=0,\tag{2.1.64}$$

$$\mathbf{H}\left(\vec{r},t\right)=\mathbf{H}_0\exp\left[i\left(\vec{k}\cdot\vec{r}-\omega t\right)\right],\ \ \vec{k}\cdot\mathbf{H}_0=0,\tag{2.1.65}$$

where $\mathbf{E}_0$ and $\mathbf{H}_0$ are constant amplitudes of the plane waves, $\left(\vec{k}\cdot\vec{r}-\omega t\right)$ is the phase of the plane waves, and $\vec{k}$ is called the wave vector. The direction of the wave vector denotes the propagation direction of the plane waves, and the magnitude of the wave vector is

$$k=\frac{\omega}{v}=\omega\sqrt{\mu\varepsilon}=\frac{2\pi}{\lambda}, \tag{2.1.66}$$

where $\lambda$ is the wavelength of the plane wave in the medium. Also, the plane waves are transverse, that is, the oscillation directions of the electromagnetic field are perpendicular to its propagation direction. Moreover, the amplitudes of the electromagnetic field using Eqs. (2.1.62), (2.1.63) have the following relations:

$$\mathbf{H}_0=\frac{\vec{k}\times\mathbf{E}_0}{\omega\mu},\quad \mathbf{E}_0=-\frac{\vec{k}\times\mathbf{H}_0}{\omega\mu} \tag{2.1.67}$$

Consequently, $\mathbf{E}$ and $\mathbf{H}$ are perpendicular to each other, and $\mathbf{E}\times\mathbf{H}$ is along the propagation direction. The time average of the energy density and the Poynting vector for a plane wave can be given as

$$\langle W\rangle=\frac{1}{2}\varepsilon|\mathbf{E}_0|^2=\frac{1}{2}\mu|\mathbf{H}_0|^2, \tag{2.1.68}$$

$$\langle\mathbf{S}\rangle=\frac{1}{2}\sqrt{\frac{\varepsilon}{\mu}}|\mathbf{E}_0|^2\hat{k}=\frac{1}{2}\sqrt{\frac{\mu}{\varepsilon}}|\mathbf{H}_0|^2\hat{k}, \tag{2.1.69}$$

where $\hat{k}=\vec{k}/k$. The permittivity $\varepsilon$ and permeability $\mu$ are assumed to be real in Eqs. (2.1.68), (2.1.69). The direction of the energy flow for the plane wave is indeed along the propagation direction, as shown in Eq. (2.1.69).

### 2.1.4 Polarization, Jones vector, Stokes vector, Jones matrix, and Mueller matrix

Because the oscillation direction of a transverse electromagnetic wave is perpendicular to the propagation direction, we can define two mutual orthogonal unit directions $(\hat{e}_1,\hat{e}_2)$ perpendicular to the propagation direction $\hat{k}$, where the unit vectors obey $\hat{e}_1\times\hat{e}_2=\hat{k}$. We will use the electric field as an example to explain the polarizations. The electric field can be written in a component form:

$$\begin{aligned}\mathbf{E}\left(\vec{r},t\right)&=\mathbf{E}_0\exp\left[i\left(\vec{k}\cdot\vec{r}-\omega t\right)\right]\\&=(\mathrm{E}_{02}\hat{e}_2+\mathrm{E}_{01}\hat{e}_1)\exp\left[i\left(\vec{k}\cdot\vec{r}-\omega t\right)\right],\end{aligned} \tag{2.1.70}$$

and

$$\mathrm{E}_{02} = a_2 \exp(i\delta_2), \quad \mathrm{E}_{01} = a_1 \exp(i\delta_1), \tag{2.1.71}$$

where $a_1$ and $a_2$ are real and denote the magnitudes for two components, $\delta_1$ and $\delta_2$ are real, and $\exp(i\delta_1)$ and $\exp(i\delta_2)$ are the phase factors for the two components. If $\delta_1 = \delta_2$, the electric field is linearly polarized, as shown in Fig. 2.2. At a fixed point, the oscillation direction of the electric field with linear polarization is fixed. The linear polarization angle is defined according to direction $\hat{e}_2$, and the clockwise direction facing into the propagation direction is defined as positive, as shown in Fig. 2.2. However, if $\delta_1 \neq \delta_2$, the electric field is elliptically polarized. The oscillation direction of the electric field sweeps all directions perpendicular to the propagation direction with frequency $\omega$, and the magnitudes at different directions construct an ellipse. The electric field is circularly polarized if the amplitudes have the same magnitude $a_1 = a_2 = a$ with a $\pi/2$ phase difference. The magnitudes are the same and construct a circle as shown in Fig. 2.3. The electric field with circular polarization becomes

$$\mathbf{E}\left(\vec{r}, t\right) = a(\hat{e}_2 \pm i\hat{e}_1)\exp\left[i\left(\vec{k} \cdot \vec{r} - \omega t\right)\right]. \tag{2.1.72}$$

A positive sign in Eq. (2.1.72) ($\hat{e}_2 + i\hat{e}_1$) indicates that the rotation is clockwise if the observer faces the incoming wave. Following this convention, the field is called

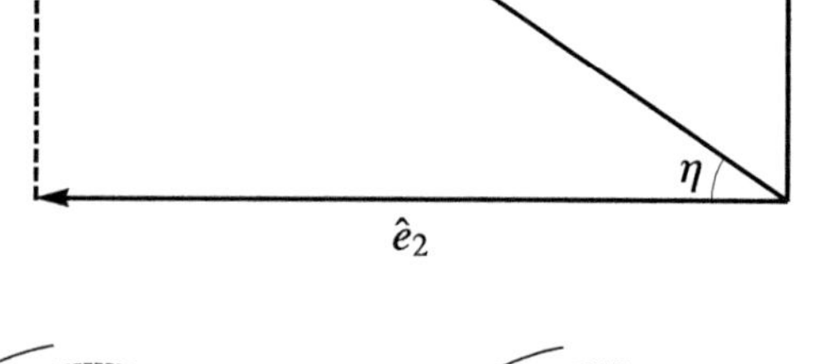


**Fig. 2.2** Linear polarized wave. The linearly polarized angle $\eta$ is defined relative to the direction $\hat{e}_2$, and a positive angle $\eta$ is in the clockwise direction if the observer faces the incoming wave.

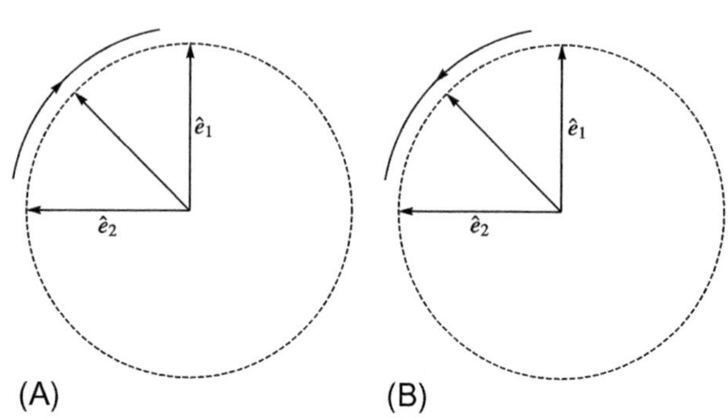


**Fig. 2.3** (A) Right-circularly polarized wave and (B) left-circularly polarized wave.

right-circularly polarized although the rotation direction with the propagation direction obeys the left-hand rule. A minus sign in Eq. (2.1.72) $(\hat{e}_2 - i\hat{e}_1)$ indicates a counterclockwise rotation. Accordingly, the field is called left-circularly polarized.

Similarly to the definition of the two orthogonal unit vectors $(\hat{e}_1, \hat{e}_2)$, two other vectors can be defined to describe the polarizations as

$$\hat{e}_\pm = \frac{1}{\sqrt{2}}(\hat{e}_2 \pm i\hat{e}_1). \tag{2.1.73}$$

Analogously, the electric field becomes

$$\mathbf{E}\left(\vec{r}, t\right) = (\mathrm{E}_+\hat{e}_+ + \mathrm{E}_-\hat{e}_-)\exp\left[i\left(\vec{k}\cdot\vec{r} - \omega t\right)\right], \tag{2.1.74}$$

and

$$\mathrm{E}_+ = a_+\exp(i\delta_+), \quad \mathrm{E}_- = a_-\exp(i\delta_-), \tag{2.1.75}$$

where $a_\pm$ denotes the magnitude and $\exp(i\delta_\pm)$ the phase factor.

The polarization state can be expressed in terms of a column vector with two elements, where the first element is the component along direction $\hat{e}_2$ and the second one is the component along direction $\hat{e}_1$. The two-element column vector describing the polarization state is also called the *Jones vector*. Table 2.2 shows different polarization states using the Jones vector, where the vector is normalized.

The Jones vector depends on the choice of the basis vectors as in Fig. 2.4. If the rotation angle of the basis vectors is $\varphi$, the Jones vector with respect to the new basis vectors becomes

$$\begin{pmatrix} \mathrm{E}_2' \\ \mathrm{E}_1' \end{pmatrix} = \begin{pmatrix} \cos\varphi & \sin\varphi \\ -\sin\varphi & \cos\varphi \end{pmatrix}\begin{pmatrix} \mathrm{E}_2 \\ \mathrm{E}_1 \end{pmatrix}, \tag{2.1.76}$$

where $\{\mathrm{E}_i\}$ are the magnitude components along the basis vectors and the primed quantities use primed basis vectors.

**Table 2.2** Jones vectors of different polarization states.

| | | | |
|---|---|---|---|
| Linearly polarized | $\begin{pmatrix} \cos\eta \\ \sin\eta \end{pmatrix}$ | $\eta = 0° \leftrightarrow \begin{pmatrix} 1 \\ 0 \end{pmatrix}$ | $\eta = 90° \leftrightarrow \begin{pmatrix} 0 \\ 1 \end{pmatrix}$ |
| | | $\eta = 45° \leftrightarrow \frac{1}{\sqrt{2}}\begin{pmatrix} 1 \\ 1 \end{pmatrix}$ | $\eta = -45° \leftrightarrow \frac{1}{\sqrt{2}}\begin{pmatrix} 1 \\ -1 \end{pmatrix}$ |
| Right-circularly polarized | $\frac{1}{\sqrt{2}}\begin{pmatrix} 1 \\ i \end{pmatrix}$ | Left-circularly polarized | $\frac{1}{\sqrt{2}}\begin{pmatrix} 1 \\ -i \end{pmatrix}$ |

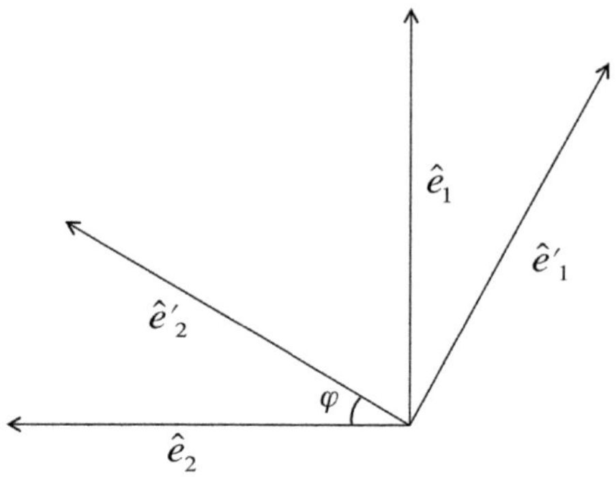


**Fig. 2.4** Rotation of the basis vectors, where $\hat{e}'_1 \times \hat{e}'_2 = \hat{e}_1 \times \hat{e}_2$.

The field intensity can be measured to determine the polarization components. The Stokes vector proposed by G.G. Stokes in 1852 is defined based on measurable quantities to describe the polarization state of light as follows ( Jackson, 1998):

$$\mathrm{I} = \mathrm{E}_2\mathrm{E}_2^* + \mathrm{E}_1\mathrm{E}_1^* = a_2^2 + a_1^2, \tag{2.1.77a}$$

$$\mathrm{Q} = \mathrm{E}_2\mathrm{E}_2^* - \mathrm{E}_1\mathrm{E}_1^* = a_2^2 - a_1^2, \tag{2.1.77b}$$

$$\mathrm{U} = \mathrm{E}_2\mathrm{E}_1^* + \mathrm{E}_1\mathrm{E}_2^* = 2a_2a_1\cos(\delta_2 - \delta_1), \tag{2.1.77c}$$

$$\mathrm{V} = i\left(\mathrm{E}_2\mathrm{E}_1^* - \mathrm{E}_1\mathrm{E}_2^*\right) = 2a_2a_1\sin(\delta_2 - \delta_1). \tag{2.1.77d}$$

The four parameters can be written as a column vector as $(\mathrm{I}, \mathrm{Q}, \mathrm{U}, \mathrm{V})^T$, which is called the Stokes vector. Comparing Eq. (2.1.77a)–(2.1.77d) with Eq. (2.1.69), the constant factor $\frac{1}{2}\sqrt{\frac{\varepsilon}{\mu}}$ is not important because relative quantities are usually measured (Bohren and Huffman, 1983). Since Eq. (2.1.77a)–(2.1.77d) have three independent variables, $a_2$, $a_1$, and $\delta_2 - \delta_1$, the four parameters for polarized light are not independent and satisfy the relation:

$$\mathrm{I}^2 = \mathrm{Q}^2 + \mathrm{U}^2 + \mathrm{V}^2. \tag{2.1.78}$$

Table 2.3 shows different polarization states using the Stokes vector, where parameter I is normalized to unity because of Eq. (2.1.78).

It is apparent that the linear polarization state depends on the choice of the orthogonal basis vectors, but the circular polarization state does not. If the basis vectors rotate at a clockwise angle $\varphi$ as in Fig. 2.4, the Stokes vector with respect to the new basis vectors can be given as

$$\begin{pmatrix} \mathrm{I}' \\ \mathrm{Q}' \\ \mathrm{U}' \\ \mathrm{V}' \end{pmatrix} = \begin{pmatrix} 1 & 0 & 0 & 0 \\ 0 & \cos 2\varphi & \sin 2\varphi & 0 \\ 0 & -\sin 2\varphi & \cos 2\varphi & 0 \\ 0 & 0 & 0 & 1 \end{pmatrix} \begin{pmatrix} \mathrm{I} \\ \mathrm{Q} \\ \mathrm{U} \\ \mathrm{V} \end{pmatrix}. \tag{2.1.79}$$

**Table 2.3** Stokes vectors of different polarization states.

| | | | |
|---|---|---|---|
| Linear polarization | $\begin{pmatrix} 1 \\ \cos 2\eta \\ \sin 2\eta \\ 0 \end{pmatrix}$ | $\eta = 0° \Leftrightarrow (1,1,0,0)^T$<br>$\eta = 45° \Leftrightarrow (1,0,1,0)^T$ | $\eta = 90° \Leftrightarrow (1,-1,0,0)^T$<br>$\eta = -45° \Leftrightarrow (1,0,-1,0)^T$ |
| Right-circular polarization | $(1,0,0,1)^T$ | Left-circular polarization | $(1,0,0,-1)^T$ |

The primed Stokes vector uses the primed basis vectors as in Fig. 2.4. The relation in Eq. (2.1.78) still holds for the primed Stokes vector. Through Eq. (2.1.79), it is evident that three quantities, I, V, and $Q^2 + U^2$, are invariant under the rotation of basis vectors (Bohren and Huffman, 1983).

The polarization description of the electric field earlier applies to a time-harmonic or monochromatic condition. For a quasimonochromatic field, the Stokes parameters are defined by the time average over the measurable quantities as follows (Bohren and Huffman, 1983; Mishchenko et al., 2002):

$$\mathrm{I} = \langle \mathrm{E}_2\mathrm{E}_2^* \rangle + \langle \mathrm{E}_1\mathrm{E}_1^* \rangle = \langle a_2^2 \rangle + \langle a_1^2 \rangle, \tag{2.1.80a}$$

$$\mathrm{Q} = \langle \mathrm{E}_2\mathrm{E}_2^* \rangle - \langle \mathrm{E}_1\mathrm{E}_1^* \rangle = \langle a_2^2 \rangle - \langle a_1^2 \rangle, \tag{2.1.80b}$$

$$\mathrm{U} = \langle \mathrm{E}_2\mathrm{E}_1^* + \mathrm{E}_1\mathrm{E}_2^* \rangle = 2\langle a_2 a_1 \cos(\delta_2 - \delta_1) \rangle, \tag{2.1.80c}$$

$$\mathrm{V} = \langle i(\mathrm{E}_2\mathrm{E}_1^* - \mathrm{E}_1\mathrm{E}_2^*) \rangle = 2\langle a_2 a_1 \sin(\delta_2 - \delta_1) \rangle. \tag{2.1.80d}$$

where $\langle\rangle$ denotes the time average over the period of the fluctuation. Eq. (2.1.78) in the quasimonochromatic case becomes (Bohren and Huffman, 1983):

$$\mathrm{I}^2 - \mathrm{Q}^2 + \mathrm{U}^2 + \mathrm{V}^2 = 4\left( \langle a_2^2 \rangle \langle a_1^2 \rangle - \langle a_2 a_1 \cos(\delta_2 - \delta_1) \rangle^2 - \langle a_2 a_1 \sin(\delta_2 - \delta_1) \rangle^2 \right) \geq 0, \tag{2.1.81}$$

where the equality only holds for polarized light. For unpolarized light, $\mathrm{Q} = \mathrm{U} = \mathrm{V} = 0$. A quasimonochromatic field is usually partially polarized, and the corresponding Stokes vector can be decomposed into the sum of Stokes vectors of the polarized field and of the unpolarized field as follows (Mishchenko et al., 2002):

$$\begin{pmatrix} \mathrm{I} \\ \mathrm{Q} \\ \mathrm{U} \\ \mathrm{V} \end{pmatrix} = \begin{pmatrix} \sqrt{\mathrm{Q}^2 + \mathrm{U}^2 + \mathrm{V}^2} \\ \mathrm{Q} \\ \mathrm{U} \\ \mathrm{V} \end{pmatrix} + \begin{pmatrix} \mathrm{I} - \sqrt{\mathrm{Q}^2 + \mathrm{U}^2 + \mathrm{V}^2} \\ 0 \\ 0 \\ 0 \end{pmatrix}. \tag{2.1.82}$$

For a quasimonochromatic field, the degree of polarization (DOP) can be defined as

$$\mathrm{DOP} = \frac{\sqrt{\mathrm{Q}^2 + \mathrm{U}^2 + \mathrm{V}^2}}{\mathrm{I}}. \tag{2.1.83}$$

Using definitions in Table 2.3, the degree of linear polarization (DOLP) and the degree of circular polarization (DOCP) can be defined as

$$\mathrm{DOLP} = \frac{\sqrt{\mathrm{Q}^2 + \mathrm{U}^2}}{\mathrm{I}}, \tag{2.1.84}$$

$$\mathrm{DOCP} = \frac{\mathrm{V}}{\mathrm{I}}. \tag{2.1.85}$$

It is evident that $\mathrm{DOP} \geq 0$ and $\mathrm{DOLP} \geq 0$. A positive DOCP indicates a right-circularly polarized component, while a negative DOCP indicates a left-circularly polarized component.

When a polarizer or a scatterer is applied to polarized light, the polarization state of the light is changed after the interaction. A mathematical theory was developed by Jones (1941, 1947) and Mueller (1948) to replace the polarization change with a linear transformation. The linear transformation between the incident and the transmitted Jones vectors is called the *Jones matrix* ( Jones, 1941, 1947). The linear transformation between the incident Stokes vector and the transmitted Stokes vector is called the Mueller matrix (Perrin, 1942; Mueller, 1948; Parke, 1948). In this subsection, using **J** (not current density) as the Jones matrix and **M** as the corresponding Mueller matrix, **M** can be expressed using **J** as follows (Parke III, 1949; Barakat, 1981):

$$\mathbf{M} = \mathbf{A}(\mathbf{J} \otimes \mathbf{J}^*)\mathbf{A}^{-1}, \tag{2.1.86}$$

where "⊗" indicates the tensor product and **A** is the following constant matrix:

$$\mathbf{A} = \begin{pmatrix} 1 & 0 & 0 & 1 \\ 1 & 0 & 0 & -1 \\ 0 & 1 & 1 & 0 \\ 0 & i & -i & 0 \end{pmatrix}, \tag{2.1.87}$$

$\mathbf{A}^{-1} = \frac{1}{2}\mathbf{A}^{\dagger}$, and the dagger † indicates the complex conjugate and transpose operation.

A rotation of basis vectors gives a new Jones vector and a new Stokes vector, and the transformation matrices in Eqs. (2.1.76), (2.1.79) can be treated as a Jones matrix and a Mueller matrix, respectively, with respect to the rotation of basis vectors. It is straightforward to verify that the Jones and the Mueller matrices under the rotation of basis vectors satisfy Eq. (2.1.86). We can use a polarizer as an example to discuss the Jones matrix and the Mueller matrix. Fig. 2.5 describes a polarizer applied to polarized light with basis vectors $(\hat{e}_1, \hat{e}_2)$, where the polarization direction of the polarizer has an included angle $\alpha$ with the basis vector $\hat{e}_2$. The subscripts "$i$" and "$t$" are used to denote

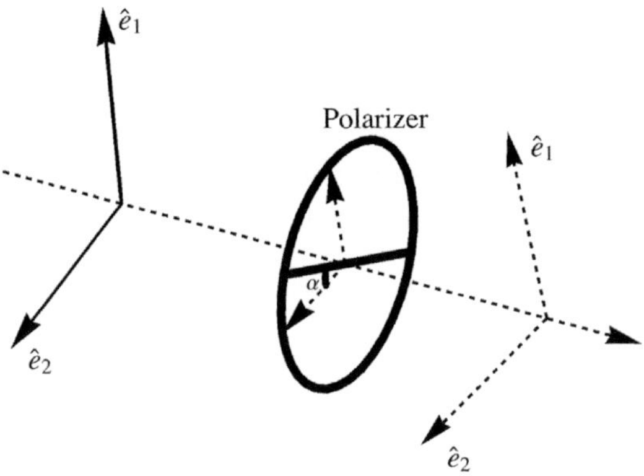


**Fig. 2.5** A polarizer is applied to the incident polarized light. The thick line inside the polarizer is the polarized direction, and angle $\alpha$ is the included angle between the polarized direction and the basis vector $\hat{e}_2$.

the incident and transmitted quantities, respectively. The Jones matrix transforming the incident polarization state to the transmitted polarization state can be given as

$$\begin{pmatrix} E_{2t} \\ E_{1t} \end{pmatrix} = \begin{pmatrix} \cos^2\alpha & \sin\alpha\cos\alpha \\ \sin\alpha\cos\alpha & \sin^2\alpha \end{pmatrix} \begin{pmatrix} E_{2i} \\ E_{1i} \end{pmatrix}, \tag{2.1.88}$$

Using the relation of Mueller-Jones matrices as in Eq. (2.1.86), the Mueller matrix transforming the incident Stokes vector to the transmitted Stokes vector can be given as (Bohren and Huffman, 1983):

$$\begin{pmatrix} I_t \\ Q_t \\ U_t \\ V_t \end{pmatrix} = \frac{1}{2} \begin{pmatrix} 1 & \cos 2\alpha & \sin 2\alpha & 0 \\ \cos 2\alpha & \cos^2 2\alpha & \cos 2\alpha \sin 2\alpha & 0 \\ \sin 2\alpha & \cos 2\alpha \sin 2\alpha & \sin^2 2\alpha & 0 \\ 0 & 0 & 0 & 0 \end{pmatrix} \begin{pmatrix} I_i \\ Q_i \\ U_i \\ V_i \end{pmatrix}. \tag{2.1.89}$$

Note that the Jones and Mueller matrices are useful because they are independent of the incident polarization state. Consequently, these matrices can be used to study a series of polarization changes.

Fig. 2.6 uses three interposed polarizers as an example to describe the effect of a series of "optical instruments" (polarizers or scatterers) on polarized light. If there is a

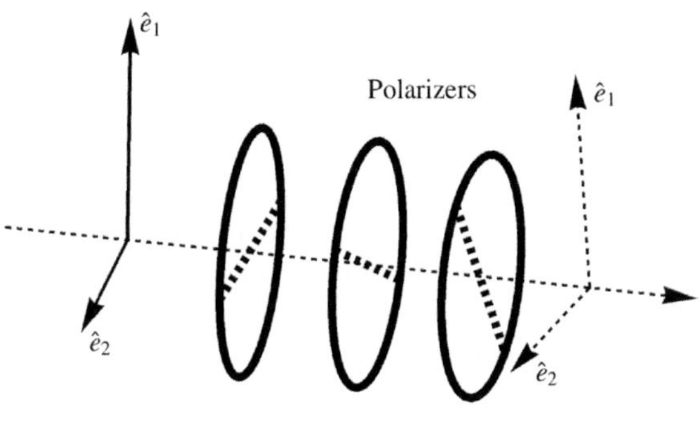


**Fig. 2.6** Polarizers applied to polarized light. The thick dashed lines inside the polarizers are the corresponding polarized directions.

series of optical instruments, the total Jones matrix is the product of the individual Jones matrices in the order of the optical instruments because Maxwell's equations are linear. If there are $N$ optical instruments interposed in a light beam, the Jones matrices and Mueller matrices are, respectively, denoted as $\{\mathbf{J}_i\}$, $i=1, \ldots, N$ and $\{\mathbf{M}_i\}$, $i=1, \ldots, N$ according to the order of the instruments applied to the light. The total Jones matrix $\mathbf{J}$ with respect to the optical instrument system is

$$\mathbf{J}=\mathbf{J}_N\mathbf{J}_{N-1}\cdots\mathbf{J}_2\mathbf{J}_1. \tag{2.1.90}$$

Accordingly, the total Mueller matrix $\mathbf{M}$ of the system can be obtained using Eq. (2.1.86) as follows:

$$\begin{aligned}\mathbf{M}&=\mathbf{A}(\mathbf{J}_N\mathbf{J}_{N-1}\cdots\mathbf{J}_2\mathbf{J}_1)\otimes(\mathbf{J}_N^*\mathbf{J}_{N-1}^*\cdots\mathbf{J}_2^*\mathbf{J}_1^*)\mathbf{A}^{-1}\\&=\mathbf{A}\left(\mathbf{J}_N\otimes\mathbf{J}_N^*\right)\left(\mathbf{J}_{N-1}\otimes\mathbf{J}_{N-1}^*\right)\cdots\left(\mathbf{J}_2\otimes\mathbf{J}_2^*\right)\left(\mathbf{J}_1\otimes\mathbf{J}_1^*\right)\mathbf{A}^{-1}\\&=\mathbf{M}_N\mathbf{M}_{N-1}\cdots\mathbf{M}_2\mathbf{M}_1,\end{aligned} \tag{2.1.91}$$

where the mixed-product property is used as follows:

$$(\mathbf{AB})\otimes(\mathbf{CD})=(\mathbf{A}\otimes\mathbf{C})(\mathbf{B}\otimes\mathbf{D}). \tag{2.1.92}$$

For a series of optical instruments, the collective effect can be described by the products of Jones matrices or the products of Mueller matrices of all instruments in the order of the instruments.

### 2.1.5 SI and Gaussian units

SI units are always assumed in the formulas in this book. In practice, Gaussian units are usually used in electromagnetism because of convenience and succinctness. Especially, electromagnetic expressions in Gaussian units do not involve the use of permittivity $\varepsilon_0$ or permeability $\mu_0$. Gaussian units use the light speed $c$ in vacuum instead, where $c=1/\sqrt{\varepsilon_0\mu_0}$. A conversion Table 2.4 between the SI and Gaussian units is given so the formulas in Gaussian units can be easily obtained by relevant substitutions following Jackson (1998, p. 782).

## 2.2 Basic physical quantities for quantifying light scattering

In Section 2.1.4, the interaction between polarized light and optical instruments is described using the Jones and Mueller matrices, which indicate the linear transformations between the incident and transmitted quantities. In addition, linear polarizers are used as examples to give explicit expressions of the Jones matrix and Mueller matrix. Particulate matter suspended in the air and the ocean acts as "optical instruments" in sunlight. However, the corresponding Jones and Mueller matrices are complex and depend on the morphology, size, refractive index, and orientation of the particles. This

**Table 2.4** Conversion table between SI and Gaussian units.

| Quantity | Conversion between SI (unprimed) and Gaussian (primed) units |
|---|---|
| Electric field | $\mathbf{E}=\frac{1}{\sqrt{4\pi\varepsilon_0}}\mathbf{E}'$ |
| Electric displacement | $\mathbf{D}=\sqrt{\frac{\varepsilon_0}{4\pi}}\mathbf{D}'$ |
| Magnetic induction | $\mathbf{B}=\sqrt{\frac{\mu_0}{4\pi}}\mathbf{B}'$ |
| Magnetic field | $\mathbf{H}=\frac{1}{\sqrt{4\pi\mu_0}}\mathbf{H}'$ |
| Electric current density | $\mathbf{J}=\sqrt{4\pi\varepsilon_0}\mathbf{J}'$ |
| Electric charge density | $\rho=\sqrt{4\pi\varepsilon_0}\rho'$ |
| Conductivity | $\sigma=4\pi\varepsilon_0\sigma'$ |
| Electric polarization | $\mathbf{P}=\sqrt{4\pi\varepsilon_0}\mathbf{P}'$ |
| Magnetization | $\mathbf{M}=\sqrt{\frac{4\pi}{\mu_0}}\mathbf{M}'$ |
| Electric susceptibility | $\chi_e=4\pi\chi_e'$ |
| Magnetic susceptibility | $\chi_m=4\pi\chi_m'$ |
| Permittivity | $\varepsilon=\varepsilon_0\varepsilon'$ |
| Permeability | $\mu=\mu_0\mu'$ |
| Light speed in vacuum | $\frac{1}{\sqrt{\varepsilon_0\mu_0}}=c$ |

section derives expressions for physical quantities associated with single-particle scattering and also briefly introduces the bulk single-scattering properties of a collection of particles. In this section, the surrounding medium is assumed to be nonabsorptive.

### 2.2.1 Amplitude scattering matrix and scattering matrix in the linear polarization representation

In the example of light scattering by a single particle shown in Fig. 2.7, a plane wave is considered as the incident wave and can be expressed as

$$\mathbf{E}^{\text{inc}}\left(\vec{r}\right)=\mathbf{E}_0\exp\left(ik\hat{e}_i\cdot\vec{r}\right), \tag{2.2.1}$$

where $k$ is the wave number in the surrounding medium, $\hat{e}_i$ is the propagation direction, and $\mathbf{E}_0$ is the amplitude of the incident wave. As shown in Fig. 2.7, the incident wave exists everywhere, and also, there is a scattered field with respect to the particle corresponding to the incident wave. The scattered wave is a transverse spherical wave in the far field, where $kr\rightarrow\infty$ and can be given as ( Jackson, 1998)

$$\mathbf{E}^{\text{sca}}\left(\vec{r}\right)\Big|_{kr\rightarrow\infty}=\frac{\exp(ikr)}{-ikr}\mathbf{E}_1^{\text{sca}}(\hat{r}), \tag{2.2.2}$$

where $\mathbf{E}_1^{\text{sca}}(\hat{r})$ is the scattering amplitude and $\hat{r}$ is the scattering direction.

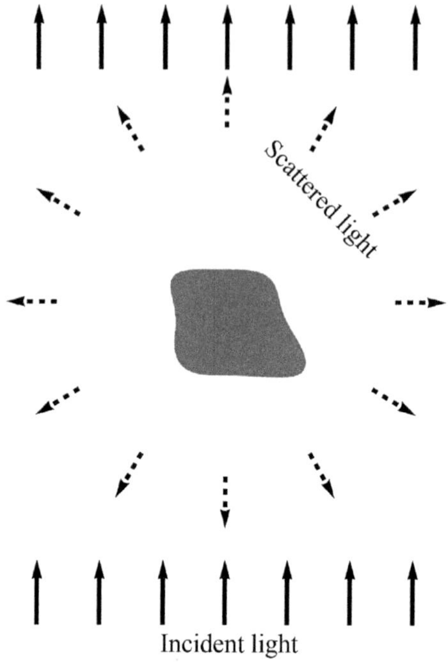


**Fig. 2.7** Light scattering by a dielectric particle (shaded area). The solid arrows denote the incident light beam, and the dashed arrows denote the scattered light. The incident light is everywhere, though not shown in the figure.

As described in Section 2.1.4, the characteristics of the polarization of the incident and scattered fields can be described by the Jones and Mueller matrices with respect to orthogonal basis vectors. Fig. 2.8 shows a laboratory frame of reference $o$-$xyz$, where the incident direction $\hat{e}_i$ is along the $z$-axis. The scattering direction $\hat{r}$ is described by the zenith and azimuthal angles $(\theta, \varphi)$. The plane spanned by the incident direction and the scattering direction is called the scattering plane as marked in Fig. 2.8. Accordingly, the incident and scattered amplitudes can be decomposed into the perpendicular and parallel components relative to the scattering plane as follows:

$$\hat{e}_\perp = \frac{\hat{r}\times\hat{e}_i}{|\hat{r}\times\hat{e}_i|}, \quad \hat{e}_\perp \times \hat{e}_{\|i} = \hat{e}_i, \quad \hat{e}_\perp \times \hat{e}_{\|s} = \hat{r}, \tag{2.2.3a}$$

$$\hat{e}_\perp = -\hat{\varphi}, \quad \hat{e}_{\|s} = \hat{\theta}, \tag{2.2.3b}$$

where as shown in Fig. 2.8, $\hat{e}_\perp$ is the unit vector perpendicular to the scattering plane and $\hat{e}_{\|i}$ and $\hat{e}_{\|s}$ are the unit vectors parallel to the scattering plane corresponding to the incident and scattering directions, respectively; $\hat{\theta}$ and $\hat{\varphi}$ are the unit vectors of the zenith and azimuthal directions of the scattering direction in spherical coordinates.

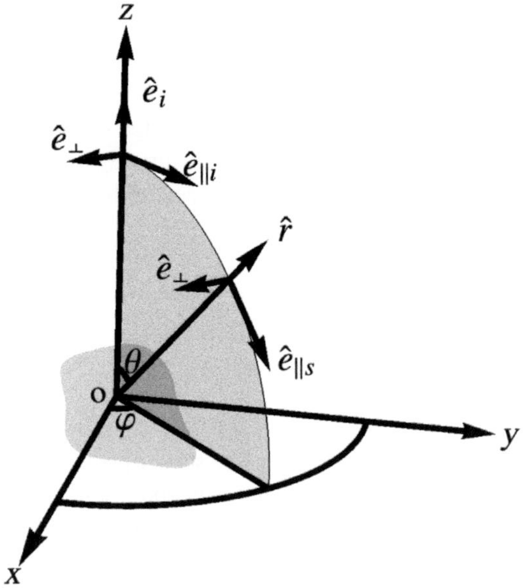


**Fig. 2.8** Scattering plane and the perpendicular and parallel components of the incident and scattering directions with respect to the scattering plane. The shaded area centered at the origin denotes the scattering particle.

If the scattering direction is in the exact forward or the exact backward directions, where $\hat{r} = \pm\hat{e}_i$, any plane containing the $z$-axis can be taken as a scattering plane.

Corresponding to the basis vectors defined in Section 2.1.4, the perpendicular ($\hat{e}_\perp$) and the parallel ($\hat{e}_\|$) components are analogs of components $\hat{e}_1$ and $\hat{e}_2$, respectively. Accordingly, the polarization states with respect to the parallel and perpendicular components become

$$\mathbf{E}_\| = \hat{e}_\| \cdot \mathbf{E}, \quad \mathbf{E}_\perp = \hat{e}_\perp \cdot \mathbf{E}, \tag{2.2.4}$$

and the Stokes parameters with respect to the parallel and perpendicular components become

$$\mathrm{I} = \mathrm{E}_\| \mathrm{E}_\|^* + \mathrm{E}_\perp \mathrm{E}_\perp^*, \tag{2.2.5}$$

$$\mathrm{Q} = \mathrm{E}_\| \mathrm{E}_\|^* - \mathrm{E}_\perp \mathrm{E}_\perp^*, \tag{2.2.6}$$

$$\mathrm{U} = \mathrm{E}_\| \mathrm{E}_\perp^* + \mathrm{E}_\perp \mathrm{E}_\|^*, \tag{2.2.7}$$

$$\mathrm{V} = i\left(\mathrm{E}_\| \mathrm{E}_\perp^* - \mathrm{E}_\perp \mathrm{E}_\|^*\right), \tag{2.2.8}$$

where the components and parameters are valid for both incident and scattered fields. Analogous to the general Jones matrix in Eq. (2.1.76), the amplitude scattering matrix **S** with respect to light scattering in the far field can be given as

$$\begin{pmatrix} E_{\|s} \\ E_{\perp s} \end{pmatrix} = \frac{\exp(ikr)}{-ikr} \begin{pmatrix} S_{11} & S_{12} \\ S_{21} & S_{22} \end{pmatrix} \begin{pmatrix} E_{\|i} \\ E_{\perp i} \end{pmatrix}, \tag{2.2.9}$$

where an explicit dependence on the incident or scattered fields is denoted using the subscripts "$i$" and "$s$," respectively. Note that here **S** is the amplitude scattering matrix, not the Poynting vector. Moreover, the scattering matrix **F** with respect to light scattering in the far field can be given as

$$\begin{pmatrix} I_s \\ Q_s \\ U_s \\ V_s \end{pmatrix} = \frac{1}{(kr)^2} \begin{pmatrix} F_{11} & F_{12} & F_{13} & F_{14} \\ F_{21} & F_{22} & F_{23} & F_{24} \\ F_{31} & F_{32} & F_{33} & F_{34} \\ F_{41} & F_{42} & F_{43} & F_{44} \end{pmatrix} \begin{pmatrix} I_i \\ Q_i \\ U_i \\ V_i \end{pmatrix}, \tag{2.2.10}$$

where the incident and scattering dependences are denoted by the subscripts "$i$" and "$s$." The scattering matrix **F** can be obtained using the corresponding amplitude scattering matrix **S** as in Eq. (2.1.86), and the formula is restated here (Parke III, 1949; Barakat, 1981):

$$\mathbf{F} = \mathbf{A}(\mathbf{S} \otimes \mathbf{S}^*)\mathbf{A}^{-1}, \tag{2.2.11}$$

where the constant matrix **A** is given in Eq. (2.1.87). Accordingly, the elements of the scattering matrix can be explicitly given by the elements of the amplitude scattering matrix as follows:

$$\begin{pmatrix} F_{11} & F_{12} \\ F_{21} & F_{22} \end{pmatrix} = \begin{pmatrix} \frac{1}{2}(|S_{11}|^2 + |S_{12}|^2 + |S_{21}|^2 + |S_{22}|^2) & \frac{1}{2}(|S_{11}|^2 - |S_{12}|^2 + |S_{21}|^2 - |S_{22}|^2) \\ \frac{1}{2}(|S_{11}|^2 + |S_{12}|^2 - |S_{21}|^2 - |S_{22}|^2) & \frac{1}{2}(|S_{11}|^2 - |S_{12}|^2 - |S_{21}|^2 + |S_{22}|^2) \end{pmatrix}, \tag{2.2.12a}$$

$$\begin{pmatrix} F_{13} & F_{14} \\ F_{23} & F_{24} \end{pmatrix} = \begin{pmatrix} Re\left[S_{11}S_{12}^* + S_{21}S_{22}^*\right] & Im\left[S_{11}S_{12}^* + S_{21}S_{22}^*\right] \\ Re\left[S_{11}S_{12}^* - S_{21}S_{22}^*\right] & Im\left[S_{11}S_{12}^* - S_{21}S_{22}^*\right] \end{pmatrix}, \tag{2.2.12b}$$

$$\begin{pmatrix} F_{31} & F_{32} \\ F_{41} & F_{42} \end{pmatrix} = \begin{pmatrix} Re\left[S_{11}S_{21}^* + S_{12}S_{22}^*\right] & Re\left[S_{11}S_{21}^* - S_{12}S_{22}^*\right] \\ -Im\left[S_{11}S_{21}^* + S_{12}S_{22}^*\right] & -Im\left[S_{11}S_{21}^* - S_{12}S_{22}^*\right] \end{pmatrix}, \tag{2.2.12c}$$

$$\begin{pmatrix} F_{33} & F_{34} \\ F_{43} & F_{44} \end{pmatrix} = \begin{pmatrix} Re\left[S_{11}S_{22}^* + S_{12}S_{21}^*\right] & Im\left[S_{11}S_{22}^* - S_{12}S_{21}^*\right] \\ -Im\left[S_{11}S_{22}^* + S_{12}S_{21}^*\right] & Re\left[S_{11}S_{22}^* - S_{12}S_{21}^*\right] \end{pmatrix}. \tag{2.2.12d}$$

The elements of the amplitude scattering matrix **S** are complex, while the elements of the scattering matrix **F** are real. There are eight independent parameters in the amplitude scattering matrix, and only the relative phase of the elements of the amplitude scattering matrix in Eqs. (2.2.12a)–(2.2.12d) is used. Accordingly, the 16 elements of the scattering matrix are not independent so there must be nine different relations connecting them (Abhyankar and Fymat, 1969; Fry and Kattawar, 1981). Element $F_{11}$ indicates the energy distribution after scattering. All elements except $F_{11}$ have the relations $|F_{ij}|/F_{11} \leq 1$, where $i, j = 1, 2, 3, 4$. With respect to the scattering plane determined by the incident direction and the scattering direction, all components and elements are given earlier in Eqs. (2.2.12a)–(2.2.12d). The scattering plane representation is usually used because the scattering matrix can show some similar characteristics.

The aforementioned quantities are described in terms of the scattering plane. The scattering plane representation is comprehensively used and accepted by the light-scattering community. A meridional plane representation is another way to describe light scattering. The incident and scattering meridional planes are, respectively, constructed by the $z$-axis and the incident direction and the $z$-axis and the scattering direction, where the directions starting from the origin are shown in Fig. 2.9. If the incident or the scattering direction is aligned with the $z$-axis, the meridional plane can be any plane containing the $z$-axis. The meridional plane representation is thoroughly introduced by Mishchenko et al. (2002). Although the component definitions are slightly different from the ones in Eq. (2.1.76), we still follow the definitions given by Mishchenko et al. (2002), but we shall point out the differences. When the incident direction is along the $z$-axis as shown in Fig. 2.9, the scattering plane and the meridional planes are set up to be overlapped, and the direct relations of the quantities associated with the two representations are pointed out. If the incident direction and the scattering direction are arbitrary, the components are given as follows:

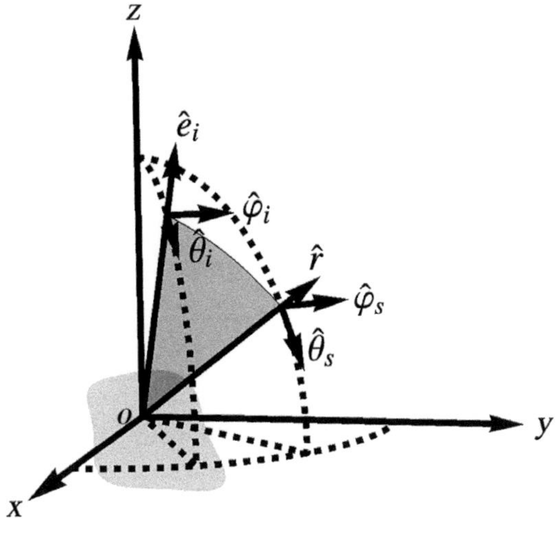


**Fig. 2.9** The zenith and azimuthal components of the incident and scattering directions with respect to the corresponding meridional planes. The scattering plane spanned by the incident and scattering directions is given in the figure. The shaded area centered at the origin denotes the scattering particle.

$$\mathrm{E}_\theta = \hat{\theta}\cdot\mathbf{E},\ \ \mathrm{E}_\varphi = \hat{\varphi}\cdot\mathbf{E}, \tag{2.2.13}$$

where $\hat{\theta}$ and $\hat{\varphi}$ are the unit vectors of the zenith and azimuthal directions in spherical coordinates, and consequently, $\hat{\theta}_i \times \hat{\varphi}_i = \hat{e}_i$, and $\hat{\theta}_s \times \hat{\varphi}_s = \hat{r}$. The scattering angle is the angle between the incident and the scattering directions, and the cosine of the scattering angle can be given as the dot product of two unit vectors pointing along the incident and the scattering directions, that is, $\cos\Theta = \hat{e}_i \cdot \hat{r}$, where $\Theta$ denotes the scattering angle. As shown in Fig. 2.9, the amplitude scattering matrix $\mathbb{S}$ in the far field with respect to the meridional planes of the incident and scattering directions can be given as

$$\left.\begin{pmatrix}\mathrm{E}_{\theta s}\\ \mathrm{E}_{\varphi s}\end{pmatrix}\right|_{kr\to\infty} = \frac{\exp(ikr)}{-ikr}\begin{pmatrix}\mathbb{S}_{11} & \mathbb{S}_{12}\\ \mathbb{S}_{21} & \mathbb{S}_{22}\end{pmatrix}\begin{pmatrix}\mathrm{E}_{\theta i}\\ \mathrm{E}_{\varphi i}\end{pmatrix}. \tag{2.2.14}$$

Note that in this chapter, we use the amplitude scattering matrix notation $\mathbb{S}$ with respect to the meridional planes instead of notation $\mathbf{S}$ used for the scattering plane, while in following chapters notation $\mathbf{S}$ is used for both representations. To distinguish the quantities with respect to the meridional planes from the quantities with respect to the scattering planes, blackboard bold letters are used in this chapter to denote the quantities with respect to the meridional planes.

If we define one of the components along $-\hat{\varphi}$ as done in Eq. (2.2.3b), the definition of the Jones matrix can be fitted into the general definition in Eq. (2.1.76) as

$$\hat{e}_1 = -\hat{\varphi},\ \ \hat{e}_2 = \hat{\theta}. \tag{2.2.15}$$

It is apparent that

$$\begin{pmatrix}\mathrm{E}_\theta\\ \mathrm{E}_{-\varphi}\end{pmatrix} = \mathbb{S}_0\begin{pmatrix}\mathrm{E}_\theta\\ \mathrm{E}_\varphi\end{pmatrix}, \tag{2.2.16a}$$

$$\mathbb{S}_0 = \begin{pmatrix}1 & 0\\ 0 & -1\end{pmatrix}. \tag{2.2.16b}$$

Using Eqs. (2.2.15), (2.1.77a)–(2.1.77d), the Stokes parameters with respect to the meridional planes become

$$\mathbb{I} = \mathrm{E}_\theta\mathrm{E}_\theta^* + \mathrm{E}_\varphi\mathrm{E}_\varphi^*, \tag{2.2.17}$$

$$\mathbb{Q} = \mathrm{E}_\theta\mathrm{E}_\theta^* - \mathrm{E}_\varphi\mathrm{E}_\varphi^*, \tag{2.2.18}$$

$$\mathbb{U} = -\mathrm{E}_\theta\mathrm{E}_\varphi^* - \mathrm{E}_\varphi\mathrm{E}_\theta^*, \tag{2.2.19}$$

$$\mathbb{V} = i\left(-\mathrm{E}_\theta\mathrm{E}_\varphi^* + \mathrm{E}_\varphi\mathrm{E}_\theta^*\right). \tag{2.2.20}$$

When the incident wave direction is along the $z$-axis, the meridional planes both overlap with the scattering plane. Using Eqs. (2.2.16a), (2.2.16b), the amplitude scattering matrices $\mathbf{S}$ and $\mathbb{S}$ have the following relations:

$$\mathbf{S} = \begin{pmatrix}1 & 0\\ 0 & -1\end{pmatrix}\mathbb{S}\begin{pmatrix}1 & 0\\ 0 & -1\end{pmatrix} = \begin{pmatrix}\mathbb{S}_{11} & -\mathbb{S}_{12}\\ -\mathbb{S}_{21} & \mathbb{S}_{22}\end{pmatrix}. \tag{2.2.21}$$

Moreover, the definitions of the Stokes parameters with respect to the meridional planes in Eqs. (2.2.17)–(2.2.20) are consistent with those with respect to the scattering plane in Eqs. (2.2.5)–(2.2.8).

The scattering matrix $\mathbb{F}$ with respect to the meridional planes becomes

$$\begin{pmatrix} \mathbb{I}_s \\ \mathbb{Q}_s \\ \mathbb{U}_s \\ \mathbb{V}_s \end{pmatrix} = \frac{1}{(kr)^2}\mathbb{F}\begin{pmatrix} \mathbb{I}_i \\ \mathbb{Q}_i \\ \mathbb{U}_i \\ \mathbb{V}_i \end{pmatrix}. \tag{2.2.22}$$

Consequently, the scattering matrix $\mathbb{F}$ and the amplitude scattering matrix $\mathbb{S}$ have the following relation:

$$\mathbb{F} = \mathbf{A}\left(\mathbb{S}_0\mathbb{S}\mathbb{S}_0^{-1}\otimes\left(\mathbb{S}_0\mathbb{S}\mathbb{S}_0^{-1}\right)^*\right)\mathbf{A}^{-1} = \mathbb{A}(\mathbb{S}\otimes\mathbb{S}^*)\mathbb{A}^{-1}, \tag{2.2.23a}$$

where

$$\mathbb{A} = \begin{pmatrix} 1 & 0 & 0 & 1 \\ 1 & 0 & 0 & -1 \\ 0 & -1 & -1 & 0 \\ 0 & -i & i & 0 \end{pmatrix}, \quad \mathbb{A}^{-1} = \frac{1}{2}\mathbb{A}^{\dagger}. \tag{2.2.23b}$$

## 2.2.2 Amplitude scattering matrix and scattering matrix in the circular polarization representation

The polarization state can also be described by using the circular polarization basis vectors in Eq. (2.1.73). Similar to the treatment in Section 2.2.1, the quantities with respect to the scattering plane and the meridional planes are, respectively, discussed.

The basis vectors with respect to the scattering plane are given by

$$\hat{e}_{\pm} = \frac{1}{\sqrt{2}}\left(\hat{e}_{\parallel} \pm i\hat{e}_{\perp}\right). \tag{2.2.24}$$

Consequently, the component forms with respect to the scattering plane are

$$\mathrm{E}_- = \hat{e}_- \cdot \mathbf{E}, \quad \mathrm{E}_+ = \hat{e}_+ \cdot \mathbf{E}. \tag{2.2.25}$$

The amplitude scattering matrix $\mathbf{C}$ in the far field in the circular polarization representation with respect to the scattering plane can be given by

$$\begin{pmatrix} \mathrm{E}_{-s} \\ \mathrm{E}_{+s} \end{pmatrix} = \frac{\exp(ikr)}{-ikr}\begin{pmatrix} \mathrm{C}_{--} & \mathrm{C}_{-+} \\ \mathrm{C}_{+-} & \mathrm{C}_{++} \end{pmatrix}\begin{pmatrix} \mathrm{E}_{-i} \\ \mathrm{E}_{+i} \end{pmatrix}. \tag{2.2.26}$$

The transformation of the components from the linear to circular polarization representations is

$$\begin{pmatrix} E_- \\ E_+ \end{pmatrix} = \mathbf{t}_s \begin{pmatrix} E_\parallel \\ E_\perp \end{pmatrix}, \tag{2.2.27}$$

where the transformation matrix $\mathbf{t}_s$ is a unitary matrix

$$\mathbf{t}_s = \frac{1}{\sqrt{2}} \begin{pmatrix} 1 & -i \\ 1 & i \end{pmatrix}, \quad \mathbf{t}_s \mathbf{t}_s^\dagger = \mathbf{I}. \tag{2.2.28}$$

The "†" denotes the complex conjugate and transpose, and **I** is a unit matrix. The transformation of the amplitude scattering matrix can be obtained by

$$\begin{aligned} \begin{pmatrix} C_{--} & C_{-+} \\ C_{+-} & C_{++} \end{pmatrix} &= \mathbf{t}_s \begin{pmatrix} S_{11} & S_{12} \\ S_{21} & S_{22} \end{pmatrix} \mathbf{t}_s^{-1} \\ &= \frac{1}{2} \begin{pmatrix} S_{11} + S_{12}i - S_{21}i + S_{22} & S_{11} - S_{12}i - S_{21}i - S_{22} \\ S_{11} + S_{12}i + S_{21}i - S_{22} & S_{11} - S_{12}i + S_{21}i + S_{22} \end{pmatrix}. \end{aligned} \tag{2.2.29}$$

The complex circularly polarized Stokes parameters $\mathbf{I}^{CP}$ with respect to the scattering plane are defined as

$$\mathbf{I}^{CP} = \begin{pmatrix} I_2 \\ I_0 \\ I_{-0} \\ I_{-2} \end{pmatrix} = \begin{pmatrix} E_+E_-^* \\ E_-E_-^* \\ E_+E_+^* \\ E_-E_+^* \end{pmatrix} = \frac{1}{2} \begin{pmatrix} Q + iU \\ I + V \\ I - V \\ Q - iU \end{pmatrix} = \mathbf{B}\mathbf{I}^{LP}, \tag{2.2.30}$$

where

$$\mathbf{B} = \frac{1}{2} \begin{pmatrix} 0 & 1 & i & 0 \\ 1 & 0 & 0 & 1 \\ 1 & 0 & 0 & -1 \\ 0 & 1 & -i & 0 \end{pmatrix}, \tag{2.2.31a}$$

$$\mathbf{B}^{-1} = \begin{pmatrix} 0 & 1 & 1 & 0 \\ 1 & 0 & 0 & 1 \\ -i & 0 & 0 & i \\ 0 & 1 & -1 & 0 \end{pmatrix}, \tag{2.2.31b}$$

and $E_-$ and $E_+$ are components with respect to the scattering plane and are defined in Eqs. (2.2.24), (2.2.25); $\mathbf{I}^{LP}$ is the Stokes vector with respect to the scattering plane defined in Eqs. (2.2.5)–(2.2.8); and $\mathbf{I}^{CP}$ is complex and different from conventional real Stokes vectors. Similarly, one can still define the complex scattering matrix as

$$\mathbf{I}_s^{CP} = \frac{1}{(kr)^2} \mathbf{F}^{CP} \mathbf{I}_i^{CP}. \tag{2.2.32}$$

Using Eq. (2.2.30), the complex scattering matrix can be given in terms of the scattering matrix in the linear polarization representation as

$$\mathbf{F}^{\mathrm{CP}} = \mathbf{BFB}^{-1} = \mathbf{A}_{\mathrm{CP}}(\mathbf{S}\otimes\mathbf{S}^{*})\mathbf{A}_{\mathrm{CP}}^{-1}, \tag{2.2.33}$$

where

$$\mathbf{A}_{\mathrm{CP}} = \mathbf{BA} = \frac{1}{2}\begin{pmatrix} 1 & i & i & -1 \\ 1 & i & -i & 1 \\ 1 & -i & i & 1 \\ 1 & -i & -i & -1 \end{pmatrix}, \tag{2.2.34a}$$

$$\mathbf{A}_{\mathrm{CP}}^{-1} = (\mathbf{BA})^{-1} = \mathbf{A}_{\mathrm{CP}}^{\dagger}. \tag{2.2.34b}$$

The constant matrix $\mathbf{A}_{\mathrm{CP}}$ is a unitary matrix. Using Eqs. (2.2.26), (2.2.30), the complex scattering matrix $\mathbf{F}^{\mathrm{CP}}$ can be directly given in terms of the amplitude scattering matrix $\mathbf{C}$ in the circular polarization representation with respect to the scattering plane as follows:

$$\mathbf{F}^{\mathrm{CP}} = \begin{pmatrix} \mathrm{C}_{++}\mathrm{C}_{--}^{*} & \mathrm{C}_{+-}\mathrm{C}_{--}^{*} & \mathrm{C}_{++}\mathrm{C}_{-+}^{*} & \mathrm{C}_{+-}\mathrm{C}_{-+}^{*} \\ \mathrm{C}_{-+}\mathrm{C}_{--}^{*} & \mathrm{C}_{--}\mathrm{C}_{--}^{*} & \mathrm{C}_{-+}\mathrm{C}_{-+}^{*} & \mathrm{C}_{--}\mathrm{C}_{-+}^{*} \\ \mathrm{C}_{++}\mathrm{C}_{+-}^{*} & \mathrm{C}_{+-}\mathrm{C}_{+-}^{*} & \mathrm{C}_{++}\mathrm{C}_{++}^{*} & \mathrm{C}_{+-}\mathrm{C}_{++}^{*} \\ \mathrm{C}_{-+}\mathrm{C}_{+-}^{*} & \mathrm{C}_{--}\mathrm{C}_{+-}^{*} & \mathrm{C}_{-+}\mathrm{C}_{++}^{*} & \mathrm{C}_{--}\mathrm{C}_{++}^{*} \end{pmatrix}. \tag{2.2.35}$$

In the circular polarization representation, the basis vectors and the component forms with respect to the meridional plane are given by

$$\hat{e}_{\pm} = \frac{1}{\sqrt{2}}\left(\hat{e}_{\theta} \mp i\hat{e}_{\varphi}\right), \tag{2.2.36}$$

$$\mathrm{E}_{-} = \hat{e}_{-}\cdot\mathbf{E}, \quad \mathrm{E}_{+} = \hat{e}_{+}\cdot\mathbf{E}, \tag{2.2.37}$$

where the same notation is used for the basis vectors and components but with respect to the meridional plane. The amplitude scattering matrix $\mathbb{C}$ in the far field in circular polarization representation with respect to the meridional planes can be given by

$$\begin{pmatrix} \mathrm{E}_{-s} \\ \mathrm{E}_{+s} \end{pmatrix} = \frac{\exp(ikr)}{-ikr}\begin{pmatrix} \mathbb{C}_{--} & \mathbb{C}_{-+} \\ \mathbb{C}_{+-} & \mathbb{C}_{++} \end{pmatrix}\begin{pmatrix} \mathrm{E}_{-i} \\ \mathrm{E}_{+i} \end{pmatrix}. \tag{2.2.38}$$

Correspondingly, the relation of the amplitude scattering matrix with respect to the meridional planes in circular polarization and linear polarization representations is (Mishchenko et al., 2002)

$$\begin{pmatrix} \mathbb{C}_{--} & \mathbb{C}_{-+} \\ \mathbb{C}_{+-} & \mathbb{C}_{++} \end{pmatrix} = \frac{1}{2}\begin{pmatrix} \mathbb{S}_{11} - \mathbb{S}_{12}i + \mathbb{S}_{21}i + \mathbb{S}_{22} & \mathbb{S}_{11} + \mathbb{S}_{12}i + \mathbb{S}_{21}i - \mathbb{S}_{22} \\ \mathbb{S}_{11} - \mathbb{S}_{12}i - \mathbb{S}_{21}i - \mathbb{S}_{22} & \mathbb{S}_{11} + \mathbb{S}_{12}i - \mathbb{S}_{21}i + \mathbb{S}_{22} \end{pmatrix}. \tag{2.2.39}$$

The complex circular polarized Stokes parameters with respect to the meridional planes are defined as (Mishchenko et al., 2002)

$$\mathbb{I}^{CP} = \begin{pmatrix} \mathbb{I}_2 \\ \mathbb{I}_0 \\ \mathbb{I}_{-0} \\ \mathbb{I}_{-2} \end{pmatrix} = \begin{pmatrix} E_+E_-^* \\ E_-E_-^* \\ E_+E_+^* \\ E_-E_+^* \end{pmatrix} = \frac{1}{2}\begin{pmatrix} \mathbb{Q}+i\mathbb{U} \\ \mathbb{I}+\mathbb{V} \\ \mathbb{I}-\mathbb{V} \\ \mathbb{Q}-i\mathbb{U} \end{pmatrix} = \mathbf{B}\mathbb{I}^{LP}, \tag{2.2.40}$$

where $E_-$ and $E_+$ components with respect to the meridional planes are defined in Eqs. (2.2.36), (2.2.37), $\mathbb{I}^{LP}$ is the Stokes vector with respect to the scattering plane defined in Eqs. (2.2.17)–(2.2.20), and matrix $\mathbf{B}$ and its inversion are defined in Eqs. (2.2.31a), (2.2.31b).When the incident direction is along the $z$-axis, Eq. (2.2.36) is the same as Eq. (2.2.24). Correspondingly, the amplitude scattering matrix and complex circular polarized Stokes parameters in circular polarization with respect to the meridional planes given in Eqs. (2.2.39), (2.2.40) are the same as the one with respect to the meridional planes given in Eqs. (2.2.29), (2.2.30), respectively. Note that the definitions for components $E_-$ and $E_+$ are reversed between the present expressions and those defined by Mishchenko et al. (2002), that is, the component $E_+$ here is the component $E_-$ in Mishchenko et al. (2002) and vice versa. The complex scattering matrix with respect to the meridional planes is defined as

$$\mathbb{I}_s^{CP} = \frac{1}{(kr)^2}\mathbb{F}^{CP}\mathbb{I}_i^{CP}. \tag{2.2.41}$$

The complex scattering matrix can be given in terms of the corresponding scattering matrix in the linear polarization representation as

$$\mathbb{F}^{CP} = \mathbf{B}\mathbb{F}\mathbf{B}^{-1} = \mathbb{A}_{CP}(\mathbb{S}\otimes\mathbb{S}^*)\mathbb{A}_{CP}^{-1}, \tag{2.2.42}$$

where

$$\mathbb{A}_{CP} = \mathbf{B}\mathbb{A} = \frac{1}{2}\begin{pmatrix} 1 & -i & -i & -1 \\ 1 & -i & i & 1 \\ 1 & i & -i & 1 \\ 1 & i & i & -1 \end{pmatrix}, \tag{2.2.43a}$$

$$\mathbb{A}_{CP}^{-1} = (\mathbf{B}\mathbb{A})^{-1} = \mathbb{A}_{CP}^{\dagger}. \tag{2.2.43b}$$

The complex scattering matrix can be directly given in terms of the amplitude scattering matrix $\mathbb{C}$ and is stated here as (Mishchenko et al., 2002)

$$\mathbb{F}^{CP} = \begin{pmatrix} \mathbb{C}_{++}\mathbb{C}_{--}^* & \mathbb{C}_{+-}\mathbb{C}_{--}^* & \mathbb{C}_{++}\mathbb{C}_{-+}^* & \mathbb{C}_{+-}\mathbb{C}_{-+}^* \\ \mathbb{C}_{-+}\mathbb{C}_{--}^* & \mathbb{C}_{--}\mathbb{C}_{--}^* & \mathbb{C}_{-+}\mathbb{C}_{-+}^* & \mathbb{C}_{--}\mathbb{C}_{-+}^* \\ \mathbb{C}_{++}\mathbb{C}_{+-}^* & \mathbb{C}_{+-}\mathbb{C}_{+-}^* & \mathbb{C}_{++}\mathbb{C}_{++}^* & \mathbb{C}_{+-}\mathbb{C}_{++}^* \\ \mathbb{C}_{-+}\mathbb{C}_{+-}^* & \mathbb{C}_{--}\mathbb{C}_{+-}^* & \mathbb{C}_{-+}\mathbb{C}_{++}^* & \mathbb{C}_{--}\mathbb{C}_{++}^* \end{pmatrix}. \tag{2.2.44}$$

Although the scattering plane and meridional plane representations are both given in this subsection, the scattering plane representation is mostly used in the following sections to discuss the scattering-related quantities. The quantities in corresponding meridional plane representation can be similarly given and the reader can also refer to Mishchenko et al. (2002).

### 2.2.3 Translation of the frame of reference

As shown in Fig. 2.10, for a translation of the frame of reference by a distance shown as vector $\vec{r}_0$, the vector from the origin to the field point p is $\vec{r}_1$ in the frame of reference with origin at $o_1$ and $\vec{r}_2$ in the frame of reference with origin at $o_2$. The field point is in the far-field region, that is, $r_1, r_2 \gg r_0$ and $\hat{r}_1 = \hat{r}_2 = \hat{r}$. The incident field and the scattered field in the far-field zone in Eqs. (2.2.1), (2.2.2) become

$$\mathbf{E}^{\text{inc}}\left(\vec{r}_1\right) = \mathbf{E}_0 \exp\left(ik\hat{e}_i \cdot \vec{r}_1\right) = \exp\left(ik\hat{e}_i \cdot \vec{r}_0\right)\mathbf{E}_0 \exp\left(ik\hat{e}_i \cdot \vec{r}_2\right), \tag{2.2.45}$$

$$\mathbf{E}^{\text{sca}}\left(\vec{r}_1\right)\Big|_{kr_1\to\infty} = \frac{\exp(ikr_1)}{-ikr_1}\mathbf{E}_1^{\text{sca}}(\hat{r}) = \exp(ik\hat{r}\cdot\hat{r}_0)\frac{\exp(ikr_2)}{-ikr_2}\mathbf{E}_1^{\text{sca}}(\hat{r}), \tag{2.2.46}$$

where $\mathbf{E}_1^{\text{sca}}$ is the scattering amplitude in the far-field zone. After the translation, the amplitude scattering matrix in Eq. (2.2.9) has the following phase difference:

$$\mathbf{S}_1 = \exp\left[ik(\hat{e}_i - \hat{r})\cdot\vec{r}_0\right]\mathbf{S}_2, \tag{2.2.47}$$

where $\mathbf{S}_1$ and $\mathbf{S}_2$ are the amplitude scattering matrices in the frames of reference at $o_1$ and $o_2$, respectively. The phase difference depends on the incident and scattering directions. The phase difference vanishes in the exact forward direction. The scattering matrix, however, is invariant under a translation of the frame of reference because

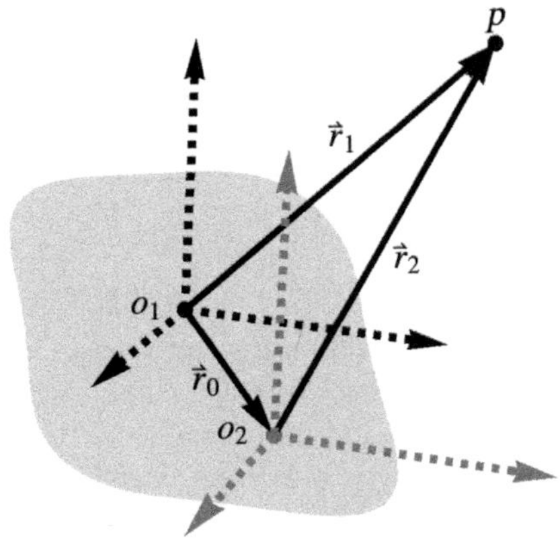


**Fig. 2.10** The translation of the frame of reference. The irregular shaded area is the scattering particle.

the phase difference is canceled by the complex conjugate operations in the computation in Eqs. (2.2.12a)–(2.2.12d).

### 2.2.4 Scattering dyad

In view of the definitions of the Jones matrix in Eqs. (2.2.9), (2.2.14), the scattering dyad in the far field can be defined as

$$\mathbf{E}^{\text{sca}}\left(\vec{r}\right)\Big|_{kr\to\infty} = \frac{\exp(ikr)}{-ikr}\overleftrightarrow{\mathrm{I}}_s \cdot \overleftrightarrow{\mathcal{A}}(\hat{r}, \hat{e}_i) \cdot \overleftrightarrow{\mathrm{I}}_i \cdot \mathbf{E}_0, \tag{2.2.48}$$

where $\overleftrightarrow{\mathcal{A}}(\hat{r}, \hat{e}_i)$ is the 2-D scattering dyad and $\overleftrightarrow{\mathrm{I}}$ is the 2-D unit dyad. The explicit expressions for $\overleftrightarrow{\mathrm{I}}$ depend on the basis vectors. For instance, $\overleftrightarrow{\mathrm{I}} = \hat{e}_{\parallel}\hat{e}_{\parallel} + \hat{e}_{\perp}\hat{e}_{\perp}$ as in Eq. (2.2.9), and $\overleftrightarrow{\mathrm{I}} = \hat{e}_{\theta}\hat{e}_{\theta} + \hat{e}_{\varphi}\hat{e}_{\varphi}$ as in Eq. (2.2.14). Since the incident field and the scattered field in the far-field zone are transverse, one can define that

$$\hat{r} \cdot \overleftrightarrow{\mathcal{A}}(\hat{r}, \hat{e}_i) = \overleftrightarrow{\mathcal{A}}(\hat{r}, \hat{e}_i) \cdot \hat{e}_i = 0. \tag{2.2.49}$$

The elements of the amplitude scattering matrix with respect to the scattering plane and the meridional planes and the scattering dyad have the following relations (Mishchenko et al., 2002):

$$\begin{pmatrix} \mathrm{S}_{11} & \mathrm{S}_{12} \\ \mathrm{S}_{21} & \mathrm{S}_{22} \end{pmatrix} = \begin{pmatrix} \hat{e}_{\parallel s} \cdot \overleftrightarrow{\mathcal{A}}(\hat{r}, \hat{e}_i) \cdot \hat{e}_{\parallel i} & \hat{e}_{\parallel s} \cdot \overleftrightarrow{\mathcal{A}}(\hat{r}, \hat{e}_i) \cdot \hat{e}_{\perp} \\ \hat{e}_{\perp} \cdot \overleftrightarrow{\mathcal{A}}(\hat{r}, \hat{e}_i) \cdot \hat{e}_{\parallel i} & \hat{e}_{\perp} \cdot \overleftrightarrow{\mathcal{A}}(\hat{r}, \hat{e}_i) \cdot \hat{e}_{\perp} \end{pmatrix}, \tag{2.2.50a}$$

$$\begin{pmatrix} \mathbb{S}_{11} & \mathbb{S}_{12} \\ \mathbb{S}_{21} & \mathbb{S}_{22} \end{pmatrix} = \begin{pmatrix} \hat{\theta}_s \cdot \overleftrightarrow{\mathcal{A}}(\hat{r}, \hat{e}_i) \cdot \hat{\theta}_i & \hat{\theta}_s \cdot \overleftrightarrow{\mathcal{A}}(\hat{r}, \hat{e}_i) \cdot \hat{\varphi}_i \\ \hat{\varphi}_s \cdot \overleftrightarrow{\mathcal{A}}(\hat{r}, \hat{e}_i) \cdot \hat{\theta}_i & \hat{\varphi}_s \cdot \overleftrightarrow{\mathcal{A}}(\hat{r}, \hat{e}_i) \cdot \hat{\varphi}_i \end{pmatrix}, \tag{2.2.50b}$$

where the relations correspond to the amplitude scattering matrix in Eqs. (2.2.9), (2.2.14), respectively.

### 2.2.5 Reciprocity and symmetry

#### 2.2.5.1 Reciprocity

Reciprocity in the electromagnetic field results from the time reversal symmetry of Maxwell's equations, which is an intrinsic symmetry and is independent of the characteristics of a particle. Reciprocity of electromagnetic field is valid in a time-invariant linear medium. Reciprocity is presented as applicable to arbitrary particles since only linear media are discussed throughout this book. Reciprocity associated with electromagnetic scattering is comprehensively presented (Blatt and Weisskopf, 1952; Morse and Feshbach, 1953; Silver, 1964; Saxon, 1955; Tsang et al., 2000;

Mishchenko et al., 2002). Saxon (1955) uses a tensor scattering matrix to illustrate the reciprocity in the electromagnetic field. The description of the reciprocity of the amplitude scattering matrix in this subsection follows the methods presented by Saxon (1955) and Mishchenko et al. (2002).

In the far-field distant from a scattering particle, where $kr \to \infty$, the electric field can be decomposed into incoming and outgoing transverse spherical waves as follows:

$$\mathbf{E}\left(\vec{r}\right)\Big|_{kr\to\infty} = \frac{\exp(ikr)}{-ikr}\mathbf{E}_1(\hat{r}) + \frac{\exp(-ikr)}{ikr}\mathbf{E}_2(\hat{r}), \tag{2.2.51}$$

where $\mathbf{E}_1(\hat{r})$ and $\mathbf{E}_2(\hat{r})$ are the amplitudes of the outgoing and incoming spherical waves, respectively, where $\hat{r}\cdot\mathbf{E}_1(\hat{r}) = \hat{r}\cdot\mathbf{E}_2(\hat{r}) = 0$. Since Maxwell's equations are linear, the tensor scattering matrix is defined to describe the linear relation between the amplitudes of the incoming and outgoing spherical waves as follows:

$$\mathbf{E}_1(\hat{r}) = \int_{4\pi} d\Omega' \overleftrightarrow{S}(\hat{r},\hat{r}')\cdot\mathbf{E}_2(-\hat{r}'), \tag{2.2.52}$$

where $d\Omega' = \sin\theta'\, d\theta'\, d\varphi$. Since the amplitudes are transverse, the tensor scattering matrix can be defined as

$$\hat{r}\cdot\overleftrightarrow{S}(\hat{r},\hat{r}') = \overleftrightarrow{S}(\hat{r},\hat{r}')\cdot\hat{r}' = 0. \tag{2.2.53}$$

The free space Maxwell's equations with time dependence $\exp(-i\omega t)$ have two sets of solutions $(\mathbf{E},\mathbf{H})$ and $(\mathbf{E}',\mathbf{H}')$, where primed fields just denote another set of solutions. Using the free space Maxwell's equations, one can get

$$\nabla\cdot(\mathbf{E}'\times\mathbf{H} - \mathbf{E}\times\mathbf{H}') = 0, \tag{2.2.54}$$

which is valid everywhere in free space. Integrating Eq. (2.2.54) over all space and using Gauss's theorem, one can straightforwardly obtain

$$r^2\int_{4\pi} d\Omega\hat{r}\cdot\left\{\mathbf{E}'\left(\vec{r}\right)\times\left[\nabla\times\mathbf{E}\left(\vec{r}\right)\right] - \mathbf{E}\left(\vec{r}\right)\times\left[\nabla\times\mathbf{E}'\left(\vec{r}\right)\right]\right\}\Big|_{kr\to\infty} = 0. \tag{2.2.55}$$

Applying the decompositions in Eq. (2.2.51) to primed and unprimed fields, one can obtain

$$\int_{4\pi} d\Omega\left[\mathbf{E}_2'(\hat{r})\cdot\mathbf{E}_1(\hat{r}) - \mathbf{E}_2(\hat{r})\cdot\mathbf{E}_1'(\hat{r})\right] = 0. \tag{2.2.56}$$

Substituting Eq. (2.2.52) into Eq. (2.2.56), one can get

$$\int_{4\pi} d\Omega\int_{4\pi} d\Omega'\left\{\mathbf{E}_2'(\hat{r})\cdot\left[\overleftrightarrow{S}(\hat{r},\hat{r}') - \overleftrightarrow{S}^T(-\hat{r}', -\hat{r})\right]\cdot\mathbf{E}_2(-\hat{r}')\right\} = 0. \tag{2.2.57}$$

Note that the prime on amplitude $\mathbf{E}_2'$ denotes one of the two solutions from the free space Maxwell's equations, while the prime on direction $\hat{r}'$ denotes the integral elements. Since the two amplitudes $\mathbf{E}_2'(\hat{r})$ and $\mathbf{E}_2(-\hat{r}')$ are arbitrary, the tensor scattering matrix must satisfy (Saxon, 1955; Mishchenko et al., 2002)

$$\overleftrightarrow{\mathcal{S}}(\hat{r}, \hat{r}') = \overleftrightarrow{\mathcal{S}}^T(-\hat{r}', -\hat{r}). \tag{2.2.58}$$

A plane wave in the far field can be decomposed into incoming and outgoing spherical waves as follows (e.g., Mishchenko et al., 2002):

$$\exp(ikr\hat{e}_i \cdot \hat{r})|_{kr\to\infty} = -2\pi\left[\delta(\hat{e}_i + \hat{r})\frac{\exp(-ikr)}{ikr} + \delta(\hat{e}_i - \hat{r})\frac{\exp(ikr)}{-ikr}\right]. \tag{2.2.59}$$

For an incident plane wave, the electric field at $kr \to \infty$ can be given in terms of the scattering dyad:

$$\mathbf{E}(r\hat{r})|_{kr\to\infty} = \frac{\exp(-ikr)}{ikr}\{-2\pi\delta(\hat{e}_i + \hat{r})\mathbf{E}_0\} + \frac{\exp(ikr)}{-ikr}\left[\overleftrightarrow{\mathcal{A}}(\hat{r}, \hat{e}_i) - 2\pi\delta(\hat{e}_i - \hat{r})\overleftrightarrow{\mathrm{I}}_i\right] \cdot \mathbf{E}_0, \tag{2.2.60}$$

where $\overleftrightarrow{\mathrm{I}}_i$ is the 2-D unit dyad. Comparing Eq. (2.2.60) with Eq. (2.2.51), one can show

$$\mathbf{E}_2(\hat{r}) = -2\pi\delta(\hat{e}_i + \hat{r})\mathbf{E}_0, \tag{2.2.61}$$

$$\mathbf{E}_1(\hat{r}) = \left[\overleftrightarrow{\mathcal{A}}(\hat{r}, \hat{e}_i) - 2\pi\delta(\hat{e}_i - \hat{r})\overleftrightarrow{I}_i\right] \cdot \mathbf{E}_0. \tag{2.2.62}$$

Using the definition of the tensor scattering matrix in Eq. (2.2.52), the tensor scattering matrix and the scattering dyad are related as follows:

$$\overleftrightarrow{\mathcal{S}}(\hat{r}, \hat{e}_i) = \delta(\hat{e}_i - \hat{r})\overleftrightarrow{I}_i - \overleftrightarrow{\mathcal{A}}(\hat{r}, \hat{e}_i). \tag{2.2.63}$$

When there is no scattering, the tensor scattering matrix is reduced to a unit operator. It is evident that the scattering dyad has the same reciprocity relation as the tensor scattering matrix (Mishchenko et al., 2002):

$$\overleftrightarrow{\mathcal{A}}(\hat{r}, \hat{e}_i) = \overleftrightarrow{\mathcal{A}}^T(-\hat{e}_i, -\hat{r}). \tag{2.2.64}$$

Using Eqs. (2.2.50a), (2.2.50b), (2.2.64), due to a reciprocal symmetry, the amplitude scattering matrix becomes (Mishchenko et al., 2002)

$$\begin{pmatrix} \mathrm{S}_{11} & \mathrm{S}_{12} \\ \mathrm{S}_{21} & \mathrm{S}_{22} \end{pmatrix} \to \begin{pmatrix} \mathrm{S}_{11} & -\mathrm{S}_{21} \\ -\mathrm{S}_{12} & \mathrm{S}_{22} \end{pmatrix}, \tag{2.2.65a}$$

$$\begin{pmatrix} \mathbb{S}_{11} & \mathbb{S}_{12} \\ \mathbb{S}_{21} & \mathbb{S}_{22} \end{pmatrix} \to \begin{pmatrix} \mathbb{S}_{11} & -\mathbb{S}_{21} \\ -\mathbb{S}_{12} & \mathbb{S}_{22} \end{pmatrix}. \tag{2.2.65b}$$

### 2.2.5.2 Symmetry and random orientation

Before introducing the relation between symmetry and random orientation, we first explain the physical meaning of random orientation. Polarization of a quasimonochromatic beam is discussed in Section 2.1.4, where the degree of polarization (DOP) can be defined. If DOP = 1, the beam is called completely coherent; otherwise, it is partially coherent or incoherent. If two or more quasimonochromatic beams are incoherent, the total Stokes parameter of the beams is just the sum of the individual Stokes parameters:

$$\mathbf{I} = \sum_n \mathbf{I}_n, \tag{2.2.66}$$

where $\mathbf{I}$ denotes the Stokes vector. If a quasimonochromatic beam illuminates a collection of identical particles with different orientations, the emergent beams are incoherent provided that the mean free path of the particles is larger than the coherent length of the quasimonochromatic beam. Accordingly, the collective scattered Stokes vector is merely the sum of the individual scattered Stokes vectors, and each incident Stokes vector is the same as the Stokes vector of the beam:

$$\mathbf{I}_s = \frac{1}{n}\sum_n \mathbf{I}_{ns}, \quad \{\mathbf{I}_{ni}\} = \mathbf{I}_i, \tag{2.2.67}$$

where the subscripts "$s$" and "$i$" denote the scattered and incident Stokes vectors and index $n$ denotes the particle orientation. Similarly, the total scattering matrix can be given by the sum of the individual scattering matrices:

$$\mathbf{F} = \frac{1}{n}\sum_n \mathbf{F}_n. \tag{2.2.68}$$

As discussed after Eqs. (2.2.12a)–(2.2.12d), the single-scattering matrix $\mathbf{F}_n$ must have nine quadratic relations since it is calculated by four complex elements of the amplitude scattering matrix and the relative phase has no effect on the result. After summing the scattering matrices from different orientations, the quadratic relations are lost, so the total scattering matrix $\mathbf{F}$ will in general have 16 independent elements. It is shown that the independent elements of the scattering matrix $\mathbf{F}$ of a particle in random orientation are reduced due to the symmetry.

The symmetry of the scattering matrix in random orientation stems directly from the symmetry of the amplitude scattering matrix. Consequently, the symmetry of the amplitude scattering matrix is first given, and then, the scattering matrix can be straightforwardly obtained.

For an arbitrary particle with different orientations, the symmetry of the scattering matrix is discussed in this section. Here, it is assumed that the scattering direction is not aligned with the incident direction (implying the exact forward or backward scattering). As known, reciprocity is an intrinsic symmetry, so an arbitrary particle has both its original and reciprocal orientations. As shown in Eqs. (2.2.65a), (2.2.65b), the reciprocal orientation gives the similar amplitude scattering matrix elements only

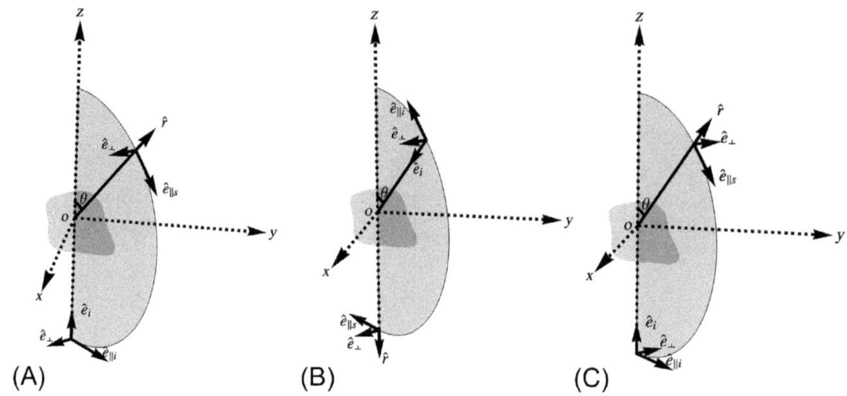


**Fig. 2.11** Light scattering in the original (A), reciprocal (B), and mirror (C) representations. The irregular shaded area denotes the scattering particle. The shaded plane including the incident and the scattering directions illustrates the scattering plane.

with sign and order differences compared with the original orientation. Other symmetries are related to the particle geometry. If a particle has a mirror symmetry, the mirror orientation can also give the similar amplitude scattering matrix elements except for sign differences.

Fig. 2.11 illustrates the original, reciprocal, and mirror representations in terms of the incident and scattering directions and also the perpendicular and parallel components of the electromagnetic field. For clarity of the derivation of the scattering matrix, the amplitude scattering matrix in the original representation as shown in Fig. 2.11A is denoted as

$$\mathbf{S}_o = \begin{pmatrix} S_{11} & S_{12} \\ S_{21} & S_{22} \end{pmatrix}. \tag{2.2.69}$$

The corresponding scattering matrix in original representation can be given using Eqs. (2.2.12a)–(2.2.12d) and denoted here as

$$\mathbf{F}_o = \begin{pmatrix} F_{11} & F_{12} & F_{13} & F_{14} \\ F_{21} & F_{22} & F_{23} & F_{24} \\ F_{31} & F_{32} & F_{33} & F_{34} \\ F_{41} & F_{42} & F_{43} & F_{44} \end{pmatrix}. \tag{2.2.70}$$

Corresponding to the original representation, the amplitude scattering matrix in the reciprocal representation, as shown in Fig. 2.11B, has been derived and given in Eqs. (2.2.65a), (2.2.65b). We restate it here as follows:

$$\mathbf{S}_r = \begin{pmatrix} S_{11} & -S_{21} \\ -S_{12} & S_{22} \end{pmatrix}. \tag{2.2.71}$$

Using Eqs. (2.2.12a)–(2.2.12d), (2.2.71), the corresponding scattering matrix in a reciprocal representation using the same elements as in Eq. (2.2.70) can be given as

$$\mathbf{F}_r = \begin{pmatrix} F_{11} & F_{21} & -F_{31} & F_{41} \\ F_{12} & F_{22} & -F_{32} & F_{42} \\ -F_{13} & -F_{23} & F_{33} & -F_{43} \\ F_{14} & F_{24} & -F_{34} & F_{44} \end{pmatrix}. \tag{2.2.72}$$

The reciprocity is an intrinsic symmetry and valid for any particle. Using Eqs. (2.2.68), (2.2.70), (2.2.72), the scattering matrix for a particle without any geometric symmetry in random orientation can be given as (van de Hulst, 1957)

$$\mathbf{F}_{\mathrm{IS}} = \begin{pmatrix} F_{11} & F_{12} & F_{13} & F_{14} \\ F_{12} & F_{22} & F_{23} & F_{24} \\ -F_{13} & -F_{23} & F_{33} & F_{34} \\ F_{14} & F_{24} & -F_{34} & F_{44} \end{pmatrix}, \tag{2.2.73}$$

where the subscript "IS" represents the results with only intrinsic symmetry and there are only 10 independent elements. Compared with the original representation as shown in Fig. 2.11A, Fig. 2.11C shows the mirror representation. Note that although Fig. 2.11C is only a schematic description about the mirror representation, where the scattering plane is overlapped with the mirror plane, the following statement is still valid for a general case where the scattering plane is different from the mirror plane. The amplitude scattering matrix in Fig. 2.11C should be the same as Eq. (2.2.69) only if using the shown parallel and perpendicular components since the effect of the mirror symmetry is canceled by the mirror symmetry of the perpendicular components. The perpendicular component, the parallel component, and the propagation direction are given in Eqs. (2.2.3a), (2.2.3b) and shown in Fig. 2.8. The parallel component should have a sign difference as follows:

$$\begin{pmatrix} E^m_{\parallel} \\ E^m_{\perp} \end{pmatrix} = \begin{pmatrix} -E^o_{\parallel} \\ E^o_{\perp} \end{pmatrix} = \begin{pmatrix} -1 & 0 \\ 0 & 1 \end{pmatrix} \begin{pmatrix} E^o_{\parallel} \\ E^o_{\perp} \end{pmatrix}, \tag{2.2.74}$$

where the superscripts "$m$" and "$o$" represent the components in the mirror and original representations, respectively. Correspondingly, the amplitude scattering matrix in the mirror representation can be obtained as

$$\mathbf{S}_m = \begin{pmatrix} -1 & 0 \\ 0 & 1 \end{pmatrix} \mathbf{S}_o \begin{pmatrix} -1 & 0 \\ 0 & 1 \end{pmatrix} = \begin{pmatrix} S_{11} & -S_{12} \\ -S_{21} & S_{22} \end{pmatrix}. \tag{2.2.75}$$

Using Eqs. (2.2.12a)–(2.2.12d), (2.2.75), the corresponding scattering matrix in mirror representation using the same elements as in Eq. (2.2.70) can be given as

$$\mathbf{F}_m = \begin{pmatrix} F_{11} & F_{12} & -F_{13} & -F_{14} \\ F_{21} & F_{22} & -F_{23} & -F_{24} \\ -F_{31} & -F_{32} & F_{33} & F_{34} \\ -F_{41} & -F_{42} & F_{43} & F_{44} \end{pmatrix}. \tag{2.2.76}$$

Using Eqs. (2.2.70), (2.2.72), (2.2.76), it is straightforward to show the scattering matrix with mirror-reciprocal symmetry as follows:

$$\mathbf{F}_{mr} = \begin{pmatrix} F_{11} & F_{21} & F_{31} & -F_{41} \\ F_{12} & F_{22} & F_{32} & -F_{42} \\ F_{13} & F_{23} & F_{33} & -F_{43} \\ -F_{14} & -F_{24} & -F_{34} & F_{44} \end{pmatrix}. \tag{2.2.77}$$

For a particle with a mirror symmetry, the scattering matrix in random orientation includes the original orientation, the mirror orientation, and their reciprocal orientations. Using Eqs. (2.2.70), (2.2.72), (2.2.76), (2.2.77), the scattering matrix for a particle with mirror symmetry under the random orientation condition can be given as (van de Hulst, 1957)

$$\mathbf{F}_{IS+MS} = \begin{pmatrix} F_{11} & F_{12} & 0 & 0 \\ F_{12} & F_{22} & 0 & 0 \\ 0 & 0 & F_{33} & F_{34} \\ 0 & 0 & -F_{34} & F_{44} \end{pmatrix}, \tag{2.2.78}$$

where the subscript "IS + MS" represents the results with intrinsic and mirror symmetry, and the matrix is block diagonal and only has six independent elements.

### 2.2.5.3 Symmetry in the exact forward and backward directions

Symmetry was given in the last subsection for an arbitrary particle shape with or without mirror symmetry when the scattering directions are not aligned with the incident direction or the scattering angle $0<\Theta<\pi$. If the scattering direction is exactly along the incident direction or $\Theta=0$, it is called exact forward scattering, while if the scattering direction is exactly opposite to the incident direction, or $\Theta=\pi$, it is called exact backward scattering. Symmetries for exact forward and backward scattering should be specifically investigated due to the nonuniqueness of scattering plane.

Although the scattering plane is not unique, a scattering plane with a certain azimuth angle can be selected as the reference plane. The reference plane can be defined as the original representation similar to Fig. 2.11A. The amplitude scattering matrix in the original representation with respect to the reference plane can be given as a reference matrix:

$$\mathbf{S}_o = \begin{pmatrix} S_{11} & S_{12} \\ S_{21} & S_{22} \end{pmatrix}. \tag{2.2.79}$$

Under intrinsic and mirror symmetries, the amplitude scattering matrix elements can be represented using the same elements given in Eq. (2.2.79) but possibly with sign differences. The symmetric relations given in Section 2.2.5.2 are still valid for exact forward and backward scattering. On the other hand, for an infinite number of orientations obtained by rotations of arbitrary angles $\varphi$ $(0<\varphi<2\pi)$ around the axis, the corresponding amplitude scattering matrix can be given in terms of the elements given in Eq. (2.2.79) and the rotation angle $\varphi$. The rotation angle $\varphi$ of the particle around the axis relative to the reference plane is set to be positive with a counterclockwise rotation facing the incident direction. The rotation is equivalent to keeping the particle fixed and rotating the reference plane the same angle $\varphi$ but in the reverse direction. The idea to reveal the symmetry of the scattering matrix, as done in Section 2.2.5.2, is to find pairs of relations, where the pair uses same elements but possibly different signs or orders. For instance, Eq. (2.2.73) is given by the pair of original and reciprocal orientations, Eq. (2.2.78) is given by two pairs, original and reciprocal orientations, and mirror and mirror-reciprocal orientations.

For exact forward and backward scattering, the same operation is used to find symmetry of exact forward and backward scattering. The symmetries for exact forward or backward scattering are different, so we present them separately.

After a rotation of the particle by an angle $\varphi$ in one direction, or of the reference plane in the reverse direction, the amplitude scattering matrix in the exact forward scattering direction can be given as (van de Hulst, 1957)

$$\mathrm{S}_{\text{for}}(\varphi) = \begin{pmatrix} c^2\mathrm{S}_{11} + cs\mathrm{S}_{12} + cs\mathrm{S}_{21} + s^2\mathrm{S}_{22} & -cs\mathrm{S}_{11} + c^2\mathrm{S}_{12} - s^2\mathrm{S}_{21} + cs\mathrm{S}_{22} \\ -cs\mathrm{S}_{11} - s^2\mathrm{S}_{12} + c^2\mathrm{S}_{21} + cs\mathrm{S}_{22} & s^2\mathrm{S}_{11} - cs\mathrm{S}_{12} - cs\mathrm{S}_{21} + c^2\mathrm{S}_{22} \end{pmatrix}, \tag{2.2.80}$$

where subscript "for" denotes exact forward scattering and $s = \sin\varphi$ and $c = \cos\varphi$. The purpose is to find symmetry relations with respect to the scattering matrix at certain azimuth angles. The experiences in intrinsic and mirror symmetry show us that the amplitude scattering matrix can be expressed with the same elements but with sign or order differences. We can generalize the experiences and speculate that the symmetry may happen at certain azimuth angles, where the element coefficients ($c^2$, $cs$, or $s^2$) should at least be rational numbers. Due to the $2\pi$ period of trigonometric functions, the satisfied azimuth angles $\varphi$ are $\pi/4$, $\pi/2$, and $3\pi/4$. Moreover, the elements with azimuth angle $3\pi/4$ are the same as the elements with azimuth angle $\pi/4$. Consequently, only azimuth angles $\pi/4$ and $\pi/2$ are used together with the original angle to explore the symmetry. Using Eq. (2.2.80), the following matrices are straightforwardly given

$$\mathrm{S}_{\text{for}}\left(\frac{\pi}{4}\right) = \frac{1}{2}\begin{pmatrix} \mathrm{S}_{11} + \mathrm{S}_{12} + \mathrm{S}_{21} + \mathrm{S}_{22} & -\mathrm{S}_{11} + \mathrm{S}_{12} - \mathrm{S}_{21} + \mathrm{S}_{22} \\ -\mathrm{S}_{11} - \mathrm{S}_{12} + \mathrm{S}_{21} + \mathrm{S}_{22} & \mathrm{S}_{11} - \mathrm{S}_{12} - \mathrm{S}_{21} + \mathrm{S}_{22} \end{pmatrix}, \tag{2.2.81}$$

$$\mathrm{S}_{\text{for}}\left(\frac{\pi}{2}\right) = \begin{pmatrix} \mathrm{S}_{22} & -\mathrm{S}_{21} \\ -\mathrm{S}_{12} & \mathrm{S}_{11} \end{pmatrix}. \tag{2.2.82}$$

Since every element in Eq. (2.2.81) is a sum of four original elements, the direct scattering matrix corresponding to Eq. (2.2.81) is tedious. We will directly study the symmetric elements and ignore the nonsymmetric elements. Using Eqs. (2.2.12a), (2.2.12d), one can get a general expression for the scattering matrix elements associated with Eq. (2.2.79) as (Mishchenko et al., 2002)

$$[\mathrm{F}_{22} - \mathrm{F}_{33}]_{\mathrm{S}_o} = \frac{1}{2}\left[|\mathrm{S}_{11} - \mathrm{S}_{22}|^2 - |\mathrm{S}_{12} + \mathrm{S}_{21}|^2\right]. \tag{2.2.83}$$

Using Eq. (2.2.81), one can give the following expression:

$$[\mathrm{F}_{22}-\mathrm{F}_{33}]_{\mathbf{S}_{\mathrm{for}}(\pi/4)}=\frac{1}{2}\left[|\mathrm{S}_{12}+\mathrm{S}_{21}|^{2}-|\mathrm{S}_{11}-\mathrm{S}_{22}|^{2}\right]. \tag{2.2.84}$$

Consequently, the original orientation and the orientation with azimuth angle $\pi/4$ in the exact forward scattering direction gives the symmetry as (Mishchenko et al., 2002)

$$\mathrm{F}_{22}=\mathrm{F}_{33}. \tag{2.2.85}$$

Using Eqs. (2.2.12a)–(2.2.12d), (2.2.82), the corresponding scattering matrix at azimuth angle $\pi/2$ can be given as

$$\mathbf{F}_{\mathrm{for}}\left(\frac{\pi}{2}\right)=\begin{pmatrix} \mathrm{F}_{11} & -\mathrm{F}_{12} & -\mathrm{F}_{13} & \mathrm{F}_{14} \\ -\mathrm{F}_{21} & \mathrm{F}_{22} & \mathrm{F}_{23} & -\mathrm{F}_{24} \\ -\mathrm{F}_{31} & \mathrm{F}_{32} & \mathrm{F}_{33} & -\mathrm{F}_{34} \\ \mathrm{F}_{41} & -\mathrm{F}_{42} & -\mathrm{F}_{43} & \mathrm{F}_{44} \end{pmatrix}. \tag{2.2.86}$$

Consequently, the original orientation and the orientation with azimuth angle $\pi/2$ give the symmetry as

$$\mathrm{F}_{12}=\mathrm{F}_{21}=\mathrm{F}_{13}=\mathrm{F}_{31}=\mathrm{F}_{24}=\mathrm{F}_{42}=\mathrm{F}_{34}=\mathrm{F}_{43}=0. \tag{2.2.87}$$

The symmetries given in Eqs. (2.2.85), (2.2.87) are general and valid for any particles. Combined Eqs. (2.2.85), (2.2.87) to the intrinsic symmetry given in Eq. (2.2.73), the scattering matrix for a particle without any geometric symmetry in random orientation for exact forward scattering can be given as (van de Hulst, 1957)

$$\mathbf{F}_{\mathrm{IS}}=\begin{pmatrix} \mathrm{F}_{11} & 0 & 0 & \mathrm{F}_{14} \\ 0 & \mathrm{F}_{22} & \mathrm{F}_{23} & 0 \\ 0 & -\mathrm{F}_{23} & \mathrm{F}_{22} & 0 \\ \mathrm{F}_{14} & 0 & 0 & \mathrm{F}_{44} \end{pmatrix}, \tag{2.2.88}$$

where there are only five independent elements. Combined Eqs. (2.2.85), (2.2.87) to the symmetry given in Eq. (2.2.78), the scattering matrix for a particle with mirror symmetry in random orientation for exact forward scattering can be given as (van de Hulst, 1957)

$$\mathbf{F}_{\mathrm{IS+MS}}=\begin{pmatrix} \mathrm{F}_{11} & 0 & 0 & 0 \\ 0 & \mathrm{F}_{22} & 0 & 0 \\ 0 & 0 & \mathrm{F}_{22} & 0 \\ 0 & 0 & 0 & \mathrm{F}_{44} \end{pmatrix}, \tag{2.2.89}$$

where there are only three independent elements.

The specialty for exact backward scattering is that its reciprocal orientation is the same with its original orientation. Using the expressions with respect to the original and the reciprocal orientations given in Eqs. (2.2.69), (2.2.71), the following element relation can be immediately given as

$$S_{12} + S_{21} = 0. \tag{2.2.90}$$

Eq. (2.2.90) is a general relation and is valid for an arbitrary amplitude scattering matrix in the exact backward scattering direction. A general expression using Eqs. (2.2.12a), (2.2.12d) can also be obtained (Mishchenko et al., 2002):

$$[F_{11} - F_{44}]_{S_o} = \frac{1}{2}\left[|S_{11} - S_{22}|^2 + |S_{12} + S_{21}|^2\right]. \tag{2.2.91}$$

Using another general expression given in Eq. (2.2.83) and Eqs. (2.2.90), (2.2.91), a general relation for the scattering matrix in exact backward scattering is (van de Hulst, 1957)

$$F_{11} - F_{44} = F_{22} - F_{33}. \tag{2.2.92}$$

Note that Eq. (2.2.92) is valid for any scattering matrix in the exact backward scattering direction, which means that it does not require the random orientation condition.

Using the same setup as with exact forward scattering, the amplitude scattering matrix after an angle $\varphi$ rotation can be expressed in terms of the matrix elements in the reference plane as (van de Hulst, 1957)

$$\mathbf{S}_{\text{bac}}(\varphi) = \begin{pmatrix} c^2S_{11} + csS_{12} - csS_{21} - s^2S_{22} & -csS_{11} + c^2S_{12} + s^2S_{21} - csS_{22} \\ csS_{11} + s^2S_{12} + c^2S_{21} + csS_{22} & -s^2S_{11} + csS_{12} - csS_{21} + c^2S_{22} \end{pmatrix}, \tag{2.2.93}$$

where subscript "bac" denotes exact backward scattering. Using the same strategy described for exact forward scattering, azimuth angles $\pi/4$ and $\pi/2$ are used to further explore the symmetry in exact backward scattering. Eq. (2.2.93) can give the following specific matrices:

$$\mathbf{S}_{\text{bac}}\left(\frac{\pi}{4}\right) = \frac{1}{2}\begin{pmatrix} S_{11} + S_{12} - S_{21} - S_{22} & -S_{11} + S_{12} + S_{21} - S_{22} \\ S_{11} + S_{12} + S_{21} + S_{22} & -S_{11} + S_{12} - S_{21} + S_{22} \end{pmatrix}, \tag{2.2.94}$$

$$\mathbf{S}_{\text{bac}}\left(\frac{\pi}{2}\right) = \begin{pmatrix} -S_{22} & S_{21} \\ S_{12} & -S_{11} \end{pmatrix}. \tag{2.2.95}$$

Due to the complexity of scattering matrix elements associated with Eq. (2.2.94), we directly go to the symmetry elements. Using Eqs. (2.2.12a), (2.2.12d), one can get a general expression as follows (Mishchenko et al., 2002):

$$[F_{22} + F_{33}]_{S_o} = \frac{1}{2}\left[|S_{11} + S_{22}|^2 - |S_{12} - S_{21}|^2\right]. \tag{2.2.96}$$

Using Eq. (2.2.94), Eq. (2.2.96) becomes

$$[\mathrm{F}_{22}+\mathrm{F}_{33}]_{\mathbf{S}_{\mathrm{bac}}(\pi/4)}=\frac{1}{2}\left[|\mathrm{S}_{12}-\mathrm{S}_{21}|^2-|\mathrm{S}_{11}+\mathrm{S}_{22}|^2\right]. \tag{2.2.97}$$

Consequently, the original orientation and the orientation with azimuth angle $\pi/4$ for exact backward scattering shows the symmetry as (Mishchenko et al., 2002)

$$\mathrm{F}_{22}+\mathrm{F}_{33}=0. \tag{2.2.98}$$

Comparing Eq. (2.2.82) and Eq. (2.2.95), one can get

$$\mathbf{S}_{\mathrm{bac}}\left(\frac{\pi}{2}\right)=-\mathbf{S}_{\mathrm{for}}\left(\frac{\pi}{2}\right). \tag{2.2.99}$$

Using the mixed tensor product property given in Eq. (2.1.92) and the relation between the amplitude scattering matrix and the scattering matrix given in Eq. (2.2.11), one can immediately obtain that the scattering matrix associated with the amplitude scattering matrix given in Eq. (2.2.95) is the same as Eq. (2.2.86). Accordingly, the original orientation and the orientation with azimuth angle $\pi/2$ give the same symmetry as

$$\mathrm{F}_{12}=\mathrm{F}_{21}=\mathrm{F}_{13}=\mathrm{F}_{31}=\mathrm{F}_{24}=\mathrm{F}_{42}=\mathrm{F}_{34}=\mathrm{F}_{43}=0. \tag{2.2.100}$$

The symmetries given in Eqs. (2.2.92), (2.2.98), (2.2.100) are general for exact backward scattering. Combining these general symmetries with the intrinsic symmetry given in Eq. (2.2.73), one can show the scattering matrix for a particle without any geometric symmetry in random orientation for exact backward scattering is (van de Hulst, 1957)

$$\mathbf{F}_{\mathrm{IS}}=\begin{pmatrix}\mathrm{F}_{11} & 0 & 0 & \mathrm{F}_{14}\\ 0 & \mathrm{F}_{22} & \mathrm{F}_{23} & 0\\ 0 & -\mathrm{F}_{23} & -\mathrm{F}_{22} & 0\\ \mathrm{F}_{14} & 0 & 0 & \mathrm{F}_{11}-2\mathrm{F}_{22}\end{pmatrix}, \tag{2.2.101}$$

where there are only four independent elements. Combining these general symmetries and the symmetry given in Eq. (2.2.78), the scattering matrix for a particle with mirror symmetry in random orientation for exact backward scattering is (van de Hulst, 1957).

$$\mathbf{F}_{\mathrm{IS+MS}}=\begin{pmatrix}\mathrm{F}_{11} & 0 & 0 & 0\\ 0 & \mathrm{F}_{22} & 0 & 0\\ 0 & 0 & -\mathrm{F}_{22} & 0\\ 0 & 0 & 0 & \mathrm{F}_{11}-2\mathrm{F}_{22}\end{pmatrix}, \tag{2.2.102}$$

where there are only two independent elements.

Further discussions about the symmetry properties for exact forward and backward scattering can be found in the literature (e.g., van de Hulst, 1957; Hu et al., 1987; Mishchenko et al., 2002).

### 2.2.6 Integrated scattering properties

The amplitude scattering matrix and scattering matrix are introduced and discussed in the sections earlier. For a specific incident direction, these matrices depend on the scattering directions and indicate actual differential scattering properties. The integrated single-scattering properties including the extinction, scattering and absorption cross sections, the asymmetry factor, and other related quantities are introduced in this subsection.

As shown in Fig. 2.12, a scattering particle surrounded by a nonabsorptive medium is illuminated by an incident plane wave. The total electric and magnetic fields in the surrounding medium are denoted as **E** and **H**, which consist of the incident and scattered fields. Using Eq. (2.1.40), the time-averaged Poynting vector $\langle \mathbf{S} \rangle$ can be decomposed into three terms (Bohren and Huffman, 1983; Mishchenko et al., 2002):

$$\begin{aligned}\langle \mathbf{S} \rangle &= \frac{1}{2} Re[\mathbf{E}^* \times \mathbf{H}] = \langle \mathbf{S}_{\text{inc}} \rangle + \langle \mathbf{S}_{\text{sca}} \rangle + \langle \mathbf{S}_{\text{ext}} \rangle \\ &= \frac{1}{2} Re\left[ \left( \mathbf{E}^{\text{inc}} \right)^* \times \left( \mathbf{H}^{\text{inc}} \right) \right] + \frac{1}{2} Re[(\mathbf{E}^{\text{sca}})^* \times (\mathbf{H}^{\text{sca}})] \\ &+ \frac{1}{2} Re\left[ \left( \mathbf{E}^{\text{inc}} \right)^* \times (\mathbf{H}^{\text{sca}}) + (\mathbf{E}^{\text{sca}})^* \times \left( \mathbf{H}^{\text{inc}} \right) \right],\end{aligned} \tag{2.2.103}$$

where the first decomposed term associated with only the incident electromagnetic field is the Poynting vector of the incident field, the second decomposed term associated with only the scattered electromagnetic field is the Poynting vector of the scattered field, and the left ones are cross terms between the incident and scattered electromagnetic fields. An illusory sphere is built to be large enough so it can enclose

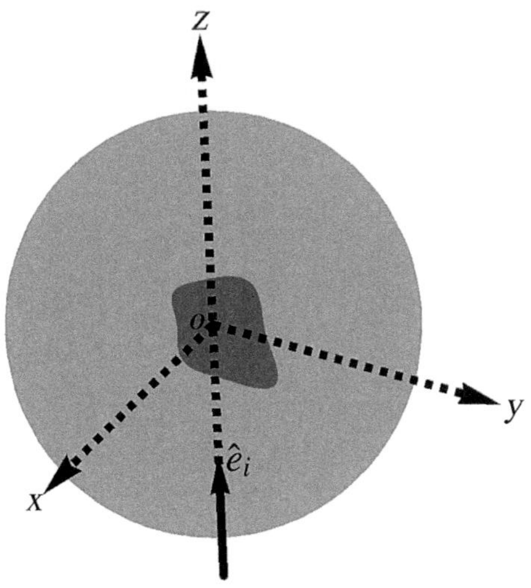


**Fig. 2.12** Extinction by an arbitrary scattering particle. The illusory sphere (gray circle) encloses the scattering particle, which is shown as an irregular darker shape.

the scattering particle. Since the surrounding medium is assumed to be nonabsorptive, the rate of incident energy absorbed by the scattering particle, or the absorption flux, according to the physical meaning of the Poynting vector, can be obtained by (Bohren and Huffman, 1983)

$$\mathrm{W}_{\mathrm{abs}} = -\oint_{\Sigma} \langle \mathbf{S} \rangle \cdot d\vec{\sigma} = -\oint_{\Sigma} \langle \mathbf{S}_{\mathrm{inc}} \rangle \cdot d\vec{\sigma} - \oint_{\Sigma} \langle \mathbf{S}_{\mathrm{sca}} \rangle \cdot d\vec{\sigma} - \oint_{\Sigma} \langle \mathbf{S}_{\mathrm{ext}} \rangle \cdot d\vec{\sigma}, \tag{2.2.104}$$

where $\Sigma$ indicates integration over the spherical surface and $d\vec{\sigma}$ is the area element vector. The first term of the right-hand-side $\oint_{\Sigma} \langle \mathbf{S}_{\mathrm{inc}} \rangle \cdot d\vec{\sigma}$ is interpreted as the absorption rate in the absence of the scattering particle and is equal to 0 with a nonabsorptive medium. The second term $\oint_{\Sigma} \langle \mathbf{S}_{\mathrm{sca}} \rangle \cdot d\vec{\sigma}$ is the rate of incident energy scattered by the scattering particle and is interpreted as the scattering flux $\mathrm{W}_{\mathrm{sca}}$. The third term $-\oint_{\Sigma} \langle \mathbf{S}_{\mathrm{ext}} \rangle \cdot d\vec{\sigma}$ is equal to the sum of the rates absorbed and scattered by the particle and reveals the rate of incident energy extinction, because of the presence of the scattering particle, interpreted as the extinction flux $\mathrm{W}_{\mathrm{ext}}$. Eq. (2.2.103) becomes

$$\mathrm{W}_{\mathrm{ext}} = \mathrm{W}_{\mathrm{sca}} + \mathrm{W}_{\mathrm{abs}}. \tag{2.2.105}$$

Due to linearity of Maxwell's equations, the rates of energy associated with absorption, scattering, and extinction are apparently proportional to the incident flux density (or irradiance). Accordingly, the normalized quantities can be defined by absorption, scattering, and extinction fluxes divided by the incident flux density. Since the dimension of each variable after normalization becomes area, the quantities are named absorption, scattering, and extinction cross sections. The illusory sphere in Fig. 2.12 can be assumed in the far field, where its radius $r$ is much larger than the characteristic length of the scattering particle so the scattered electric and magnetic fields become transverse fields. Using Eqs. (2.2.2), (2.1.58), one can give (Mishchenko et al., 2002)

$$\mathbf{E}^{\mathrm{sca}}\left(\vec{r}\right)\Big|_{kr\to\infty} = \frac{\exp(ikr)}{-ikr}\mathbf{E}_1^{\mathrm{sca}}(\hat{r}), \tag{2.2.106}$$

$$\mathbf{H}^{\mathrm{sca}}\left(\vec{r}\right)\Big|_{kr\to\infty} = \sqrt{\frac{\varepsilon}{\mu_0}}\hat{r} \times \mathbf{E}^{\mathrm{sca}}\left(\vec{r}\right)\Big|_{kr\to\infty}, \tag{2.2.107}$$

where $\varepsilon$ is the permittivity of the surrounding medium. Accordingly, one can explicitly give the expression for the scattering flux:

$$\mathrm{W}_{\mathrm{sca}}\big|_{kr\to\infty} = \frac{1}{k^2}\frac{1}{2}\sqrt{\frac{\varepsilon}{\mu_0}}\oint_{4\pi} \left|\mathbf{E}_1^{\mathrm{sca}}(\hat{r})\right|^2 d\Omega. \tag{2.2.108}$$

The incident fields in the far field can be decomposed into the incoming and outgoing spherical waves and can be given as follows (Mishchenko et al., 2002)

$$\mathbf{E}^{\mathrm{inc}}\left(\vec{r}\right)\Big|_{kr\to\infty} = -2\pi\left[\delta(\hat{e}_i+\hat{r})\frac{\exp(-ikr)}{ikr} + \delta(\hat{e}_i-\hat{r})\frac{\exp(ikr)}{-ikr}\right]\mathbf{E}_0, \tag{2.2.109}$$

where $\mathbf{E}_0$ is the amplitude of the incident field so it is still transverse in the far field. Subsequently, the magnetic field using Eq. (2.1.58) is

$$\mathbf{H}^{\mathrm{inc}}\left(\vec{r}\right)\Big|_{kr\to\infty} = \sqrt{\frac{\varepsilon}{\mu_0}}\hat{e}_i \times \mathbf{E}^{\mathrm{inc}}\left(\vec{r}\right)\Big|_{kr\to\infty}. \tag{2.2.110}$$

After some algebra, one can obtain

$$\mathrm{W}_{\mathrm{ext}}\big|_{kr\to\infty} = \frac{4\pi}{k^2}\frac{1}{2}\sqrt{\frac{\varepsilon}{\mu_0}}Re\left[\mathbf{E}_0^* \cdot \mathbf{E}_1^{\mathrm{sca}}(\hat{e}_i)\right]. \tag{2.2.111}$$

The extinction flux only depends on the forward scattering amplitude. Using Eq. (2.1.69), the time-averaged incident flux density of the incident plane wave can be given as $\frac{1}{2}\sqrt{\frac{\varepsilon}{\mu_0}}|\mathbf{E}_0|^2$. Using Eqs. (2.2.108), (2.2.111), the scattering and extinction cross sections can be obtained as

$$C_{\mathrm{sca}} = \frac{\mathrm{W}_{\mathrm{sca}}\big|_{kr\to\infty}}{\frac{1}{2}\sqrt{\frac{\varepsilon}{\mu_0}}|\mathbf{E}_0|^2} = \frac{1}{k^2}\frac{\oint_{4\pi}\left|\mathbf{E}_1^{\mathrm{sca}}(\hat{r})\right|^2 d\Omega}{|\mathbf{E}_0|^2}, \tag{2.2.112}$$

$$C_{\mathrm{ext}} = \frac{\mathrm{W}_{\mathrm{ext}}\big|_{kr\to\infty}}{\frac{1}{2}\sqrt{\frac{\varepsilon}{\mu_0}}|\mathbf{E}_0|^2} = \frac{4\pi Re\left[\mathbf{E}_0^* \cdot \mathbf{E}_1^{\mathrm{sca}}(\hat{e}_i)\right]}{k^2 |\mathbf{E}_0|^2}, \tag{2.2.113}$$

where the area dimension is explicitly in the term $1/k^2$. Eqs. (2.2.111), (2.2.113) are called the *optical theorem*. Actually, the optical theorem states that extinction only depends on the forward scattering amplitude and is a general rule of scattering theory, which is valid not only in classical scattering such as acoustic and electromagnetic wave scattering but also in quantum scattering such as elementary particle scattering. Using Eq. (2.2.105), the absorption cross section is

$$C_{\mathrm{abs}} = \frac{\mathrm{W}_{\mathrm{abs}}}{\frac{1}{2}\sqrt{\frac{\varepsilon}{\mu_0}}|\mathbf{E}_0|^2} = C_{\mathrm{ext}} - C_{\mathrm{sca}}. \tag{2.2.114}$$

Since the surrounding medium is nonabsorptive, energy absorption only occurs inside the scattering particle and is not related to the radius of the illusory sphere in Fig. 2.12. The *single-scattering albedo* is the ratio of the scattering cross section to the extinction cross section:

$$\mathrm{SSA} = \frac{C_{\mathrm{sca}}}{C_{\mathrm{ext}}}. \tag{2.2.115}$$

Since the dimensions of the cross sections are all area, the cross sections can be further normalized by the geometric cross section of a scattering particle, which is defined as projected cross section of the scattering particle over the plane perpendicular to the incident direction. For instance, the geometric cross section of a sphere is a circular disk with area of $\pi$ times the square of the radius. The further normalized quantity is termed as efficiency. The corresponding scattering, absorption, and extinction efficiencies are defined as follows:

$$Q_{\text{sca}} = C_{\text{sca}}/A, \quad Q_{\text{abs}} = C_{\text{abs}}/A, \quad Q_{\text{ext}} = C_{\text{ext}}/A, \tag{2.2.116}$$

where $A$ represents the geometric cross section. The extinction efficiency approaches 2 when the scattering particle becomes large, which is called the extinction dilemma and explained by van de Hulst (1957).

The scattering cross section given in Eq. (2.2.112) is an integrated scattering property. Formally, $dC_{\text{sca}}/d\Omega$ is used to denote the differential scattering cross section but cannot be considered as the derivative with respect to solid angle $\Omega$. It can be represented as

$$\frac{dC_{\text{sca}}}{d\Omega} = \frac{\left|\mathbf{E}_1^{\text{sca}}(\hat{r})\right|^2}{k^2 |\mathrm{E}_0|^2} \tag{2.2.117}$$

According to Eqs. (2.2.9), (2.2.10), the differential scattering cross section is related to the scattering matrix and the incident Stokes vector. For unpolarized light, the differential scattering cross section becomes

$$\frac{dC_{\text{sca}}}{d\Omega} = \frac{\mathrm{F}_{11}}{k^2} = \frac{C_{\text{sca}}\mathrm{P}_{11}}{4\pi}, \tag{2.2.118}$$

The element $\mathrm{F}_{11}$ indicates the distribution of scattered energy and is related to the differential scattering cross section. As shown in Eq. (2.2.118), integrating element $\mathrm{F}_{11}$ over $4\pi$ gives $k^2 C_{\text{sca}}$ (van de Hulst, 1957). Accordingly, the normalized scattering matrix denoted as **P** and named as the scattering phase matrix is described as

$$\mathbf{P} = \frac{4\pi}{k^2 C_{\text{sca}}}\mathbf{F}, \tag{2.2.119}$$

where the element $\mathrm{P}_{11}$ of the scattering phase matrix is called the *scattering phase function*. Consequently, the integration of the scattering phase function is unity:

$$\frac{1}{4\pi}\int_{4\pi} d\Omega \mathrm{P}_{11} = 1. \tag{2.2.120}$$

If the scattering phase function is azimuthally symmetric, Eq. (2.2.120) becomes

$$\frac{1}{2}\int_0^{\pi} \mathrm{P}_{11} \sin\Theta d\Theta = 1, \tag{2.2.121}$$

where $\Theta$ is the scattering angle. The *asymmetry factor* $g$ is defined as the mean cosine of the scattering angle, averaged over the phase function, as

$$g = \frac{1}{4\pi} \int_{4\pi} \mathrm{P}_{11} \cos\Theta d\Omega, \tag{2.2.122}$$

where the cosine of the scattered angle $\cos\Theta = \hat{e}_i \cdot \hat{r}$. The asymmetry factor indicates the tendency for forward or backward scattering. Here, forward scattering is associated with scattering angles from 0° to 90°, while backward scattering is associated with scattering angles from 90° to 180°. If $g=0$, the scattering is symmetric around the forward and backward scattering hemispheres, such as Rayleigh scattering; scattering is dominated by forward scattering if $0<g<1$, while scattering is dominated by backward scattering if $-1<g<0$; for $g=1$ or $g=-1$, all scattering is only in the exact forward or backward direction, respectively.

## 2.2.7 Bulk optical properties

The particle in random orientation introduced in Section 2.2.5.2 deals with a situation, where identical particles are distributed in all possible orientations and the emergent light is incoherent. For a macroscopically small volume containing many individual particles with a particle number-size distribution, bulk properties can be obtained by an ensemble average of the optical properties of the particles over the distribution if the emergent scattered waves with respect to the particles are incoherent. Since only a chromatic or monochromatic wave is considered, all properties are spectral.

$n(D)dD$ is the number of particles per unit volume within the size interval $(D, D+dD)$, where $n(D)$ is defined as the particle number-size distribution and $D$ is the particle diameter (for spherical particles) or the particle equivalent diameter (for nonspherical particles). For nonspherical particles, the effective diameter can be defined as the diameter of a sphere with the same volume or surface area as the nonspherical particle, although other definitions of the equivalent diameter can be used. The unit for $n(D)$ is $\mathrm{m}^{-4}$ or $\mathrm{m}^{-3}\,\mu\mathrm{m}^{-1}$. Since the particle size is usually microscopic with $D$ from much smaller than a micron to thousands of microns, the unit for the particle diameter usually uses micrometer (Mobley, 1994).

As long as the particle number-size distribution $n(D)$ is provided, the bulk optical properties of ensemble particles can be given by integration over the distribution. The *effective diameter* can be defined as (Foot, 1988; Wendisch and Yang, 2012)

$$D_{\mathrm{eff}} = \frac{3V_{\mathrm{eff}}}{2A_{\mathrm{eff}}} = \frac{3}{2} \frac{\int_{D_1}^{D_2} V(D)n(D)dD}{\int_{D_1}^{D_2} A(D)n(D)dD}, \tag{2.2.123}$$

where $V(D)$ and $A(D)$ are the volume and the projected area on a plane perpendicular to the incident direction with respect to a particle with diameter $D$, respectively; the lower limit $D_1$ and the upper limit $D_2$ are determined according to $n(D)$. The coefficient 3/2 in Eq. (2.2.123) is given to be consistent with the definition for spherical particles. For an ensemble of spherical particles, the effective diameter can be explicitly given by

$$D_{\text{eff}} = \frac{\int_{D_1}^{D_2} D^3 n(D) dD}{\int_{D_1}^{D_2} D^2 n(D) dD}. \tag{2.2.124}$$

The bulk integrated scattering properties of the particles can be obtained by integrating the physical quantities of individual particles over the number-size distribution $n(D)$. The integrations of the extinction, scattering, and absorption cross sections provide the corresponding coefficients associated with volumetric particles. The conventional notations $a$, $b$, and $c$ are, respectively, used to denote the absorption, scattering, and extinction coefficients (Mobley, 1994). The bulk extinction, scattering, and absorption coefficients can be given as follows:

$$c = \int_{D_1}^{D_2} C_{\text{ext}}(D) n(D) dD, \tag{2.2.125}$$

$$b = \int_{D_1}^{D_2} C_{\text{sca}}(D) n(D) dD, \tag{2.2.126}$$

$$a = \int_{D_1}^{D_2} C_{\text{abs}}(D) n(D) dD, \tag{2.2.127}$$

where the units for the coefficients are $\text{m}^{-1}$. It is evident that the coefficients satisfy $c = b + a$. The dimensionless bulk single-scattering albedo is given by

$$\langle \text{SSA} \rangle = \frac{b}{c}, \tag{2.2.128}$$

where the symbol $\langle ... \rangle$ represents the ensemble average over the distribution, which is distinguished from the same symbol given in Eq. (2.1.27) that indicates a time average. The bulk scattering phase function gives the *volume scattering function* as

$$\beta = \int_{D_1}^{D_2} \text{P}_{11}(\Omega, D) C_{\text{sca}}(D) n(D) dD, \tag{2.2.129}$$

where the volume scattering function is actually the differential scattering coefficient and the unit for the volume scattering function is $\text{m}^{-1}\text{sr}^{-1}$. The *bulk scattering phase function* for the ensemble volume can be given by

$$\langle \mathrm{P}_{11}(\Omega) \rangle = \frac{\beta}{b}, \tag{2.2.130}$$

where the unit for the effective phase function is $sr^{-1}$. The *bulk asymmetry factor* is given by

$$\langle g \rangle = \frac{\int_{D_1}^{D_2} g(D) C_{sca}(D) n(D) dD}{b}. \tag{2.2.131}$$

The scattering phase matrix is defined in Eq. (2.2.119). The *bulk scattering phase matrix* is then given by

$$\langle \mathbf{P} \rangle = \frac{\int_{D_1}^{D_2} \mathbf{P}(\Omega, D) C_{sca}(D) n(D) dD}{b}. \tag{2.2.132}$$

The total particle number density (unit $m^{-3}$) can be given by

$$N_{total} = \int_{D_1}^{D_2} n(D) dD. \tag{2.2.133}$$

Then the particle size distribution can be defined as

$$f(D) = \frac{n(D)}{N_{total}}, \tag{2.2.134}$$

where the unit for $f(D)$ is $m^{-1}$ or $\mu m^{-1}$ and it is evident that

$$\int_{D_1}^{D_2} f(D) dD = 1. \tag{2.2.135}$$

Since the bulk properties, such as $D_{eff}$, $\langle SSA \rangle$, $\langle P_{11} \rangle$, $\langle g \rangle$ and $\langle \mathbf{P} \rangle$, which are obtained using one integration divided by another integration over the distribution, do not depend on the number density $N_{total}$ of the particles, they can also be computed by replacing the particle size distribution $f(D)$ for $n(D)$ in those equations.

However, the particle size distribution does not contain the number density information, so the extinction, scattering, and absorption coefficients and the volume scattering function cannot be computed using only the particle size distribution. Instead one can obtain the bulk extinction, scattering, and absorption cross sections and the bulk differential scattering cross section as follows:

$$\langle C_{ext} \rangle = \int_{D_1}^{D_2} C_{ext}(D) f(D) dD, \tag{2.2.136}$$

$$\langle C_{\text{sca}} \rangle = \int_{D_1}^{D_2} C_{\text{sca}}(D) f(D) dD, \tag{2.2.137}$$

$$\langle C_{\text{abs}} \rangle = \int_{D_1}^{D_2} C_{\text{abs}}(D) f(D) dD, \tag{2.2.138}$$

$$\begin{aligned} \left\langle \frac{dC_{\text{sca}}}{d\Omega}(\Omega) \right\rangle &= \frac{1}{4\pi} \int_{D_1}^{D_2} p(\Omega, D) C_{\text{sca}}(D) f(D) dD \\ &= \int_{D_1}^{D_2} \frac{dC_{\text{sca}}}{d\Omega}(\Omega, D) f(D) dD. \end{aligned} \tag{2.2.139}$$

## 2.3 A brief summary of rigorous methods for light scattering by nonspherical and inhomogeneous particles

For a given nonspherical or inhomogeneous particle, represented by a distribution of the refractive index, a number of computational methods have been developed to compute the single-scattering properties by solving Maxwell's equations or their mathematical equivalents. In this section, we briefly discuss the features of popular rigorous numerical methods. By "rigorous methods," we refer to the methods to solve Maxwell's equations with controllable numerical errors within well-defined applicable regimes.

The separation of variables method (SOVM), as applied in the Lorenz–Mie theory, can be extended to spheroids and infinite circular cylinders because spheroidal and cylindrical coordinate systems can be used to conform to the particle geometry so that the boundary conditions can be specified. Similar to a multilayered sphere, the SOVM can also deal with multilayered spheroids and infinite cylinders. Although the SOVM has limited applicability in terms of the particle shape, it has been proven to be valuable in the development of computational methods for arbitrarily shaped nonspherical and inhomogeneous particles. First, the results provided by the SOVM for a specified geometry (e.g., a sphere) have high accuracy, which permits us to test the validity of other numerical methods and to trace errors in the computational program. Second, the analytical formulation of light scattering allows the possibility to obtain an accurate understanding of the scattering mechanism by developing appropriate approximations. The knowledge obtained from these efforts is useful for understanding the results from other nonspherical particles. One example is the semiclassical approximations developed based on the Lorenz–Mie theory by employing the Debye expansion and complex angular momentum theory (Nussenzveig, 1992). The insight and approximations not only are useful to understand the scattering by a sphere but also have been demonstrated to be useful for understanding the semiclassical effects of light scattering by nonspherical particles. Third, the techniques developed in the Lorenz-Mie theory can also be incorporated into other techniques to form a new hybrid method to compute the optical properties of nonspherical and inhomogeneous particles. For example, for a multiple-sphere cluster, the separation of variable method is employed to treat multiple spheres (Mackowski and Mishchenko, 2011). As will be discussed in Chapter 4, the Lorenz-Mie theory is used to compute the T-matrix of an inscribed sphere of a nonspherical particle.

The finite-difference time-domain (FDTD) method is a popular numerical method that has no limitation with respect to the particle shape. Yee (1966) pioneered the development of the FDTD method for simulating the propagation of electromagnetic waves and scattering. Instead of solving vector Helmholtz equations in the frequency domain, the FDTD directly solves Maxwell's curl equations in the time domain (Taflove and Hagness, 2000). By imposing an absorbing boundary (Berenger, 1994), which absorbs the scattered waves from the particle without reflections that could contaminate the near field, the FDTD is suitable for simulating the electromagnetic scattering by an arbitrary particle in a truncated space. To improve the accuracy or the efficiency, a class of time-domain methods have been developed, including the pseudospectral time domain (PSTD) (Liu, 1997), the multiple-resolution time domain (MRTD) (Krumpholz and Katehi, 1996), and the discontinuous Galerkin time domain (Cockburn et al., 2000). All of these methods start with solving Maxwell's curl equations in the time domain but use different numerical schemes to compute the near field, followed by transforming the fields from the time domain to the frequency domain and then computing the single-scattering properties (the extinction efficiency, the absorption efficiency, and the phase matrix) using relevant integrals in terms of the near field.

Another class of approaches solves volume integral equations in the frequency domain. For example, the discrete dipole approximation (DDA) method solves the electric field volume integral equation in the frequency domain. The DDA method was introduced by Purcell and Pennypacker (1973) to study the optical properties of interstellar dust grains and further developed and popularized by Draine and Flatau (2003) with a widely used program. The basic idea of the DDA can be traced even earlier to DeVoe (1964) in a study of optical properties of molecular aggregates. Simply stated, the particle is assumed to be a finite group of dipoles in the DDA method. Each dipole responds to the incident field and generates scattered waves. The mutual interactions among all of the dipoles establish a final self-consistent field distribution that represents the solution of the electric field volume integral equation. In the most original form, the DDA equations are established on the basis of a hybrid of physical insights and mathematics. Recent efforts to put the DDA approach firmly within the electric field volume integral equations include the work of Yurkin and Hoekstra (2007). There are many other numerical methods, including the method of moment (Harrington, 1968) and the multilevel fast multipole algorithm (Song et al., 1997). Both the DDA and FDTD have similar applicability in terms of the particle shape. However, the two methods have different numerical performances with respect to the refractive index. The readers are referred to Yurkin et al. (2007) and Liu et al. (2012) for comparison studies. A comprehensive introduction to numerical methods in electromagnetic scattering is given by Kahnert (2010).

It should be noted that all of the previous methods compute the single-scattering properties by imposing a specified incident field. Namely, for each simulation, only the optical properties of a single orientation can be obtained. The SOVM, FDTD, and DDA all obtain the solution of light scattering by specifying an incident wave. Unlike these numerical methods, the T-matrix methods are semianalytical by expanding the incident and scattered fields in suitable vector wave functions and computing the T-matrix that transforms the coefficients of the incident field to those of the scattered field. By this approach, the T-matrix obtained does not depend on the incident field and the particle orientation. Thus, the T-matrix method is most suitable for computing the optical properties of randomly oriented particles, once the T-matrix is obtained. Different T-matrix methods use different approaches to compute the T-matrix. Each T-matrix method has its own advantages for particular particle types. The extended boundary condition method (EBCM) was initially introduced by Waterman (1965, 1971) and has been shown to be an efficient and accurate method for scattering calculations involving axially rotationally symmetric nonspherical particles, such as spheroids and circular cylinders (Mishchenko et al., 1996). The EBCM has been further improved by Kahnert (2013) for particles with finite symmetries and the null-field method with discrete sources (e.g., Doicu et al., 2006). The superposition principle that explores the use of the addition theorem is demonstrated to be particularly useful for a cluster of spherical particles and has recently been extended to an inhomogeneous sphere with multiple spherical inclusions (Mackowski and Mishchenko, 2011). It should be noted that the resolvent kernel method that solves the volume integral equation could also obtain a matrix that contains all of the information of scattering (Kattawar et al., 1987), although this method has not been further explored.

From this summary, we can conclude that different approaches have different advantages. The present invariant imbedding T-matrix method is most suitable for computing the single-scattering properties of randomly oriented irregular and inhomogeneous particles. For fixed orientations, the FDTD and DDA are more suitable because of their computational efficiency. For homogeneous and axially symmetric particles, the EBCM has been demonstrated to be an efficient tool, although the II-TM is applicable to a wider size parameter range.

# T-matrix concept

# 3

In this chapter, we present a number of fundamental subjects related to the T-matrix method. Section 3.1 gives a brief history of the T-matrix method along with its development and advantages. Section 3.2 introduces the T-matrix method in terms of the vector spherical wave functions (VSWF) as a means to expand the incident and scattered fields. Two effective methods for computing the T-matrix are briefly discussed and compared in Section 3.3. The unitarity of the T-matrix is discussed in Section 3.4.1. The reciprocity and the symmetries of the T-matrix associated with the shape of a scattered particle are discussed in Section 3.4.2. Sections 3.4.3 and 3.4.4 give the scattering phase matrix associated with the T-matrix for an arbitrarily shaped particle and for particles with various symmetries under the random orientation condition. Section 3.4.5 gives the scattering phase matrix of an arbitrarily shaped particle that has a specific orientation.

## 3.1 A brief history

In Section 2.3, we reviewed several rigorous methods for light scattering by nonspherical and inhomogeneous particles. The T-matrix method is an efficient semianalytical technique, proposed by Waterman first for acoustic scattering in 1969 and then for electromagnetic scattering in 1971 (Waterman, 1969, 1971). In the T-matrix method, the incident and scattered fields are expanded using the regular and outgoing VSWF, respectively. Since the boundary conditions are linear, the T-matrix of a particle stems from the transition matrix $\mathbf{T}$ that relates the incident expansion coefficient vector $\boldsymbol{a}$ to the scattered expansion coefficient vector $\boldsymbol{p}$, that is, $\boldsymbol{p} = \mathbf{T}\boldsymbol{a}$. The T-matrix of a particle is dependent on the particle shape, composition, and the frame of reference, but is independent of the direction and polarization state of the incident wave. The T-matrix of a particle can be used to analytically compute the scattering phase matrix of a particle under the random orientation condition.

The scattered electromagnetic field can be expressed using a surface- or volume-integral equation derived from Maxwell's equations. The T-matrix can correspondingly be formulated in terms of the surface- or volume-integral equation. The T-matrix method using the surface integral is well known as the extended boundary condition method (EBCM). Note that this method is also referred to as the null field method (Doicu and Wriedt, 2018). The EBCM was first proposed by Waterman in 1965 for perfectly conducting particles (Waterman, 1965) and in 1969 for dielectric particles (Waterman, 1969) and was used to compute the T-matrix in 1969 and 1971 (Waterman, 1969, 1971). The T-matrix method using the EBCM was improved and implemented by Mishchenko, particularly in the case of particles with axial

**Invariant Imbedding T-matrix Method for Light Scattering by Nonspherical and Inhomogeneous Particles**
**https://doi.org/10.1016/B978-0-12-818090-7.00003-6**

symmetries (e.g., Barber and Hill, 1990; Mishchenko et al., 2004). The T-matrix method using the volume integral is called the invariant imbedding T-matrix method (II-TM). The II-TM was first proposed by Johnson (1988). A state-of-the-art computational program using the II-TM was developed by Bi et al. (2013) to compute the scattering properties of arbitrary particles. Although methods other than the EBCM and the II-TM have been employed for computing the T-matrix, such as the separation-of-variable method (Schulz et al., 1998), the point-matching method (Nieminen et al., 2003), and the discrete dipole approximation (Mackowski, 2002), they are not discussed in this book. The method to compute the random orientation average single-scattering properties associated with the T-matrix of a particle with axial symmetry was originally developed by Mishchenko (1991), and then, the algorithm was generalized to compute the random-orientation averaged single-scattering properties of an arbitrary particle (Khlebtsov, 1992; Mackowski and Mishchenko, 1996; Bi and Yang, 2014).

For a homogeneous spherical particle, the T-matrix method is reduced to an analytical solution based on the Lorenz-Mie theory (LMT). For multiple separate spherical particles, the scattering solution is accomplished by combining the addition theorem associated with the VSWF (Friedman and Russek, 1954; Stein, 1961; Cruzan, 1962) and the LMT and called the multisphere T-matrix method (Bruning and Lo, 1971a, b; Fuller and Kattawar, 1988a, b; Xu, 1995, 1997; Mackowski and Mishchenko, 1996; Fuller and Mackowski, 2000). If the incident wave is not a plane wave, the LMT is generalized to deal with the scattering properties of a particle in a transversely localized incident beam, which is called the generalized Lorenz-Mie theory (GLMT) (e.g., Gouesbet and Grehan, 2011). The Lorenz-Mie coefficients of a spherical particle can be decomposed into Debye series, for example, diffraction and external reflection and refraction without internal reflection (Hovenac and Lock, 1992). The Debye series of a nonspherical particles is furthermore obtained using the EBCM (Xu et al., 2010; Xu and Lock, 2010) and used to expand the T-matrix for studying the depolarization of nearly spherical particles (Bi et al., 2018b). The component combining the diffraction and external reflection of the Debye series for the T-matrix of a nonspherical particle has been developed using the II-TM (Bi et al., 2015; Bi and Yang, 2016). The Debye series form of the T-matrix is not discussed in this book.

The T-matrix method is an important scientific legacy of Peter C. Waterman, whose biographical and scientific sketches can be found in a review article entitled *Peter Waterman and T-matrix methods* by Mishchenko and Martin (2013). Moreover, a bibliographic database of studies related to the T-matrix method has been continually updated (Mishchenko et al., 2004, 2007, 2008, 2010, 2013, 2014, 2016, 2017; Zakharova et al., 2012).

## 3.2 Expansion of the incident and scattered fields

A time-harmonic electromagnetic wave is assumed with time dependence $\exp(-i\omega t)$, where $\omega = 2\pi/T$ is the angular frequency and $T$ is the wave period. The time dependence is suppressed in this book. Since the incident and scattered fields are expanded using the VSWF, the VSWF is mathematically introduced using spherical harmonics and spherical Bessel functions.

### 3.2.1 Spherical coordinates

When a dielectric particle scatters an electromagnetic wave, the scattered wave in the far-field zone is a spherical wave that is proportional to $\exp(ikr)/r$, where $k$ is the wave number and $r$ is the radial coordinate in a spherical coordinate system. The arguments of the corresponding amplitude scattering matrix and scattering phase matrix introduced in Chapter 2 are spherical angles $(\theta, \varphi)$, where $\theta$ is the zenith angle relative to a chosen $z$-axis and $\varphi$ is the azimuth angle counterclockwise relative to the $x$-axis in the $x$-$y$ plane, when facing the positive $z$-axis as shown in Fig. 2.8. Moreover, specific axes are chosen relative to a scattering particle at the origin to take advantage of particle symmetries, as discussed in each relevant particle origin case. The spherical coordinate noted as $(r, \theta, \varphi)$ is frequently used in light scattering. In spherical coordinates, unit vectors have the following expressions in Cartesian coordinates:

$$\hat{r} = (\sin\theta\cos\varphi,\ \sin\theta\sin\varphi,\ \cos\theta), \tag{3.2.1a}$$

$$\hat{\theta} = (\cos\theta\sin\varphi,\ \cos\theta\cos\varphi,\ -\sin\theta), \tag{3.2.1b}$$

$$\hat{\varphi} = (-\sin\varphi,\ \cos\varphi, 0). \tag{3.2.1c}$$

The gradient, divergence, curl, and Laplace operator have the following forms in spherical coordinates (Morse and Feshbach, 1953; Arfken and Weber, 1995, p. 126):

$$\nabla\Psi = \frac{\partial\Psi}{\partial r}\hat{r} + \frac{1}{r}\frac{\partial\Psi}{\partial\theta}\hat{\theta} + \frac{1}{r\sin\theta}\frac{\partial\Psi}{\partial\varphi}\hat{\varphi}, \tag{3.2.2a}$$

$$\nabla\cdot\mathbf{A} = \frac{1}{r^2}\frac{\partial}{\partial r}\left(r^2 \mathrm{A}_r\right) + \frac{1}{r\sin\theta}\frac{\partial}{\partial\theta}(\sin\theta \mathrm{A}_\theta) + \frac{1}{r\sin\theta}\frac{\partial \mathrm{A}_\varphi}{\partial\varphi}, \tag{3.2.2b}$$

$$\begin{aligned}\nabla\times\mathbf{A} = {} & \frac{1}{r\sin\theta}\left[\frac{\partial}{\partial\theta}(\sin\theta \mathrm{A}_\theta) - \frac{\partial \mathrm{A}_\varphi}{\partial\varphi}\right]\hat{r} + \frac{1}{r}\left[\frac{1}{\sin\theta}\frac{\partial \mathrm{A}_\varphi}{\partial\varphi} - \frac{\partial}{\partial r}\left(r\mathrm{A}_\varphi\right)\right]\hat{\theta} \\ & + \frac{1}{r}\left[\frac{\partial}{\partial r}(r\mathrm{A}_\theta) - \frac{\partial \mathrm{A}_r}{\partial\theta}\right]\hat{\varphi},\end{aligned} \tag{3.2.2c}$$

$$\nabla^2\Psi = \frac{1}{r^2}\frac{\partial}{\partial r}\left(r^2\frac{\partial\Psi}{\partial r}\right) + \frac{1}{r^2\sin\theta}\frac{\partial}{\partial\theta}\left(\sin\theta\frac{\partial\Psi}{\partial\theta}\right) + \frac{1}{r^2\sin^2\theta}\frac{\partial^2\Psi}{\partial\varphi^2}, \tag{3.2.2d}$$

where $\Psi = \Psi(r, \theta, \varphi)$ and $\mathbf{A} = \mathbf{A}(r, \theta, \varphi)$ represent arbitrary scalar and vector functions, respectively.

### 3.2.2 Scalar and vector spherical wave functions

For traveling waves, such as sound waves and electromagnetic waves, the governing wave equation satisfies the following expression:

$$\nabla^2\psi - \frac{1}{\upsilon^2}\frac{\partial^2\psi}{\partial t^2} = 0, \tag{3.2.3a}$$

$$\nabla^2\mathbf{F} - \frac{1}{\upsilon^2}\frac{\partial^2\mathbf{F}}{\partial t^2} = 0, \tag{3.2.3b}$$

where $\upsilon$ is the wave velocity and $\psi$ and $\mathbf{F}$ are the corresponding scalar and vector wave fields, respectively. When the time dependence is time harmonic in the form of $\exp(-i\omega t)$, the wave equations become the scalar and vector Helmholtz equations:

$$\nabla^2\psi + k^2\psi = 0, \tag{3.2.4a}$$

$$\nabla^2\mathbf{F} + k^2\mathbf{F} = 0, \tag{3.2.4b}$$

where $k = \omega/\upsilon = 2\pi/\lambda$ is the wave number and $\lambda$ is the wavelength.

In spherical coordinates $(r, \theta, \varphi)$, the Laplace operator in Eq. (3.2.2d) can be represented as

$$\nabla^2 = \frac{1}{r^2}\frac{\partial}{\partial r}r^2\frac{\partial}{\partial r} + \frac{1}{r^2}\nabla^2_\Omega, \tag{3.2.5}$$

The spherical Laplace operator on the surface of a unit sphere is denoted as

$$\nabla^2_\Omega = \frac{1}{\sin^2\theta}\frac{\partial}{\partial\theta}\sin\theta\frac{\partial}{\partial\theta} + \frac{1}{\sin^2\theta}\frac{\partial^2}{\partial\varphi^2}. \tag{3.2.6}$$

Accordingly, the scalar Helmholtz equation in Eq. (3.2.4a) can be solved using the separation of variables in spherical coordinates. Using Eqs. (3.2.4), (3.2.5), one can ascertain that the wave number $k$ and radial length $r$ are always paired together to give the dimensionless quantity $kr$. Correspondingly, what really matters in the wave equation is the ratio of the particle size to the wavelength, rather than only the particle size. The scalar Helmholtz equation can be separated by writing $\psi$ as the product of a radial function and an angular function:

$$\psi = z(kr)\mathbb{Y}(\theta, \varphi). \tag{3.2.7}$$

We will show that the radial function can be expressed in terms of spherical Bessel functions and the angular function is expressed in spherical harmonics. Since the spherical harmonics used here have a factor difference with conventional spherical harmonics $\mathbf{Y}$, blackboard-bold $\mathbb{Y}$ is used to distinguish from the conventional form. Using Eq. (3.2.7), the scalar Helmholtz equation in Eq. (3.2.4) is decomposed into a spherical harmonic equation and a spherical Bessel equation as follows:

$$\nabla_{\Omega}^{2}\mathbb{Y}(\theta,\varphi) = -\Lambda\mathbb{Y}(\theta,\varphi), \tag{3.2.8a}$$

$$\left(\frac{d^2}{d(kr)^2}+\frac{2}{kr}\frac{d}{d(kr)}\right)z(kr)+\left[1-\frac{\Lambda}{(kr)^2}\right]z(kr) = 0, \tag{3.2.8b}$$

where $\Lambda$ is a constant used in the separation of variables. Next, we will briefly introduce how to obtain the solutions of Eqs. (3.2.8a) and (3.2.8b).

For Eq. (3.2.8a), further separation can be made by using

$$\mathbb{Y}(\theta,\varphi) = \Theta(\theta)\Phi(\varphi), \tag{3.2.9}$$

so

$$\frac{d^2\Phi}{d\varphi^2}+m^2\Phi = 0, \tag{3.2.10a}$$

$$\sin\theta\frac{d}{d\theta}\sin\theta\frac{d\Theta}{d\theta}+\left(\Lambda\sin^2\theta-m^2\right)\Theta = 0. \tag{3.2.10b}$$

The additional constant in the separation of variables $(\theta,\varphi)$ is $m^2$, where $m$ is an integer due to the $2\pi$ period of variable $\varphi$. The complex solution for variable $\varphi$ is $\exp(im\varphi)$. Eq. (3.2.10b) is the so-called associated Legendre equation. A power series expansion is used to solve the associated Legendre equation. The function $\Theta$ of variable $\theta$ is physically acceptable or well behaved only if constant $\Lambda$ has a discrete infinite number of nonnegative values (Morse and Feshbach, 1953; Arfken and Weber, 1995). The constant is then $n(n+1)$, $n=0$, 1, 2, ..., and then, the solutions are denoted by associated Legendre polynomials $P_n^m(\cos\theta)$. Accordingly, we have

$$\mathbb{Y}_{nm}(\theta,\varphi) = P_n^m(\cos\theta)\exp(im\varphi),\quad n=0,1,2,\ldots,\quad m=0,\pm1,\ldots,\pm n. \tag{3.2.11}$$

When $m=0$, Eq. (3.2.10b) is reduced to the Legendre equation. The solutions are Legendre polynomials, defined in the form

$$P_n(x) = \frac{1}{2^n n!}\frac{d^n}{dx^n}\left(x^2-1\right)^n,\quad n=0,1,2,\ldots, \tag{3.2.12}$$

where $x=\cos\theta$. It is evident that the parity relation is

$$P_n(-x) = (-1)^n P_n(x). \tag{3.2.13}$$

The first three Legendre polynomials are as follows:

$$P_0(x) = 1,\ \ P_1(x) = x,\ \ P_2(x) = \frac{1}{2}\left(3x^2 - 1\right). \tag{3.2.14}$$

The values for $x = 0, \pm 1$ of the Legendre polynomials are

$$P_{2n+1}(0) = 0,\ \ P_n(1) = 1,\ \ P_n(-1) = (-1)^n,\ \ n = 0,1,2,\ldots \tag{3.2.15}$$

Fig. 3.1 shows the Legendre polynomials $P_n(\cos\theta)$ in polar coordinates, where $n$ is from 0 to 11 and $\theta$ is from 0 to $2\pi$. The Legendre polynomial displays fine features when the order $n$ is increased. However, the values of the Legendre polynomials are always between $-1$ and 1. The Legendre polynomials have the following orthogonality and recurrence relations (Morse and Feshbach, 1953; Arfken and Weber, 1995):

$$\int_{-1}^{1} P_n P_{n'} dx = \frac{2}{2n+1}\delta_{nn'}. \tag{3.2.16}$$

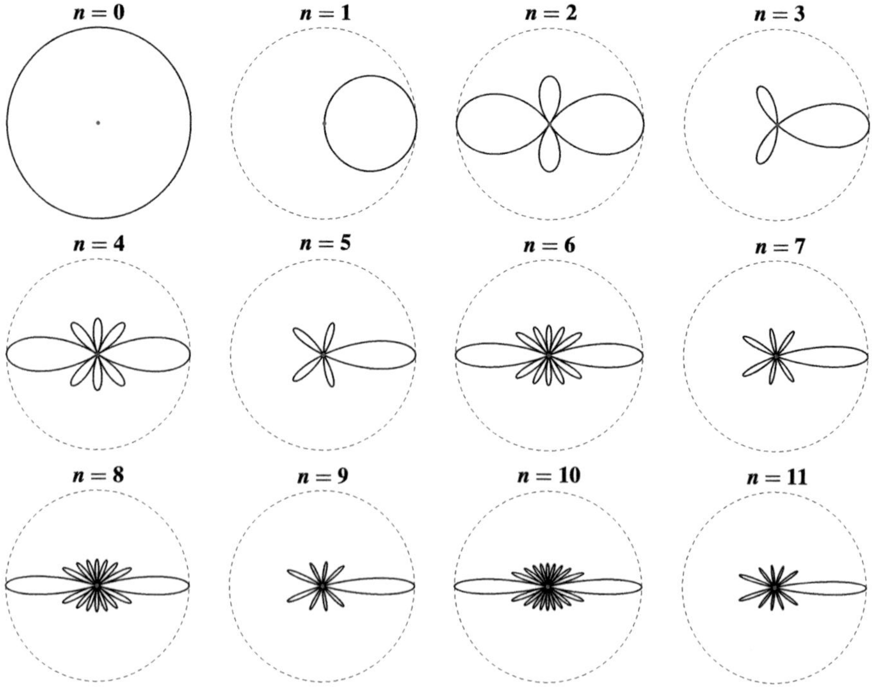


**Fig. 3.1** Legendre polynomials $P_n(\cos\theta)$, $\theta = (0, 2\pi)$ in polar coordinates, where $n = 0, \ldots, 11$. For comparison, a dashed unit circle is given in the figure, where the origin (0,0) is marked as a gray point (see the case with $n = 0$).

$$(n+1)P_{n+1} = (2n+1)xP_n - nP_{n-1}, \tag{3.2.17a}$$

$$(x^2-1)\frac{dP_n}{dx} = xnP_n - nP_{n-1}. \tag{3.2.17b}$$

The argument $x$ of the Legendre polynomials has been suppressed in the relations, and $\delta_{nn'}$ is the Kronecker delta, which is equal to 1 when $n=n'$ and 0 otherwise.

The associated Legendre polynomials can be given as the derivatives of Legendre polynomials for $m \geq 0$ (Abramowitz and Stegun 1965; Tsang et al., 2000):

$$P_n^m(x) = (-1)^m\left(1-x^2\right)^{m/2}\frac{d^m}{dx^m}P_n(x), \quad \begin{matrix} n = 0,1,2,\ldots, \\ m = 0,1,2,\ldots,n \end{matrix}. \tag{3.2.18}$$

It is evident that

$$P_n^0(x) = P_n(x). \tag{3.2.19}$$

If $m$ is negative,

$$P_n^{-m}(x) = (-1)^m\frac{(n-m)!}{(n+m)!}P_n^m(x). \tag{3.2.20}$$

Eq. (3.2.20) is valid for both positive and negative $m$ values. The parity relation for associated Legendre polynomials is

$$P_n^m(-x) = (-1)^{n+m}P_n^m(x). \tag{3.2.21}$$

The associated Legendre polynomials for $n=0$, 1, and 2 are

$$P_0^0(x) = 1, \tag{3.2.22a}$$

$$P_1^0(x) = x, \ \ P_1^1(x) = -\left(1-x^2\right)^{1/2}, \tag{3.2.22b}$$

$$P_2^0(x) = \frac{1}{2}\left(3x^2-1\right), \ \ P_2^1(x) = -3x\left(1-x^2\right)^{1/2}, \ \ P_2^2(x) = 3\left(1-x^2\right). \tag{3.2.22c}$$

Figs. 3.2 and 3.3 show the first several associated Legendre polynomials with $m=1$ and $m=2$ in polar coordinates, respectively. With increased $n$ and $m$, the associated Legendre polynomials display not only finer structures but also larger values. The maximum values of the magnitude of the associated Legendre polynomials approach

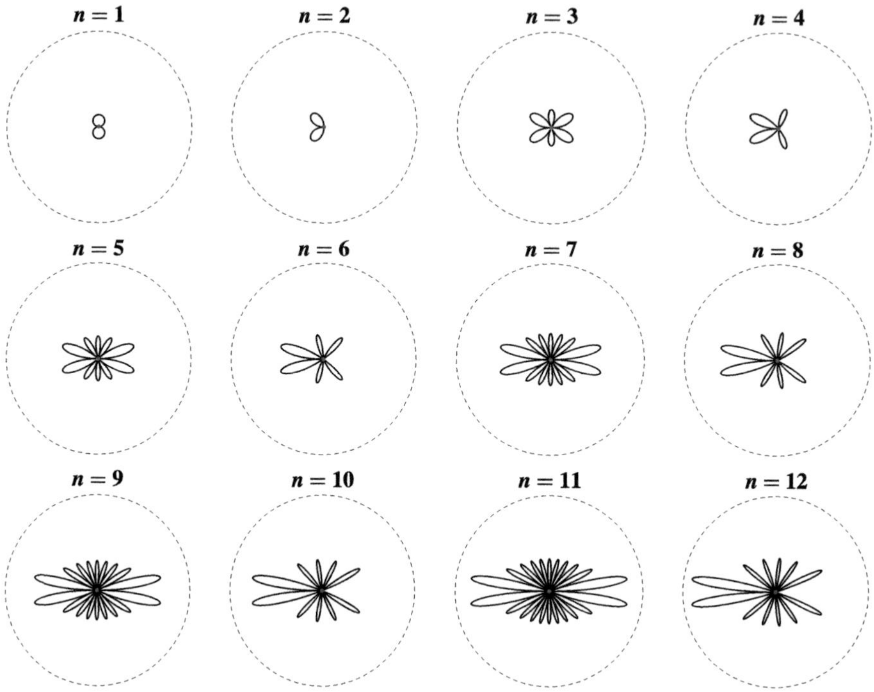


**Fig. 3.2** Associated Legendre polynomial for $P_n^1(\cos\theta)$ in polar coordinates, where $\theta = (0, 2\pi)$ and $n = 1, \ldots, 12$. A dashed circle with radius 8 is given for comparison, where the origin (0,0) is marked in gray.

infinity when $n$ approaches infinity. The values associated with $m < 0$ can be obtained using Eq. (3.2.20). The associated Legendre polynomials have the following orthogonality and recurrence relations (Stratton, 1941):

$$\int_{-1}^{1} P_n^m P_{n'}^m dx = \frac{(n+m)!}{(n-m)!}\frac{2}{2n+1}\delta_{nn'}. \tag{3.2.23}$$

$$(2n+1)xP_n^m = (n-m+1)P_{n+1}^m + (n+m)P_{n-1}^m, \tag{3.2.24a}$$

$$(2n+1)\sqrt{1-x^2}P_n^m = P_{n-1}^{m+1} - P_{n+1}^{m+1}, \tag{3.2.24b}$$

$$(2n+1)\sqrt{1-x^2}P_n^m = (n-m+2)(n-m+1)P_{n+1}^{m-1} - (n+m)(n+m-1)P_{n-1}^{m-1}, \tag{3.2.24c}$$

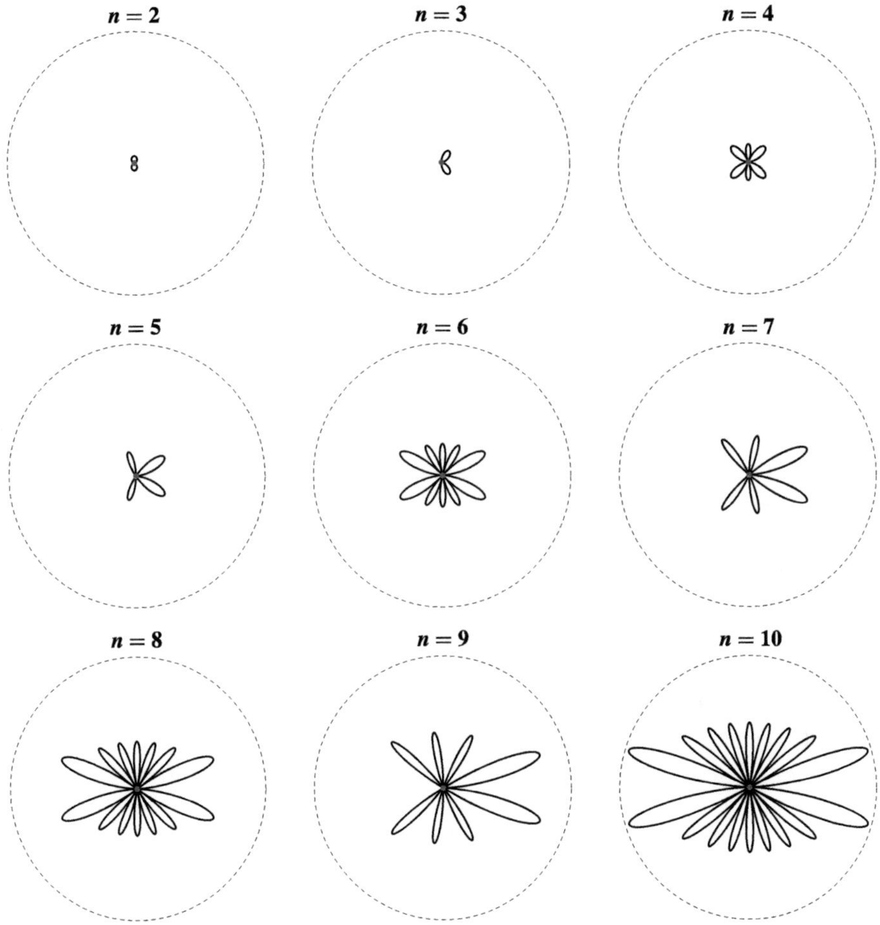


**Fig. 3.3** Associated Legendre polynomial for $P_n^2(\cos\theta)$ in polar coordinates, where $\theta = (0, 2\pi)$ and $n = 2, \ldots, 10$. A dashed circle with radius 55 is given for comparison, where the origin (0,0) is marked in gray.

$$\frac{mP_n^m}{\sqrt{1-x^2}} = -\frac{1}{2}\left[P_{n-1}^{m+1} + (n+m-1)(n+m)P_{n-1}^{m-1}\right], \tag{3.2.24d}$$

$$\frac{mP_n^m}{\sqrt{1-x^2}} = -\frac{1}{2}\left[P_{n+1}^{m+1} + (n-m+1)(n-m+2)P_{n+1}^{m-1}\right], \tag{3.2.24e}$$

$$\frac{mx}{\sqrt{1-x^2}}P_n^m = -\frac{1}{2}\left[(n-m+1)(n+m)P_n^{m-1} + P_n^{m+1}\right], \tag{3.2.24f}$$

$$\left(1-x^2\right)\frac{dP_n^m}{dx}=\frac{(n+1)(n+m)}{2n+1}P_{n-1}^m-\frac{n(n-m+1)}{2n+1}P_{n+1}^m, \tag{3.2.24g}$$

$$\sqrt{1-x^2}\frac{dP_n^m}{dx}=\frac{1}{2}\left[(n-m+1)(n+m)P_n^{m-1}-P_n^{m+1}\right], \tag{3.2.24h}$$

$$\begin{aligned}x\sqrt{1-x^2}\frac{dP_n^m}{dx}=&-\frac{1}{2}\frac{n}{2n+1}\left[P_{n+1}^{m+1}-(n-m+1)(n-m+2)P_{n+1}^{m-1}\right]\\&-\frac{1}{2}\frac{n+1}{2n+1}\left[P_{n-1}^{m+1}-(n+m-1)(n+m)P_{n-1}^{m-1}\right].\end{aligned} \tag{3.2.24i}$$

When it is possible, argument $x$ has been suppressed in the earlier relations for convenience. The earlier formulas are derived from or directly given by Stratton (1941). Note that due to the $(-1)^m$ difference in the definition of the associated Legendre polynomial in Eq. (3.2.18), some recurrence relations have opposite signs compared with those given by Stratton (1941). Accordingly, the spherical harmonics have the following orthogonal relation:

$$\frac{1}{4\pi}\int_{4\pi}\mathbb{Y}_{n'm'}^*(\theta,\varphi)\mathbb{Y}_{nm}(\theta,\varphi)\sin\theta d\theta d\varphi=\delta_{nn'}\delta_{mm'}\frac{1}{2n+1}\frac{(n+m)!}{(n-m)!}. \tag{3.2.25}$$

For unit vectors $\hat{r}_1=(\sin\theta_1\cos\varphi_1,\ \sin\theta_1\sin\varphi_1,\ \cos\theta_1)$ and $\hat{r}_2=(\sin\theta_2\cos\varphi_2,\ \sin\theta_2\sin\varphi_2,\ \cos\theta_2)$, with the relation $\cos\Theta=\hat{r}_1\cdot\hat{r}_2=\cos\theta_1\cos\theta_2+\sin\theta_1\sin\theta_2\cos(\varphi_1-\varphi_2)$, the Legendre polynomials and the associated Legendre polynomials have the following addition theorem (Arfken and Weber, 1995):

$$P_n(\cos\Theta)=\sum_{m=-n}^{n}\frac{(n-m)!}{(n+m)!}P_n^m(\cos\theta_1)P_n^m(\cos\theta_2)\exp[im(\varphi_1-\varphi_2)]. \tag{3.2.26}$$

After the spherical harmonic equation in Eq. (3.2.8a) is explicitly solved, we next consider the spherical Bessel equation. The solution of the spherical Bessel equation in Eq. (3.2.8b) with constant $\Lambda=n(n+1)$ can be rewritten as follows:

$$\left(\frac{d^2}{d(kr)^2}+\frac{2}{kr}\frac{d}{d(kr)}\right)z(kr)+\left[1-\frac{n(n+1)}{(kr)^2}\right]z(kr)=0. \tag{3.2.27}$$

Two sets of independent spherical Bessel functions exist only if their Wronskian is not zero. There are many kinds of Bessel function solutions (Morse and Feshbach, 1953; Korn and Korn, 2000; Arfken and Weber, 1995). Four types of spherical Bessel functions are frequently used for light scattering, which are the spherical Bessel function $j_n$, the spherical Neumann function $y_n$, and the spherical Hankel functions of the first and second kinds $h_n^{(1)}$ and $h_n^{(2)}$. The definitions are as follows (Morse and Feshbach, 1953; Arfken and Weber, 1995):

$$j_n(\rho) = (-\rho)^n \left(\frac{1}{\rho}\frac{d}{d\rho}\right)^n \frac{\sin\rho}{\rho}, \tag{3.2.28a}$$

$$y_n(\rho) = -(-\rho)^n \left(\frac{1}{\rho}\frac{d}{d\rho}\right)^n \frac{\cos\rho}{\rho}, \tag{3.2.28b}$$

$$h_n^{(1)}(\rho) = j_n(\rho) + iy_n(\rho), \tag{3.2.28c}$$

$$h_n^{(2)}(\rho) = j_n(\rho) - iy_n(\rho), \tag{3.2.28d}$$

where $\rho = kr$ is used in this subsection for convenience.

All spherical Bessel functions have the following recurrence relations (Morse and Feshbach, 1953; Arfken and Weber, 1995):

$$\frac{2n+1}{\rho} z_n(\rho) = z_{n-1}(\rho) + z_{n+1}(\rho), \tag{3.2.29a}$$

$$(2n+1)\frac{dz_n(\rho)}{d\rho} = nz_{n-1}(\rho) - (n+1)z_{n+1}(\rho), \tag{3.2.29b}$$

where $z_n$ represents any of the spherical Bessel functions ($j_n$, $y_n$, $h_n^{(1)}$, or $h_n^{(2)}$) and $\rho$ is the argument of the functions.

The asymptotic forms of the spherical Bessel function, the spherical Neumann function, and their product in asymptotic forms, that is, as $n \to \infty$, can be expressed as (Abramowitz and Stegun, 1965)

$$j_n(\rho) \sim \frac{1}{n \cdot 2\sqrt{2}} \left(\frac{e\rho}{2n}\right)^n, \tag{3.2.30a}$$

$$y_n(\rho) \sim -\frac{\sqrt{2}}{\rho} \left(\frac{e\rho}{2n}\right)^{-n}, \tag{3.2.30b}$$

$$j_n(\rho) \cdot y_n(\rho) \sim -\frac{1}{2n\rho}, \tag{3.2.30c}$$

where $n > \rho$ and $e$ is Euler's number. The spherical Bessel function $j_n$ and the spherical Neumann function $y_n$ for $n = 0, \ldots, 4$ are shown in Fig. 3.4. Using Eq. (3.2.30) and Fig. 3.4, the spherical Bessel function $j_n$ for a fixed argument is well behaved for asymptotic large orders, while the spherical Neumann functions for a fixed argument are divergent for asymptotic large orders. The product of the spherical Bessel function

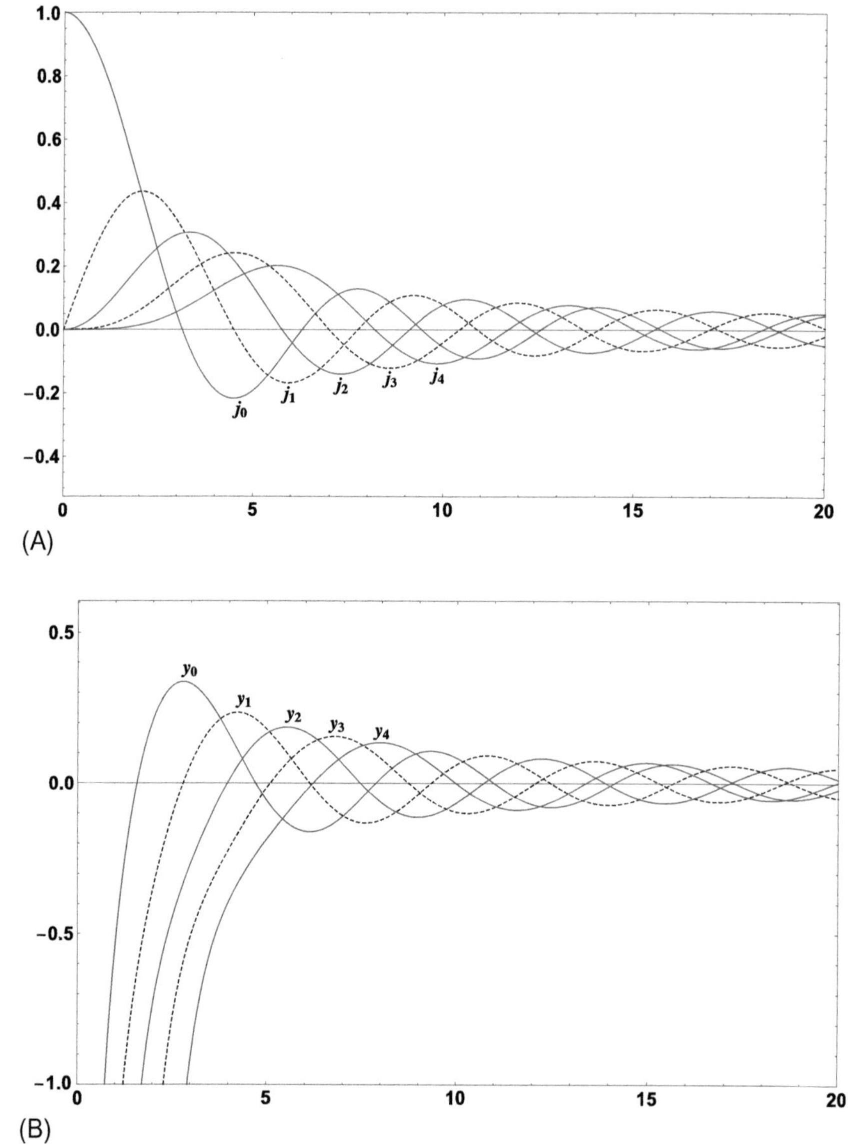


**Fig. 3.4** The spherical Bessel function $j_n$ in panel (A) and the spherical Neumann function $y_n$ in panel (B), where $n=0, \ldots, 4$.

and the spherical Neumann function of the same order, however, is well behaved when the order approaches infinity. Using Eq. (3.2.30), the recurrence relations for the spherical Bessel and spherical Neumann functions are

$$\frac{j_{n-1}(\rho)}{j_n(\rho)} \sim \frac{2n}{e\rho}, \tag{3.2.31a}$$

$$\frac{y_{n-1}(\rho)}{y_n(\rho)} \sim \frac{e\rho}{2n}, \tag{3.2.31b}$$

The value $|j_{n-1}(\rho)/j_n(\rho)|$ increases when order $n$ increases, while the value $|y_{n-1}(\rho)/y_n(\rho)|$ does the opposite. Accordingly, only the downward recurrence relation for the spherical Bessel function computation is stable (Kattawar and Plass, 1967), whereas the upward recurrence relation is stable for the spherical Neumann function.

The Wronskians for the spherical Bessel functions are as follows (Arfken and Weber, 1995):

$$\Delta(j_n(\rho), y_n(\rho)) = \begin{vmatrix} j_n(\rho) & y_n(\rho) \\ \frac{dj_n(\rho)}{d\rho} & \frac{dy_n(\rho)}{d\rho} \end{vmatrix} = \frac{1}{\rho^2}, \tag{3.2.32a}$$

$$\Delta\left(j_n(\rho), h_n^{(1)}(\rho)\right) = \begin{vmatrix} j_n(\rho) & h_n^{(1)}(\rho) \\ \frac{dj_n(\rho)}{d\rho} & \frac{dh_n^{(1)}(\rho)}{d\rho} \end{vmatrix} = \frac{i}{\rho^2}, \tag{3.2.32b}$$

$$\Delta\left(h_n^{(2)}(\rho), h_n^{(1)}(\rho)\right) = \begin{vmatrix} h_n^{(2)}(\rho) & h_n^{(1)}(\rho) \\ \frac{dh_n^{(2)}(\rho)}{d\rho} & \frac{dh_n^{(1)}(\rho)}{d\rho} \end{vmatrix} = \frac{2i}{\rho^2}. \tag{3.2.32c}$$

The first three terms for each of the four types of spherical Bessel functions are as follows:

$$j_0(\rho) = \frac{\sin\rho}{\rho}, \quad j_1(\rho) = \frac{\sin\rho}{\rho^2} - \frac{\cos\rho}{\rho}, \quad j_2(\rho) = \left(\frac{3}{\rho^2} - 1\right)\frac{\sin\rho}{\rho} - \frac{3\cos\rho}{\rho^2}, \tag{3.2.33a}$$

$$y_0(\rho) = -\frac{\cos\rho}{\rho}, \quad y_1(\rho) = -\frac{\cos\rho}{\rho^2} - \frac{\sin\rho}{\rho}, \quad y_2(\rho) = -\left(\frac{3}{\rho^2} - 1\right)\frac{\cos\rho}{\rho} - \frac{3\sin\rho}{\rho^2}, \tag{3.2.33b}$$

$$h_0^{(1)}(\rho) = \frac{\exp(i\rho)}{i\rho}, \quad h_1^{(1)}(\rho) = \left(\frac{1}{\rho} - i\right)\frac{\exp(i\rho)}{i\rho}, \quad h_2^{(1)}(\rho) = \left(\frac{3}{\rho^2} + \frac{3}{i\rho} - 1\right)\frac{\exp(i\rho)}{i\rho}, \tag{3.2.33c}$$

$$h_0^{(2)}(\rho) = \frac{\exp(-i\rho)}{-i\rho}, \quad h_1^{(2)}(\rho) = \left(\frac{1}{\rho} + i\right)\frac{\exp(-i\rho)}{-i\rho}, \quad h_2^{(2)}(\rho) = \left(\frac{3}{\rho^2} - \frac{3}{i\rho} - 1\right)\frac{\exp(-i\rho)}{-i\rho}. \tag{3.2.33d}$$

The four types of spherical Bessel functions behave at the origin and infinity as follows (Morse and Feshbach, 1953; Arfken and Weber, 1995):

$$j_n(\rho)\big|_{\rho\to 0} = \frac{\rho^n}{(2n+1)!!}, \quad j_n(\rho)\big|_{\rho\to\infty} = \frac{\sin(\rho - n\pi/2)}{\rho}, \tag{3.2.34a}$$

$$y_n(\rho)\big|_{\rho\to 0} = -\frac{(2n-1)!!}{\rho^{n+1}}, \quad y_n(\rho)\big|_{\rho\to\infty} = -\frac{\cos(\rho - n\pi/2)}{\rho}, \tag{3.2.34b}$$

$$h_n^{(1)}(\rho)\big|_{\rho\to 0} = \frac{\rho^n}{(2n+1)!!} - i\frac{(2n-1)!!}{\rho^{n+1}}, \quad h_n^{(1)}(\rho)\big|_{\rho\to\infty} = \frac{\exp(i\rho)}{i\rho}(-i)^n, \tag{3.2.34c}$$

$$h_n^{(2)}(\rho)\big|_{\rho\to 0} = \frac{\rho^n}{(2n+1)!!} + i\frac{(2n-1)!!}{\rho^{n+1}}, \quad h_n^{(2)}(\rho)\big|_{\rho\to\infty} = \frac{\exp(-i\rho)}{-i\rho}i^n. \tag{3.2.34d}$$

Consequently, the spherical Bessel function satisfies the finite condition at the origin, and the spherical Hankel function of the first kind at infinity satisfies the Sommerfeld radiation condition as follows (Schot, 1992):

$$\lim_{\rho\to\infty} \rho\left(\frac{\partial}{\partial\rho} - i\right) h_n(\rho) = 0, \tag{3.2.35}$$

where $h_n(\rho)$ denotes the spherical Hankel function of the first kind or second kind. The spherical Hankel function of the first kind behaves as an outgoing radiation field radiating from the origin, while the second kind behaves as an incoming sink field from infinity to the origin.

The solution of the spherical harmonic equation in Eq. (3.2.8a) is given in detail from Eq. (3.2.9) to Eq. (3.2.26). The solution of the spherical Bessel equation in Eq. (3.2.8b), or Eq. (3.2.27) used in light scattering, is given in detail from Eqs. (3.2.28)–(3.2.35). Accordingly, the solution of the scalar Helmholtz equation in Eq. (3.2.4a) is the scalar spherical wave function, which can be written as

$$\psi_{mn}(\rho,\theta,\varphi) = z_n(\rho)P_n^m(\cos\theta)\exp(im\varphi), \quad \begin{array}{l} n = 0, 1, 2, \ldots, \\ m = 0, \pm 1, \pm 2, \ldots, \pm n, \end{array} \tag{3.2.36}$$

where $z_n(\rho)$ represents one of the four spherical Bessel functions. The incident field is finite at the origin, and the scattered field satisfies the Sommerfeld outgoing radiation condition. Accordingly, the spherical Bessel function $j_n(\rho)$ and the spherical Hankel function of the first kind $h_n^{(1)}(\rho)$ are employed in light scattering discussions. Following conventional notation and being consistent between the scalar and the vector functions, the spherical wave functions using $h_n^{(1)}(\rho)$ and $j_n(\rho)$ are respectively noted as $\psi$ and Rg$\psi$ as follows (Tsang et al., 2000; Mishchenko et al., 2002):

$$\begin{pmatrix} \psi_{mn}(\rho,\theta,\varphi) \\ \mathrm{Rg}\psi_{mn}(\rho,\theta,\varphi) \end{pmatrix} = \begin{pmatrix} h_n^{(1)}(\rho) \\ j_n(\rho) \end{pmatrix} P_n^m(\cos\theta)\exp(im\varphi), \tag{3.2.37}$$

where "Rg" denotes the regular spherical wave functions and is associated with $j_n(\rho)$. The scalar wave functions in Eq. (3.2.37) form a complete set of solution functions for the scalar Helmholtz equation. Owing to the linearity of the scalar Helmholtz equation, a general solution can be written as a linear superposition of these spherical wave functions. The corresponding superposition coefficients can be determined by the use of the boundary conditions.

For the vector Helmholtz equation in Eq. (3.2.4b), the solution is expressed as vector spherical wave functions, which can be constructed in terms of the scalar spherical wave functions in Eq. (3.2.37). The vector field **F** in Eq. (3.2.4b) can be decomposed into one longitudinal and two transverse components. One should expect that the longitudinal component should be curl free and can be given as the gradient of the scalar spherical wave functions. Also, the two transverse components should be divergence free and can be given as the curl and the double curl of certain vector functions associated with the scalar spherical wave functions. The three sets of vector spherical wave functions can be constructed as follows (Stratton, 1941; Morse and Feshbach, 1953; Tsang et al., 2000; Mishchenko et al., 2002):

$$\begin{pmatrix} \mathbf{L}_{mn}(\rho,\theta,\varphi) \\ \mathrm{Rg}\mathbf{L}_{mn}(\rho,\theta,\varphi) \end{pmatrix} = \gamma'_{mn}\nabla_\rho \begin{pmatrix} \psi_{mn}(\rho,\theta,\varphi) \\ \mathrm{Rg}\psi_{mn}(\rho,\theta,\varphi) \end{pmatrix}$$

$$= \gamma'_{mn}\left\{ \frac{d}{d\rho}\begin{pmatrix} h_n^{(1)}(\rho) \\ j_n(\rho) \end{pmatrix} \mathbf{P}_{mn}(\theta,\varphi) + \frac{1}{\rho}\begin{pmatrix} h_n^{(1)}(\rho) \\ j_n(\rho) \end{pmatrix} \mathbf{B}_{mn}(\theta,\varphi) \right\}, \tag{3.2.38a}$$

$$\begin{pmatrix} \mathbf{M}_{mn}(\rho,\theta,\varphi) \\ \mathrm{Rg}\mathbf{M}_{mn}(\rho,\theta,\varphi) \end{pmatrix} = \gamma_{mn}\nabla_\rho \times \left[ \vec{\rho} \begin{pmatrix} \psi_{mn}(\rho,\theta,\varphi) \\ \mathrm{Rg}\psi_{mn}(\rho,\theta,\varphi) \end{pmatrix} \right]$$

$$= \gamma_{mn}\begin{pmatrix} h_n^{(1)}(\rho) \\ j_n(\rho) \end{pmatrix} \mathbf{C}_{mn}(\theta,\varphi) \tag{3.2.38b}$$

$$= \nabla_\rho \times \begin{pmatrix} \mathbf{N}_{mn}(\rho,\theta,\varphi) \\ \mathrm{Rg}\mathbf{N}_{mn}(\rho,\theta,\varphi) \end{pmatrix},$$

$$\begin{pmatrix} \mathbf{N}_{mn}(\rho,\theta,\varphi) \\ \mathrm{Rg}\mathbf{N}_{mn}(\rho,\theta,\varphi) \end{pmatrix} = \nabla_\rho \times \begin{pmatrix} \mathbf{M}_{mn}(\rho,\theta,\varphi) \\ \mathrm{Rg}\mathbf{M}_{mn}(\rho,\theta,\varphi) \end{pmatrix}$$

$$= \gamma_{mn}\left\{ \frac{n(n+1)}{\rho}\begin{pmatrix} h_n^{(1)}(\rho) \\ j_n(\rho) \end{pmatrix} \mathbf{P}_{mn}(\theta,\varphi) + \frac{1}{\rho}\frac{d}{d\rho}\left[ \rho \begin{pmatrix} h_n^{(1)}(\rho) \\ j_n(\rho) \end{pmatrix} \right] \mathbf{B}_{mn}(\theta,\varphi) \right\}, \tag{3.2.38c}$$

where $\vec{\rho} = k\vec{r}$ and $\nabla_\rho = \nabla/k$; "Rg" in the earlier equations denotes the regular vector spherical wave functions, where the spherical Bessel function $j_n(\rho)$ is used; otherwise, the equations are the outgoing vector spherical wave functions, where the spherical Hankel function of the first kind $h_n^{(1)}(\rho)$ is used. Note that $n$ starts with 0 for $\{(\mathrm{Rg})\mathbf{L}_{mn}\}$ while $n$ starts with 1 for $\{(\mathrm{Rg})\mathbf{M}_{mn}\}$ and $\{(\mathrm{Rg})\mathbf{N}_{mn}\}$. It is indeed true that $\{(\mathrm{Rg})\mathbf{L}_{mn}\}$ is curl free and $\{(\mathrm{Rg})\mathbf{M}_{mn}\}$ and $\{(\mathrm{Rg})\mathbf{N}_{mn}\}$ are divergence free. Also, the coefficients and the functions of angular arguments in Eq. (3.2.38) are (Tsang et al., 2000; Mishchenko et al., 2002)

$$\gamma'_{mn} = \sqrt{\frac{(2n+1)(n-m)!}{4\pi(n+m)!}}, \tag{3.2.39a}$$

$$\gamma_{mn} = \sqrt{\frac{(2n+1)(n-m)!}{4\pi n(n+1)(n+m)!}}, \tag{3.2.39b}$$

and

$$\mathbb{Y}_{nm}(\theta,\varphi) = P_n^m(\cos\theta)\exp(im\varphi), \tag{3.2.40}$$

$$\mathbf{V}_{mn}^{(1)}(\theta,\varphi) = \mathbf{P}_{mn}(\theta,\varphi) = \hat{r}\mathbb{Y}_{nm}(\theta,\varphi), \tag{3.2.41a}$$

$$\begin{aligned}\mathbf{V}_{mn}^{(2)}(\theta,\varphi) &= \mathbf{B}_{mn}(\theta,\varphi) = \rho\nabla_\rho\mathbb{Y}_{nm}(\theta,\varphi)\\ &= \left[\hat{\theta}\frac{dP_n^m(\cos\theta)}{d\theta} + \hat{\varphi}\frac{im}{\sin\theta}P_n^m(\cos\theta)\right]\exp(im\varphi)\\ &= \hat{r}\times\mathbf{C}_{mn}(\theta,\varphi),\end{aligned} \tag{3.2.41b}$$

$$\begin{aligned}\mathbf{V}_{mn}^{(3)}(\theta,\varphi) &= \mathbf{C}_{mn}(\theta,\varphi) = \nabla_\rho\times\left[\vec{\rho}\;\mathbb{Y}_{nm}(\theta,\varphi)\right]\\ &= \left[\hat{\theta}\frac{im}{\sin\theta}P_n^m(\cos\theta) - \hat{\varphi}\frac{dP_n^m(\cos\theta)}{d\theta}\right]\exp(im\varphi)\\ &= \mathbf{B}_{mn}(\theta,\varphi)\times\hat{r}\end{aligned} \tag{3.2.41c}$$

where $\mathbf{V}_{mn}^{(1-3)}$ are vector spherical harmonics and $\hat{r}$, $\hat{\theta}$, and $\hat{\varphi}$ are the unit vectors in spherical coordinates, and the complex conjugate of the vector spherical harmonics has the unique relation as follows (Mishchenko et al., 2002):

$$\mathbf{V}_{mn}^{(i)*}(\theta,\varphi) = (-1)^m\frac{(n+m)!}{(n-m)!}\mathbf{V}_{-mn}^{(i)}(\theta,\varphi), \quad i = 1,2,3. \tag{3.2.41d}$$

$\mathbf{P}_{mn}(\theta,\varphi)$ denotes the longitudinal vector spherical harmonics, and $\mathbf{B}_{mn}(\theta,\varphi)$ and $\mathbf{C}_{mn}(\theta,\varphi)$ are the transverse vector spherical harmonics. Their orthogonality relations are shown as (Mishchenko et al., 2002)

$$\begin{aligned}\int_{4\pi} d\Omega\mathbf{B}_{mn}(\theta,\varphi)\cdot\mathbf{C}_{m'n'}^*(\theta,\varphi) &= \int_{4\pi} d\Omega\mathbf{P}_{mn}(\theta,\varphi)\cdot\mathbf{B}_{m'n'}^*(\theta,\varphi)\\ &= \int_{4\pi} d\Omega\mathbf{C}_{mn}(\theta,\varphi)\cdot\mathbf{P}_{m'n'}^*(\theta,\varphi) = 0,\end{aligned} \tag{3.2.42a}$$

$$\int_{4\pi} d\Omega \mathbf{B}_{mn}(\theta,\varphi)\cdot\mathbf{B}^*_{m'n'}(\theta,\varphi) = \int_{4\pi} d\Omega \mathbf{C}_{mn}(\theta,\varphi)\cdot\mathbf{C}^*_{m'n'}(\theta,\varphi)$$

$$= \frac{1}{(\gamma_{mn})^2}\delta_{mm'}\delta_{nn'}, \tag{3.2.42b}$$

$$\int_{4\pi} d\Omega \mathbf{P}_{mn}(\theta,\varphi)\cdot\mathbf{P}^*_{m'n'}(\theta,\varphi) = \frac{1}{(\gamma'_{mn})^2}\delta_{mm'}\delta_{nn'}. \tag{3.2.42c}$$

Using the parity relation of the associated Legendre polynomials $P_n^m$ in Eq. (3.2.21), the parity relations of the vector spherical harmonics can be given as follows:

$$\mathbf{P}_{mn}(\pi-\theta,\pi+\varphi) = (-1)^{n+1}\mathbf{P}_{mn}(\theta,\varphi), \tag{3.2.43a}$$

$$\mathbf{B}_{mn}(\pi-\theta,\pi+\varphi) = (-1)^{n+1}\mathbf{B}_{mn}(\theta,\varphi), \tag{3.2.43b}$$

$$\mathbf{C}_{mn}(\pi-\theta,\pi+\varphi) = (-1)^{n}\mathbf{C}_{mn}(\theta,\varphi). \tag{3.2.43c}$$

Consequently, the parity relations of the (regular) vector spherical wave functions are given as

$$(\mathrm{Rg})\mathbf{L}_{mn}\left(-k\vec{r}\right) = (-1)^{n+1}(\mathrm{Rg})\mathbf{L}_{mn}\left(k\vec{r}\right), \tag{3.2.44a}$$

$$(\mathrm{Rg})\mathbf{M}_{mn}\left(-k\vec{r}\right) = (-1)^{n}(\mathrm{Rg})\mathbf{M}_{mn}\left(k\vec{r}\right), \tag{3.2.44b}$$

$$(\mathrm{Rg})\mathbf{N}_{mn}\left(-k\vec{r}\right) = (-1)^{n+1}(\mathrm{Rg})\mathbf{N}_{mn}\left(k\vec{r}\right), \tag{3.2.44c}$$

where the parity relations are valid for both vector spherical wave functions because the parity is invariant for spherical Bessel or Hankel functions.

Note that there is a constant coefficient with respect to $n$ and $m$ between the $\mathbb{Y}_{nm}(\theta,\varphi)$ and conventionally used spherical harmonics $\mathrm{Y}_{nm}(\theta,\varphi)$. They have the following relations and properties:

$$\mathrm{Y}_{nm}(\theta,\varphi) = \sqrt{\frac{(2n+1)(n-m)!}{4\pi(n+m)!}}\mathbb{Y}_{nm}(\theta,\varphi). \tag{3.2.45a}$$

$$\mathrm{Y}^*_{nm}(\theta,\varphi) = (-1)^m\mathrm{Y}_{n-m}(\theta,\varphi), \tag{3.2.45b}$$

$$\mathbb{Y}^*_{nm}(\theta,\varphi) = (-1)^m\frac{(n+m)!}{(n-m)!}\mathbb{Y}_{n-m}(\theta,\varphi). \tag{3.2.45c}$$

The "*" denotes the complex conjugate here and throughout this book.

$\mathbf{L}_{mn}$ at infinity acting as an outgoing longitudinal wave and $\mathbf{M}_{mn}$ and $\mathbf{N}_{mn}$ at infinity acting as an outgoing transverse wave can be expressed in terms of Eq. (3.2.34c) as follows (Mishchenko et al., 2002):

$$\mathbf{L}_{mn}(\rho,\theta,\varphi)\big|_{\rho\to\infty} = \frac{(-i)^n \exp(i\rho)}{\rho}\gamma'_{mn}\mathbf{P}_{mn}(\theta,\varphi), \tag{3.2.46a}$$

$$\mathbf{M}_{mn}(\rho,\theta,\varphi)\big|_{\rho\to\infty} = \frac{(-i)^{n+1} \exp(i\rho)}{\rho}\gamma_{mn}\mathbf{C}_{mn}(\theta,\varphi), \tag{3.2.46b}$$

$$\mathbf{N}_{mn}(\rho,\theta,\varphi)\big|_{\rho\to\infty} = \frac{(-i)^n \exp(i\rho)}{\rho}\gamma_{mn}\mathbf{B}_{mn}(\theta,\varphi). \tag{3.2.46c}$$

However, the regular vector spherical wave functions at infinity act as standing waves, or addition of the outgoing and the incoming waves, in terms of Eq. (3.2.34a) as follows:

$$\begin{aligned}\mathrm{Rg}\mathbf{L}_{mn}(\rho,\theta,\varphi)\big|_{\rho\to\infty} &= \frac{\cos(\rho-n\pi/2)}{\rho}\gamma'_{mn}\mathbf{P}_{mn}(\theta,\varphi)\\ &= \frac{1}{2}\left[\frac{(-i)^n\exp(i\rho)}{\rho}+\frac{i^n\exp(-i\rho)}{\rho}\right]\gamma'_{mn}\mathbf{P}_{mn}(\theta,\varphi),\end{aligned} \tag{3.2.47a}$$

$$\begin{aligned}\mathrm{Rg}\mathbf{M}_{mn}(\rho,\theta,\varphi)\big|_{\rho\to\infty} &= \frac{\sin(\rho-n\pi/2)}{\rho}\gamma_{mn}\mathbf{C}_{mn}(\theta,\varphi)\\ &= \frac{1}{2}\left[\frac{(-i)^{n+1}\exp(i\rho)}{\rho}+\frac{i^{n+1}\exp(-i\rho)}{\rho}\right]\gamma_{mn}\mathbf{C}_{mn}(\theta,\varphi),\end{aligned} \tag{3.2.47b}$$

$$\begin{aligned}\mathrm{Rg}\mathbf{N}_{mn}(\rho,\theta,\varphi)\big|_{\rho\to\infty} &= \frac{\cos(\rho-n\pi/2)}{\rho}\gamma_{mn}\mathbf{B}_{mn}(\theta,\varphi)\\ &= \frac{1}{2}\left[\frac{(-i)^n\exp(i\rho)}{\rho}+\frac{i^n\exp(-i\rho)}{\rho}\right]\gamma_{mn}\mathbf{B}_{mn}(\theta,\varphi).\end{aligned} \tag{3.2.47c}$$

The regular vector spherical wave functions can also be presented as integrals of vector spherical harmonics as follows (Tsang et al., 2000):

$$\mathrm{Rg}\mathbf{L}_{mn}(\rho,\theta,\varphi) = \frac{(-i)^{n-1}}{4\pi}\gamma'_{mn}\int_{4\pi} d\Omega' \exp\left(i\vec{\rho}\cdot\hat{r}'\right)\mathbf{P}_{mn}(\theta',\varphi'), \tag{3.2.48a}$$

$$\mathrm{Rg}\mathbf{M}_{mn}(\rho,\theta,\varphi) = \frac{(-i)^{n}}{4\pi}\gamma_{mn}\int_{4\pi} d\Omega' \exp\left(i\vec{\rho}\cdot\hat{r}'\right)\mathbf{C}_{mn}(\theta',\varphi'), \tag{3.2.48b}$$

$$\mathrm{Rg}\mathbf{N}_{mn}(\rho,\theta,\varphi)=\frac{(-i)^{n-1}}{4\pi}\gamma_{mn}\int\limits_{4\pi}d\Omega'\exp\left(i\vec{\rho}\cdot\hat{r}'\right)\mathbf{B}_{mn}(\theta',\varphi'). \tag{3.2.48c}$$

Moreover, the Dirac delta function can be expressed as a sum of the spherical harmonics ( Jackson, 1998; Arfken and Weber, 1995):

$$\delta(\cos\theta-\cos\theta')\delta(\varphi-\varphi')=\sum_{n=0}^{\infty}\sum_{m=-n}^{n}\mathrm{Y}_{nm}(\theta,\varphi)\mathrm{Y}_{nm}^{*}(\theta',\varphi'). \tag{3.2.49}$$

The left-hand side of Eq. (3.2.49) describes a Dirac delta function in spherical coordinates. The delta function is mathematically defined, but only physically meaningful when integrated. The physical meaning is that the delta function expression is zero when $\cos\theta\neq\cos\theta'$ or $\varphi\neq\varphi'$ integrates to unity over a unit sphere, that is, $\int_{4\pi}\delta(\cos\theta-\cos\theta')\delta(\varphi-\varphi')d\Omega=1$. The dyad $\overleftrightarrow{\mathrm{I}}\exp\left(\mathrm{i}\vec{p}\cdot\vec{r}\right)$ can be expressed in terms of Eqs. (3.2.48), (3.2.49) as follows (Morse and Feshbach, 1953; Tsang et al., 2000; Mishchenko et al., 2002):

$$\begin{aligned}\overleftrightarrow{\mathrm{I}}\exp\left(\mathrm{i}\vec{p}\cdot\vec{r}\right)=\sum_{n=0}^{\infty}\sum_{m=-n}^{n}(-1)^{m}\frac{(2n+1)}{n(n+1)}i^{n}\Bigg[&n(n+1)\frac{\mathbf{P}_{-mn}\left(\theta_p,\varphi_p\right)}{i\gamma'_{mn}}\mathrm{Rg}\mathbf{L}_{mn}(pr,\theta,\varphi)\\&+\frac{\mathbf{C}_{-mn}\left(\theta_p,\varphi_p\right)}{\gamma_{mn}}\mathrm{Rg}\mathbf{M}_{mn}(pr,\theta,\varphi)+\frac{\mathbf{B}_{-mn}\left(\theta_p,\varphi_p\right)}{i\gamma_{mn}}\mathrm{Rg}\mathbf{N}_{mn}(pr,\theta,\varphi)\Bigg],\end{aligned} \tag{3.2.50}$$

where $\vec{p}=\{p,\theta_p,\varphi_p\}$ in spherical coordinates and $\overleftrightarrow{\mathrm{I}}$ is a unit dyad, that is, $\mathbf{F}\cdot\overleftrightarrow{\mathrm{I}}=\overleftrightarrow{\mathrm{I}}\cdot\mathbf{F}=\mathbf{F}$, where $\mathbf{F}$ is an arbitrary vector; the juxtaposition of two vectors in the right-hand side of Eq. (3.2.50) serves as a dyad. Note that, for the three terms, the summation index $n$ starts at 0 is only valid for the first term of Eq. (3.2.50) and starts at 1 in other terms.

A plane wave is given by $\mathbf{E}_0\exp\left(ik\hat{e}_i\cdot\vec{r}\right)$, where $\mathbf{E}_0$ is the amplitude of the electric field and $\hat{e}_i$ is the propagation direction. Using Eq. (3.2.50), the plane wave can be expressed using the regular vector spherical wave functions as follows:

$$\begin{aligned}\mathbf{E}_0\exp\left(ik\hat{e}_i\cdot\vec{r}\right)=\sum_{n=1}^{\infty}\sum_{m=-n}^{n}(-1)^{m}\frac{(2n+1)}{n(n+1)}\frac{i^{n}}{\gamma_{mn}}\{&[\mathbf{E}_0\cdot\mathbf{C}_{-mn}(\theta_i,\varphi_i)]\\&\mathrm{Rg}\mathbf{M}_{mn}(kr,\theta,\varphi)-i[\mathbf{E}_0\cdot\mathbf{B}_{-mn}(\theta_i,\varphi_i)]\mathrm{Rg}\mathbf{N}_{mn}(kr,\theta,\varphi)\},\end{aligned} \tag{3.2.51}$$

where the unit direction $\hat{e}_i=(\theta_i,\varphi_i)$ and $\mathbf{E}_0\cdot\hat{e}_i=0$, and Eq. (3.2.41a) is used; the summation index $n$ starts from 1 instead of 0 because the longitudinal component disappears.

### 3.2.3 Expansion of incident and scattered fields and T-matrix

We have exploited the completeness of the vector spherical wave functions (VSWF). The linearity of the vector Helmholtz equation allows us to write the incident and scattered fields, as well as the field inside the particle, as a linear superposition of VSWF. For an incident plane wave $\mathbf{E}^{\text{inc}}\left(\vec{r}\right) = \mathbf{E}_0 \exp\left(ik\hat{e}_i \cdot \vec{r}\right)$, where $\hat{e}_i$ is the incident direction, Eq. (3.2.51) has actually given the expansion of the incident field using the regular VSWF because the incident plane wave at the origin is regular. The incident and scattered fields both satisfy the divergence-free condition. Considering the incident plane wave scattering of a dielectric particle, the scattered wave at infinity is an outgoing spherical wave. Correspondingly, the incident and scattered electric fields can be expanded as follows (Barber and Hill, 1990; Tsang et al., 2000; Mishchenko et al., 2002):

$$\mathbf{E}^{\text{inc}}\left(\vec{r}\right) = \sum_{n,m} \left[a_{mn} \text{Rg}\mathbf{M}_{mn}\left(k\vec{r}\right) + b_{mn} \text{Rg}\mathbf{N}_{mn}\left(k\vec{r}\right)\right], \tag{3.2.52a}$$

$$\mathbf{E}^{\text{sca}}\left(\vec{r}\right) = \sum_{n,m} \left[p_{mn} \mathbf{M}_{mn}\left(k\vec{r}\right) + q_{mn} \mathbf{N}_{mn}\left(k\vec{r}\right)\right], \tag{3.2.52b}$$

where the ranges of the summation indices $n$ and $m$ are the same as in Eq. (3.2.51) and are suppressed for convenience. The incident expansion coefficients $a_{mn}$ and $b_{mn}$ are known and straightforwardly given by Eq. (3.2.51). Since the boundary conditions are linear, there must be a linear relation between the expansion coefficients of the incident and the scattered fields. This relation is given by the T-matrix (Waterman, 1965; Barber and Hill, 1990; Tsang et al., 2000; Mishchenko et al., 2002):

$$\begin{pmatrix} p_{mn} \\ q_{mn} \end{pmatrix} = \sum_{n',m'} \begin{pmatrix} \mathrm{T}^{11}_{mnm'n'} & \mathrm{T}^{12}_{mnm'n'} \\ \mathrm{T}^{21}_{mnm'n'} & \mathrm{T}^{22}_{mnm'n'} \end{pmatrix} \begin{pmatrix} a_{m'n'} \\ b_{m'n'} \end{pmatrix}. \tag{3.2.53}$$

As given in Eq. (3.2.51), the expansion coefficients are determined by two indices $m$ and $n$. In a compact vector and matrix form, the indices $m$ and $n$ can be replaced by a combined index l. There is a unique correspondence between $l$ and $(n,m)$:

$$l = n(n+1) + m \Leftrightarrow \begin{cases} n = \text{int}(\sqrt{l}) \\ m = l - n(n+1) \end{cases}, \tag{3.2.54}$$

where the function "int" is the largest integer that does not exceed the value of $\sqrt{l}$. Using Eq. (3.2.54), the index $l$ and indices $(m,n)$ have unique correspondence, where $l$ is from 1 to infinity. Without causing confusion, the index $l$ is alternatively used to denote the indices $(n,m)$ when a quantity is only a notation without specific expression of the indices $(n,m)$.

The T-matrix expressions in Eq. (3.2.53) can be also given using coefficient vectors $\begin{pmatrix} \mathbf{a} \\ \mathbf{b} \end{pmatrix}$ and $\begin{pmatrix} \mathbf{p} \\ \mathbf{q} \end{pmatrix}$ and the matrix form $\begin{pmatrix} \mathbf{T}^{11} & \mathbf{T}^{12} \\ \mathbf{T}^{21} & \mathbf{T}^{22} \end{pmatrix}$. These infinite series for the vectors and the T-matrix have to be truncated in real computations. Assume that the truncation order for the index $n$ is at $N$, that is, $n_{\max}=N$. Consequently, the dimension for the vectors **p**, **q**, **a**, and **b** is

$$L=\sum_{n=1}^{N}(2n+1)=N(N+2). \tag{3.2.55}$$

The dimension of the T-matrix is $2L\times 2L$, and all the vector or matrix elements are complex.

The quantities of single scattering by a particle include the scattering phase matrix and the scattering, absorption, and extinction cross sections. The T-matrix represents the intrinsic properties of the particle, and it only depends on the refractive index, size parameter, morphology, and the origin location and orientation of the frame of reference, but not on the incident direction and polarization state. However, except for homogeneous spherical particles, the single-scattering properties for an oriented particle are related to the incident direction and polarization state. The scattering phase matrix for an oriented particle is calculated using the amplitude scattering matrix. The electromagnetic field is expanded using VSWF, expressed in the spherical coordinate system. Consequently, the amplitude scattering matrix associated with the corresponding meridional planes is more straightforward than the one associated with the corresponding scattering plane. Using Eqs. (3.2.46), (3.2.51), (3.2.52), the scattered electric field in the far field and the corresponding amplitude scattering matrix are written as

$$\mathbf{E}^{\text{sca}}\big|_{kr\to\infty}=\frac{\exp(ikr)}{-ikr}\mathbf{E}_1^{\text{sca}}, \tag{3.2.56}$$

$$\begin{pmatrix}\mathbf{E}_1^{\text{sca}}\cdot\hat{\theta}_s\\ \mathbf{E}_1^{\text{sca}}\cdot\hat{\varphi}_s\end{pmatrix}=\begin{pmatrix}S_{11}(\theta_s,\varphi_s;\theta_i,\varphi_i) & S_{12}(\theta_s,\varphi_s;\theta_i,\varphi_i)\\ S_{21}(\theta_s,\varphi_s;\theta_i,\varphi_i) & S_{22}(\theta_s,\varphi_s;\theta_i,\varphi_i)\end{pmatrix}\begin{pmatrix}\mathbf{E}^{\text{inc}}\cdot\hat{\theta}_i\\ \mathbf{E}^{\text{inc}}\cdot\hat{\varphi}_i\end{pmatrix}=\mathbf{S}\begin{pmatrix}\mathbf{E}^{\text{inc}}\cdot\hat{\theta}_i\\ \mathbf{E}^{\text{inc}}\cdot\hat{\varphi}_i\end{pmatrix}, \tag{3.2.57}$$

where the unit scattering direction $\hat{r}=(\theta_s,\varphi_s)$, the elements of the amplitude scattering matrix with respect to the incident angles $(\theta_i,\varphi_i)$, and the scattering angles $(\theta_s,\varphi_s)$ are (Mishchenko et al., 2002)

$$S_{11}=\sum_{l,l'}\alpha_{mnm'n'}\exp[i(m\varphi_s-m'\varphi_i)]\big[\mathrm{T}_{ll'}^{11}\pi_{mn}(\theta_s)\pi_{m'n'}(\theta_i)+\mathrm{T}_{ll'}^{21}\tau_{mn}(\theta_s)\pi_{m'n'}(\theta_i)$$
$$+\mathrm{T}_{ll'}^{12}\pi_{mn}(\theta_s)\tau_{m'n'}(\theta_i)+\mathrm{T}_{ll'}^{22}\tau_{mn}(\theta_s)\tau_{m'n'}(\theta_i)\big], \tag{3.2.58a}$$

$$S_{12}=-i\sum_{l,l'}\alpha_{mnm'n'}\exp[i(m\varphi_s-m'\varphi_i)]\big[\mathrm{T}_{ll'}^{11}\pi_{mn}(\theta_s)\tau_{m'n'}(\theta_i)+\mathrm{T}_{ll'}^{21}\tau_{mn}(\theta_s)\tau_{m'n'}(\theta_i)$$
$$+\mathrm{T}_{ll'}^{12}\pi_{mn}(\theta_s)\pi_{m'n'}(\theta_i)+\mathrm{T}_{ll'}^{22}\tau_{mn}(\theta_s)\pi_{m'n'}(\theta_i)\big], \tag{3.2.58b}$$

$$S_{21} = i\sum_{l,l'} \alpha_{mnm'n'} \exp\left[i(m\varphi_s - m'\varphi_i)\right]\left[\mathrm{T}_{ll'}^{11}\tau_{mn}(\theta_s)\pi_{m'n'}(\theta_i) + \mathrm{T}_{ll'}^{21}\pi_{mn}(\theta_s)\pi_{m'n'}(\theta_i)\right.$$
$$\left. + \mathrm{T}_{ll'}^{12}\tau_{mn}(\theta_s)\tau_{m'n'}(\theta_i) + \mathrm{T}_{ll'}^{22}\pi_{mn}(\theta_s)\tau_{m'n'}(\theta_i)\right], \tag{3.2.58c}$$

$$S_{22} = \sum_{l,l'} \alpha_{mnm'n'} \exp\left[i(m\varphi_s - m'\varphi_i)\right]\left[\mathrm{T}_{ll'}^{11}\tau_{mn}(\theta_s)\tau_{m'n'}(\theta_i) + \mathrm{T}_{ll'}^{21}\pi_{mn}(\theta_s)\tau_{m'n'}(\theta_i)\right.$$
$$\left. + \mathrm{T}_{ll'}^{12}\tau_{mn}(\theta_s)\pi_{m'n'}(\theta_i) + \mathrm{T}_{ll'}^{22}\pi_{mn}(\theta_s)\pi_{m'n'}(\theta_i)\right], \tag{3.2.58d}$$

where the arguments of the four elements are suppressed; the summation index $l$ is from 1 to infinity or $L$ if truncated. In addition, the following relations exist (Mishchenko et al., 2002):

$$\alpha_{mnm'n'} = i^{n'-n}(-1)^{m+m'+1}\sqrt{\frac{(2n+1)(2n'+1)}{n(n+1)n'(n'+1)}}, \tag{3.2.59a}$$

$$\pi_{mn}(\theta) = (-1)^m\sqrt{\frac{(n-m)!}{(n+m)!}}\frac{mP_n^m(\cos\theta)}{\sin\theta}, \tag{3.2.59b}$$

$$\tau_{mn}(\theta) = (-1)^m\sqrt{\frac{(n-m)!}{(n+m)!}}\frac{dP_n^m(\cos\theta)}{d\theta}, \tag{3.2.59c}$$

and

$$\pi_{-mn}(\theta) = (-1)^{m+1}\pi_{mn}(\theta), \quad \pi_{mn}(\pi-\theta) = (-1)^{m+n}\pi_{mn}(\theta), \tag{3.2.59d}$$

$$\tau_{-mn}(\theta) = (-1)^{m}\tau_{mn}(\theta), \quad \tau_{mn}(\pi-\theta) = (-1)^{m+n+1}\tau_{mn}(\theta). \tag{3.2.59e}$$

The amplitude scattering matrix elements in Eq. (3.2.58) are associated with the meridional planes. Usually, the amplitude scattering matrix associated with the scattering plane is used to compute the scattering phase matrix. The transformation of the amplitude scattering matrix from the meridional planes to the scattering plane is discussed by Mishchenko (2000). The $\mathbf{E}_1^{\text{sca}}$ and the amplitude scattering matrix $\mathbf{S}$ are dimensionless. Moreover, the length appearing in $\mathbf{E}_1^{\text{sca}}$ and $\mathbf{S}$ is always normalized by the wave number $k$, and the normalized quantity is called the size parameter. The computation of the amplitude scattering matrix or scattering matrix depends only on the size parameter of the particle. However, the extinction and scattering cross sections depend on the incident wavelength.

Using the optical theorem, the extinction cross section can be given as (Mishchenko et al., 2002)

$$C_{\text{ext}} = \frac{1}{k^2}\frac{4\pi Re\left[\mathbf{E}_1^{\text{sca}}(\theta_i,\varphi_i)\cdot\mathbf{E}_0^*\right]}{|\mathbf{E}_0|^2} = -\frac{1}{k^2}\frac{\sum_l Re\left[a_l p_l^* + b_l q_l^*\right]}{|\mathbf{E}_0|^2}. \tag{3.2.60}$$

The scattering cross section depends on the integral of the scattered field. Using the orthogonality properties of vector spherical harmonics in Eq. (3.2.42), the scattering cross section can be given as follows (Mishchenko et al., 2002):

$$C_{\mathrm{sca}} = \frac{1}{k^2} \frac{\int_{4\pi} d\Omega_s \left|\mathbf{E}_1^{\mathrm{sca}}(\theta_s, \varphi_s)\right|^2}{\left|\mathbf{E}_0\right|^2} = \frac{1}{k^2} \frac{\sum_l \left[|p_l|^2 + |q_l|^2\right]}{\left|\mathbf{E}_0\right|^2}. \tag{3.2.61}$$

The dimension of the cross section is represented by the area dimension $1/k^2$, and the remaining expressions in Eqs. (3.2.60) and (3.2.61) are dimensionless. The extinction and scattering cross sections are described using the incident and scattered expansion coefficients and evidently depend on the incident direction and polarization states. However, the cross sections are independent of the magnitude of the incident field due to the linearity of Maxwell's equations.

### 3.2.4 Rotation and translation of vector spherical wave function

As mentioned in Section 3.2.3, the T-matrix of a particle depends on the location of the origin and the orientation of the frame of reference. Usually, the T-matrix of the particle is given in the particle frame of reference, which is attached to the particle. The particle frame of reference can be chosen to efficiently utilize the particle morphology, so the T-matrix computation can be done as efficiently as possible. However, the T-matrix in another frame of reference is usually necessary, so almost any light scattering problem involves translation due to a different origin (or a different reference point) and rotation due to a different orientation associated with the T-matrix. For instance, in many-particle problems, every particle has a reference point and its corresponding T-matrix. The total scattering relative to the origin of the laboratory frame of reference involves both translation and rotation of all the T-matrices associated with the particles. Another example is a particle in random orientation, which depends on rotations of the T-matrix. Moreover, two arbitrary right-hand frames of reference can be connected by a rotational operation and a translational operation. Accordingly, rotations and translations of the T-matrix are important to the T-matrix concept. Since computation of the T-matrix is implemented in terms of the VSWF expansion in Eqs. (3.2.52)–(3.2.53), rotations and translations of the VSWF are discussed in this section.

#### 3.2.4.1 Rotation

Rotations of the VSWF are only related to angular functions, which are the vector spherical harmonics shown in Eq. (3.2.41), because the radial functions shown in Eq. (3.2.38) are independent of rotations of the frame of reference. The vector spherical harmonics are generated by spherical harmonics as shown in Eqs. (3.2.40) and (3.2.41). Accordingly, the rotations of the spherical harmonics are presented first, and then, the rotations of the vector spherical harmonics or the VSWF are discussed based on the rotations of the spherical harmonics. This section will introduce the rotation matrices—the Wigner-D and Wigner-d functions, the Clebsch-Gordan (CG) coefficient, and the corresponding symmetric properties.

For rotations from one frame of reference to another, three Euler angles $(\alpha,\beta,\gamma)$ with the "$z$-$y$-$z$" convention are always used in this book. After rotations by Euler angles $(\alpha,\beta,\gamma)$, the laboratory frame of reference L$(x_L, y_L, z_L)$ is rotated to the particle frame of reference P$(x_P, y_P, z_P)$. A positive angle denotes a right-handed counterclockwise rotation along the positive direction of the corresponding axis. The Euler angle rotations in the "$z$-$y$-$z$" convention follow three steps (Tsang et al., 2000; Mishchenko et al., 2002; Doicu et al., 2006):

- Rotate about the axis $z_L$ by angle $\alpha$, so axis $y_L$ is rotated to coincide with the line of nodes, that is, the intersection line between planes $x_Ly_L$ and $x_Py_P$. The range of angle $\alpha$ is $0 \le \alpha < 2\pi$. The frame of reference change is $(x_L, y_L, z_L) \xrightarrow{\alpha} (x', y', z_L)$.
- Rotate about the new axis $y'$ by angle $\beta$, so axis $z_L$ is rotated to coincide with axis $z_P$. The range of angle $\beta$ is $0 \le \beta \le \pi$. The frame of reference change is $(x', y', z_L) \xrightarrow{\beta} (x'', y', z_P)$.
- Rotate about the axis $z_P$ by angle $\gamma$, so axis $y'$ is rotated to coincide with axis $y_P$. The range for angle $\gamma$ is $0 \le \gamma < 2\pi$. The frame of reference change is $(x'', y', z_P) \xrightarrow{\gamma} (x_P, y_P, z_P)$.

Fig. 3.5 shows the rotation process from the laboratory frame of reference L$(x_L, y_L, z_L)$ to the particle frame of reference P$(x_P, y_P, z_P)$, where $(x,y,z)$ are the Cartesian coordinates in the corresponding frame of reference. The two frames of reference have the same origin. The dashed line aligns with the line of nodes, and $x'$, $y'$, and $x''$ are the intermediate axes during rotations.

The Wigner-D function and Wigner-d function describe the rotation of the angular momentum, and the CG coefficients describe the coupling of the angular momentum.

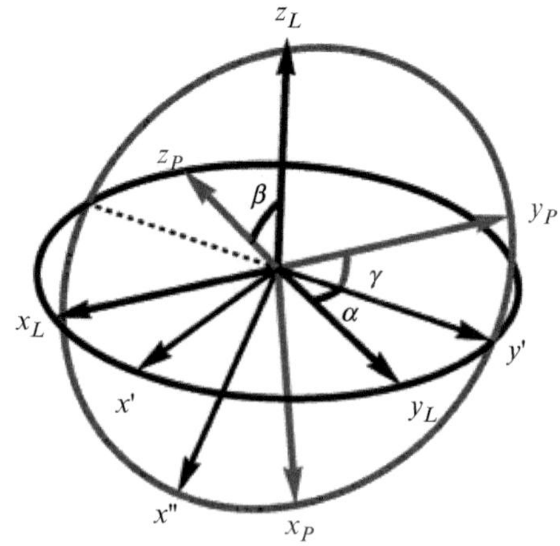


**Fig. 3.5** Euler angle rotations $(\alpha,\beta,\gamma)$ from the laboratory frame of reference L$(x_L, y_L, z_L)$ to the particle frame of reference P$(x_P, y_P, z_P)$, where $(x,y,z)$ is the set of coordinates in the corresponding frame of reference. The dashed line aligns with the line of nodes, which is the intersection line between the $x_Ly_L$ plane and the $x_Py_P$ plane.

These are discussed in this subsection. Since only orbital angular momentum is used in this book, only integer indices are involved for the CG coefficients and the Wigner functions. Classically, orbital angular momentum is defined as the cross product of displacement and momentum or $\mathbf{L} = \vec{r} \times \vec{p}$. Here, the bold **L** and $(\mathrm{L}_x, \mathrm{L}_y, \mathrm{L}_z)$ in the following subsection denote the orbital angular momentum and its components and should be distinguished from the laboratory frame of reference L.

The quantum mechanical description is used in following derivations for convenience. In going from classical mechanics to quantum mechanics, all of the classical quantities are replaced by operators. In the position representation, the operator of the displacement $\vec{r}$ is itself, while the momentum $\vec{p}$ is replaced by the operator $-i\nabla$ (the Planck constant $\hbar$ is always used as a unit). The operators of the angular momentum components and the total angular momentum can be written as follows (Morse and Feshbach, 1953; Arfken and Weber, 1995):

$$\mathrm{L}_x = -i\left(y\frac{\partial}{\partial z} - z\frac{\partial}{\partial y}\right) = i\left(\sin\varphi\frac{\partial}{\partial\theta} + \frac{\cos\varphi}{\tan\theta}\frac{\partial}{\partial\varphi}\right), \tag{3.2.62a}$$

$$\mathrm{L}_y = -i\left(z\frac{\partial}{\partial x} - x\frac{\partial}{\partial z}\right) = -i\left(\cos\varphi\frac{\partial}{\partial\theta} - \frac{\sin\varphi}{\tan\theta}\frac{\partial}{\partial\varphi}\right), \tag{3.2.62b}$$

$$\mathrm{L}_z = -i\left(x\frac{\partial}{\partial y} - y\frac{\partial}{\partial x}\right) = -i\frac{\partial}{\partial\varphi}, \tag{3.2.62c}$$

$$\mathbf{L}^2 = -\nabla_\Omega^2. \tag{3.2.63}$$

It is easy to verify that $\mathbf{L}^2$ commutes with all components, that is,

$$\left[\mathbf{L}^2, \mathrm{L}_i\right] = \mathbf{L}^2\mathrm{L}_i - \mathrm{L}_i\mathbf{L}^2 = 0, \quad i = x, y, z. \tag{3.2.64}$$

The complete simultaneous eigenfunctions of the operators $(\mathbf{L}^2, \mathrm{L}_z)$ are scalar spherical harmonics:

$$\mathbf{L}^2\mathrm{Y}_{nm}(\theta,\varphi) = n(n+1)\mathrm{Y}_{nm}(\theta,\varphi), \quad \mathrm{L}_z\mathrm{Y}_{nm}(\theta,\varphi) = m\mathrm{Y}_{nm}(\theta,\varphi). \tag{3.2.65}$$

Note that even though the definition of the spherical harmonics might have a factor difference, such as $\mathbb{Y}_{nm}(\theta,\varphi)$ and $\mathrm{Y}_{nm}(\theta,\varphi)$ as in Eqs. (3.2.40) and (3.2.44), Eq. (3.2.65) still holds. The spherical harmonics form a complete set in the orbital angular momentum space. Accordingly, an eigenfunction as in Eq. (3.2.65) can be viewed as a state vector in the orbital angular momentum space. Using Dirac symbols for convenience, the "ket" and "bra" have the following expressions:

$$|nm\rangle = \mathrm{Y}_{nm}(\theta,\varphi), \tag{3.2.66a}$$

$$\langle n'm'|\mathbf{A}|nm\rangle = \int_{4\pi} d\Omega \mathrm{Y}^*_{n'm'}(\theta,\varphi)\mathbf{A}\mathrm{Y}_{nm}(\theta,\varphi). \tag{3.2.66b}$$

$$\sum_{m=-n}^{n} |nm\rangle\langle nm| = \mathrm{I}, \tag{3.2.66c}$$

where "bra" $\langle\,|$ is the dual state of the corresponding "ket" $|\,\rangle$, $|nm\rangle\langle nm|$ is a projection operator, $\{|nm\rangle\}$, $m=-n, \ldots, n$ forms a complete set in fixed $n$ space, and I is a unit operator.

The rotational operator corresponding to the Euler angle rotations in the "$z$-$y$-$z$" convention is (Schulz et al., 1998, 1999)

$$\mathrm{U}(\alpha,\beta,\gamma) = \exp(-i\alpha\mathrm{L}_z)\exp\left(-i\beta\mathrm{L}_y\right)\exp(-i\gamma\mathrm{L}_z). \tag{3.2.67}$$

It is evident that the inverse rotational operator is

$$\mathrm{U}^{-1}(\alpha,\beta,\gamma) = \mathrm{U}(-\gamma,\,-\beta,\,-\alpha) = \exp(i\gamma\mathrm{L}_z)\exp\left(i\beta\mathrm{L}_y\right)\exp(i\alpha\mathrm{L}_z). \tag{3.2.68}$$

The Wigner-D function and the Wigner-d function are given by the matrix representations of the rotational operators as follows (e.g., Schulz et al., 1998, 1999):

$$D^n_{m'm}(\alpha,\beta,\gamma) = \langle nm'|\mathrm{U}(\alpha,\beta,\gamma)|nm\rangle, \tag{3.2.69a}$$

$$d^n_{m'm}(\beta) = \langle nm'|\exp\left(-i\beta\mathrm{L}_y\right)|nm\rangle, \tag{3.2.69b}$$

$$D^n_{m'm}(\alpha,\beta,\gamma) = \exp(-im'\alpha)d^n_{m'm}(\beta)\exp(-im\gamma), \tag{3.2.70}$$

where the operation over a fixed $n$ is due to the commutations of Eq. (3.2.64). Using the operators in Eqs. (3.2.66)–(3.2.68), the rotation and inverse rotation of the spherical harmonics can be obtained as

$$\begin{aligned}\mathrm{U}(\alpha,\beta,\gamma)|nm\rangle &= \sum_{m'=-n}^{n} |nm'\rangle\langle nm'|\mathrm{U}(\alpha,\beta,\gamma)|nm\rangle \\ &= \sum_{m'=-n}^{n} D^n_{m'm}(\alpha,\beta,\gamma)|nm'\rangle,\end{aligned} \tag{3.2.71a}$$

$$\begin{aligned}\mathrm{U}(-\gamma,\,-\beta,\,-\alpha)|nm\rangle &= \sum_{m'=-n}^{n} |nm'\rangle\langle nm'|\mathrm{U}(-\gamma,\,-\beta,\,-\alpha)|nm\rangle \\ &= \sum_{m'=-n}^{n} D^n_{m'm}(-\gamma,\,-\beta,\,-\alpha)|nm'\rangle.\end{aligned} \tag{3.2.71b}$$

Since rotations only apply to the spherical harmonics, and not on the Bessel functions of the scalar spherical wave functions, rotation over the scalar spherical wave function is the same as rotation over the spherical harmonics. The rotation operator and the curl operator are commutative (Varadan and Varadan, 1980; Khlebtsov, 1992; Schulz et al., 1999), that is, $[\mathrm{U},\nabla\times]=0$. Accordingly, rotations of the vector spherical wave functions $\{(\mathrm{Rg})\mathbf{M}_{mn}\}$ and $\{(\mathrm{Rg})\mathbf{N}_{mn}\}$ have the same rotational properties as the spherical harmonics (Tsang et al., 1985; Mishchenko et al., 2002):

$$\mathbf{F}_{mn}\left(k\vec{r}_P\right) = \mathrm{U}(\alpha,\beta,\gamma)\mathbf{F}_{mn}\left(k\vec{r}_L\right) = \sum_{m'=-n}^{n} D^n_{m'm}(\alpha,\beta,\gamma)\mathbf{F}_{m'n}\left(k\vec{r}_L\right), \quad (3.2.72a)$$

$$\begin{aligned}\mathbf{F}_{mn}\left(k\vec{r}_L\right) &= \mathrm{U}(-\gamma,-\beta,-\alpha)\mathbf{F}_{mn}\left(k\vec{r}_P\right)\\ &= \sum_{m'=-n}^{n} D^n_{m'm}(-\gamma,-\beta,-\alpha)\mathbf{F}_{m'n}\left(k\vec{r}_P\right),\end{aligned} \quad (3.2.72b)$$

where $\mathbf{F}_{mn}$ could be $\mathbf{M}_{mn}$, $\mathbf{N}_{mn}$, $\mathrm{Rg}\mathbf{M}_{mn}$, or $\mathrm{Rg}\mathbf{N}_{mn}$ and $\vec{r}_P$ and $\vec{r}_L$ are the same position vector but in the frame of references P and L, respectively. Following Eq. (3.2.52), the VSWF expansions of the incident field and the scattered field can be both implemented in frames of reference L and P:

$$\begin{aligned}\mathbf{E}^{\mathrm{inc}}\left(\vec{r}_L\right) &= \sum_{n=1}^{\infty}\sum_{m=-n}^{n}\left(\mathrm{Rg}\mathbf{M}_{mn}\left(k\vec{r}_L\right)\ \mathrm{Rg}\mathbf{N}_{mn}\left(k\vec{r}_L\right)\right)\begin{pmatrix}a^L_{mn}\\ b^L_{mn}\end{pmatrix}\\ &= \sum_{n=1}^{\infty}\sum_{m=-n}^{n}\left(\mathrm{Rg}\mathbf{M}_{mn}\left(k\vec{r}_P\right)\ \mathrm{Rg}\mathbf{N}_{mn}\left(k\vec{r}_P\right)\right)\left[\sum_{m'=-n}^{n} D^n_{mm'}(-\gamma,-\beta,-\alpha)\begin{pmatrix}a^L_{m'n}\\ b^L_{m'n}\end{pmatrix}\right],\end{aligned} \quad (3.2.73a)$$

$$\begin{aligned}\mathbf{E}^{\mathrm{sca}}\left(\vec{r}_L\right) &= \sum_{n=1}^{\infty}\sum_{m=-n}^{n}\left(\mathbf{M}_{mn}\left(k\vec{r}_L\right)\ \mathbf{N}_{mn}\left(k\vec{r}_L\right)\right)\begin{pmatrix}p^L_{mn}\\ q^L_{mn}\end{pmatrix}\\ &= \sum_{n=1}^{\infty}\sum_{m=-n}^{n}\left(\mathbf{M}_{mn}\left(k\vec{r}_P\right)\ \mathbf{N}_{mn}\left(k\vec{r}_P\right)\right)\left[\sum_{m'=-n}^{n} D^n_{mm'}(-\gamma,-\beta,-\alpha)\begin{pmatrix}p^L_{m'n}\\ q^L_{m'n}\end{pmatrix}\right],\end{aligned} \quad (3.2.73b)$$

$$\begin{aligned}\mathbf{E}^{\mathrm{inc}}\left(\vec{r}_P\right) &= \sum_{n=1}^{\infty}\sum_{m=-n}^{n}\left(\mathrm{Rg}\mathbf{M}_{mn}\left(k\vec{r}_P\right)\ \mathrm{Rg}\mathbf{N}_{mn}\left(k\vec{r}_P\right)\right)\begin{pmatrix}a^P_{mn}\\ b^P_{mn}\end{pmatrix}\\ &= \sum_{n=1}^{\infty}\sum_{m=-n}^{n}\left(\mathrm{Rg}\mathbf{M}_{mn}\left(k\vec{r}_L\right)\ \mathrm{Rg}\mathbf{N}_{mn}\left(k\vec{r}_L\right)\right)\left[\sum_{m'=-n}^{n} D^n_{mm'}(\alpha,\beta,\gamma)\begin{pmatrix}a^P_{m'n}\\ b^P_{m'n}\end{pmatrix}\right],\end{aligned} \quad (3.2.74a)$$

$$\begin{aligned}\mathbf{E}^{\mathrm{sca}}\left(\vec{r}_P\right) &= \sum_{n=1}^{\infty}\sum_{m=-n}^{n}\left(\mathbf{M}_{mn}\left(k\vec{r}_P\right)\ \mathbf{N}_{mn}\left(k\vec{r}_P\right)\right)\begin{pmatrix}p^P_{mn}\\ q^P_{mn}\end{pmatrix}\\ &= \sum_{n=1}^{\infty}\sum_{m=-n}^{n}\left(\mathbf{M}_{mn}\left(k\vec{r}_L\right)\ \mathbf{N}_{mn}\left(k\vec{r}_L\right)\right)\left[\sum_{m'=-n}^{n} D^n_{mm'}(\alpha,\beta,\gamma)\begin{pmatrix}p^P_{m'n}\\ q^P_{m'n}\end{pmatrix}\right],\end{aligned} \quad (3.2.74b)$$

where subscripts and superscripts "$L$" and "$P$" correspond to the quantities in the frames of reference L and P, respectively. Using Eq. (3.2.53) and Eqs. (3.2.73)–(3.2.74), the T-matrix has the following rotational relations (Schulz et al., 1998, 1999; Mishchenko et al., 2002):

$$\mathrm{T}^{ij}_{mnm'n'}(\mathrm{L}) = \sum_{m_1=-n}^{n}\sum_{m_2=-n'}^{n'} D^n_{mm_1}(\alpha,\beta,\gamma)\mathrm{T}^{ij}_{m_1nm_2n'}(\mathrm{P})D^{n\prime}_{m_2m'}(-\gamma,-\beta,-\alpha), i,j=1,2, \tag{3.2.75a}$$

$$\mathrm{T}^{ij}_{mnm'n'}(\mathrm{P}) = \sum_{m_1=-n}^{n}\sum_{m_2=-n'}^{n'} D^n_{mm_1}(-\gamma,-\beta,-\alpha)\mathrm{T}^{ij}_{m_1nm_2n'}(\mathrm{L})D^{n\prime}_{m_2m'}(\alpha,\beta,\gamma), i,j=1,2, \tag{3.2.75b}$$

where $L$ and $P$ in the T-matrix denote that the T-matrix is given in the frame of reference L or P. Note that the Euler angles $(\alpha,\beta,\gamma)$ are defined to rotate the frame of reference L to the frame of reference P.

The properties of the Wigner-D function and Wigner-d function are directly given without proof in the subsection. The orthogonality properties lead to the following expression (Biedenharn and Louck, 1981; Mishchenko et al., 2002):

$$\frac{1}{2}\int_0^{\pi} d^n_{mm'}(\beta)d^{n'}_{mm'}(\beta)\sin\beta d\beta = \frac{\delta_{nn'}}{2n+1}, \tag{3.2.76a}$$

$$\frac{1}{8\pi^2}\int_0^{2\pi} d\alpha\int_0^{\pi}\sin\beta d\beta\int_0^{2\pi} d\gamma\left[D^n_{mm_1}(\alpha,\beta,\gamma)\right]^* D^{n'}_{m'm'_1}(\alpha,\beta,\gamma) = \frac{\delta_{nn'}\delta_{mm'}\delta_{m_1m'_1}}{2n+1}. \tag{3.2.76b}$$

The unitary conditions can be obtained using the unit operator in Eqs. (3.2.66c) and (3.2.71):

$$\begin{aligned}\sum_{m'=-n}^{n} & D^n_{mm'}(\alpha,\beta,\gamma)D^n_{m'm_1}(-\gamma,-\beta,-\alpha)\\ &= \sum_{m'=-n}^{n}\langle nm|\mathrm{U}(\alpha,\beta,\gamma)|nm'\rangle\langle nm'|\mathrm{U}(-\gamma,-\beta,-\alpha)|nm_1\rangle\\ &= \langle nm|\, nm_1\rangle = \delta_{mm_1},\end{aligned} \tag{3.2.77a}$$

$$\sum_{m'=-n}^{n} d^n_{mm'}(\beta)d^n_{m'm_1}(-\beta) = \delta_{mm_1}, \tag{3.2.77b}$$

and the Wigner-d function has the following symmetric properties (Biedenharn and Louck, 1981; Mishchenko et al., 2002):

$$d^n_{mm'}(\beta) = (-1)^{m+m'}d^n_{m'm}(\beta) = (-1)^{m+m'}d^n_{-m-m'}(\beta) = d^n_{m'm}(-\beta), \tag{3.2.78a}$$

$$d^n_{mm'}(\pi-\beta) = (-1)^{n+m'} d^n_{-mm'}(\beta) = (-1)^{n+m} d^n_{m-m'}(\beta), \tag{3.2.78b}$$

$$d^n_{mm'}(0) = \delta_{mm'}, \quad d^n_{mm'}(\pi) = (-1)^{n+m}\delta_{m-m'}. \tag{3.2.78c}$$

The relations between the Wigner-d function $d^n_{mm'}(\beta)$, the associated Legendre polynomial $P^m_n(\cos\beta)$, and the generalized spherical function $P^n_{mm'}(\cos\beta)$, which is usually used in the expansion of the scattering phase matrix, are (Hovenier and van der Mee, 1983; Mishchenko et al., 2002)

$$d^n_{m0}(\beta) = \sqrt{\frac{(n-m)!}{(n+m)!}} P^m_n(\cos\beta), \tag{3.2.79a}$$

$$d^n_{mm'}(\beta) = i^{m'-m} P^n_{mm'}(\cos\beta). \tag{3.2.79b}$$

The generalized spherical functions are usually used in the scattering phase matrix expansion, which shall be mentioned in the computation of the scattering phase matrix of a particle under the random orientation condition.

The CG coefficients describe the coupling of two angular momentums. There are two independent orbital angular momentums $\mathbf{L}_1$ and $\mathbf{L}_2$, where the coupled angular momentum $\mathbf{L}$ and its $z$-component $\mathrm{L}_z$ are related by $\mathbf{L}=\mathbf{L}_1+\mathbf{L}_2$ and $\mathrm{L}_z=\mathrm{L}_{1z}+\mathrm{L}_{2z}$. The noncoupled operators $(\mathbf{L}_1^2, \mathrm{L}_{1z}, \mathbf{L}_2^2, \mathrm{L}_{2z})$ construct an operator complete set, and the noncoupled state vectors $|n_1 m_1\rangle$ and $|n_2 m_2\rangle$ can be jointly used as $|n_1 m_1 n_2 m_2\rangle$, an eigenfunction in the coupled angular momentum space. Since the eigenfunctions $\{|n_1 m_1 n_2 m_2\rangle\}$ form a complete set in the coupled angular momentum space, the simultaneous eigenfunction $|nm\rangle$ of the coupled operators $(\mathbf{L}^2, \mathrm{L}_z)$ in the coupled angular momentum space can be expanded using the eigenfunctions of the noncoupled operators:

$$|nm\rangle = \sum_{m_1=-n_1}^{n_1} \sum_{m_2=-n_2}^{n_2} \langle n_1 m_1 n_2 m_2 | nm\rangle |n_1 m_1 n_2 m_2\rangle, \tag{3.2.80}$$

where similarly to Eq. (3.2.66c), $\sum_{m_1=-n_1}^{n_1}\sum_{m_2=-n_2}^{n_2} |n_1 m_1 n_2 m_2\rangle\langle n_1 m_1 n_2 m_2|$ is an unit operator and the expansion coefficients $\langle n_1 m_1 n_2 m_2 | nm\rangle$ can be always made to be real and are called the CG coefficients. In quantum mechanics, the Wigner 3j-symbols are usually used as an alternative to the CG coefficients. The Wigner 3j-symbols can be given in terms of the CG coefficients as (Wigner, 1993; Mishchenko et al., 2002; Arfken and Weber, 1995)

$$\begin{pmatrix} n_1 & n_2 & n \\ m_1 & m_2 & m \end{pmatrix} = \frac{(-1)^{n_1+n_2+m}}{\sqrt{2n+1}} \langle n_1 m_1 n_2 m_2 | n-m\rangle. \tag{3.2.81}$$

The CG coefficient is expressed using $C^{nm}_{n_1 m_1 n_2 m_2}$ in this book, where $C^{nm}_{n_1 m_1 n_2 m_2} = \langle n_1 m_1 n_2 m_2 | nm\rangle$. The CG coefficients are also used in the computation of the scattering phase matrix of a particle in random orientation.

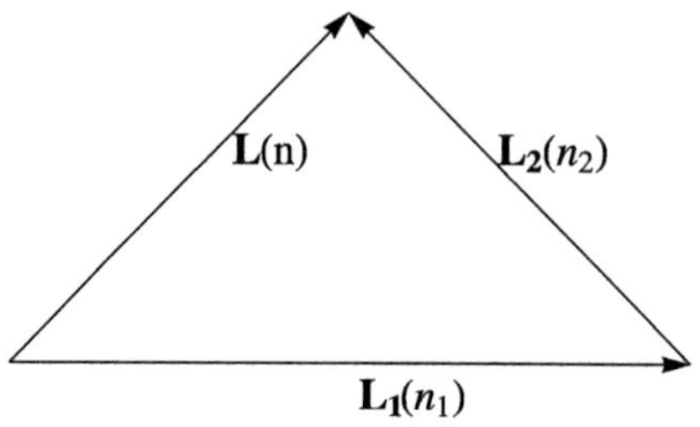


**Fig. 3.6** Coupling of angular momentums. Index $n$ of the coupled angular momentum **L** and indices $n_1$ and $n_2$ of the independent angular momentums $\mathbf{L}_1$ and $\mathbf{L}_2$ obey the triangle rule.

The index $n$ of the eigenfunction $|nm\rangle$ in the coupled angular momentum space obeys the triangle rule $|n_1-n_2|\leq n\leq(n_1+n_2)$ as in Fig. 3.6 and $-n\leq m\leq n$. The CG coefficients have the following symmetric properties (Varshalovich et al., 1988; Mishchenko et al., 2002; Arfken and Weber, 1995):

$$C^{nm}_{n_1m_1n_2m_2} = (-1)^{n_1+n_2+n}C^{n-m}_{n_1-m_1n_2-m_2} = (-1)^{n_1+n_2+n}C^{nm}_{n_2m_2n_1m_1}$$
$$= (-1)^{n_1+m_1}\sqrt{\frac{2n+1}{2n_2+1}}C^{n_2-m_2}_{n_1m_1n-m}. \tag{3.2.82}$$

The Wigner-d function and the Wigner-D function and the CG coefficients have the following relations (Varshalovich et al., 1988; Mishchenko et al., 2002):

$$d^{n_1}_{m_1m_1'}(\beta)d^{n_2}_{m_2m_2'}(\beta) = \sum_{n=|n_1-n_2|}^{n_1+n_2} C^{n,m_1+m_2}_{n_1m_1n_2m_2}d^n_{m_1+m_2,m_1'+m_2'}(\beta)C^{n,m_1'+m_2'}_{n_1m_1'n_2m_2'}, \tag{3.2.83a}$$

$$D^{n_1}_{m_1m_1'}(\alpha,\beta,\gamma)D^{n_2}_{m_2m_2'}(\alpha,\beta,\gamma) = \sum_{n=|n_1-n_2|}^{n_1+n_2} C^{n,m_1+m_2}_{n_1m_1n_2m_2}D^n_{m_1+m_2,m_1'+m_2'}(\alpha,\beta,\gamma)C^{n,m_1'+m_2'}_{n_1m_1'n_2m_2'}. \tag{3.2.83b}$$

### *3.2.4.2 Translation*

Translation of a frame of reference is illustrated in Fig. 3.7. There are two frames of reference, which are marked as frame 1 and frame 2 with the origins $O_1$ and $O_2$, respectively. The same field point P has position vectors $\vec{r}_1$ and $\vec{r}_2$ in frame 1 and frame 2, respectively. The direction vector from origin $O_1$ to origin $O_2$ is marked as $\vec{r}_0$. It is evident in Fig. 3.7 that $\vec{r}_1 = \vec{r}_0 + \vec{r}_2$. For spherical wave functions, the translation means that the spherical wave function in frame 1 can be expanded using the spherical wave functions in frame 2, and the expansion coefficients are related to the direction vector connecting the two frames. The earlier described expansion with respect to the scalar and vector spherical wave functions is called the

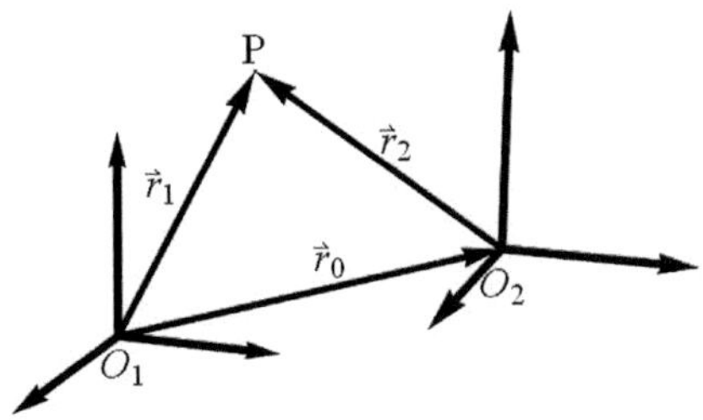


**Fig. 3.7** Translation of the frames of reference. The reference points $O_1$ and $O_2$ are the origins of two right-hand frames of reference, and the two frames are referred to as frame 1 and frame 2, respectively. P is the location of a field point. The position vectors $\vec{r}_1$ and $\vec{r}_2$ have origins at $O_1$ and $O_2$, respectively, and $\vec{r}_0$ is the direction vector from origin $O_1$ to origin $O_2$.

"addition theorem" and has been extensively exploited (Friedman and Russek, 1954; Stein, 1961; Cruzan, 1962). The addition theorem is significantly important for many-body light scattering associated with multiple preferred reference points. If the many-body system is composed entirely of spherical particles, the problem is drastically simplified because the scattering of each spherical particle has an analytical Lorenz-Mie type solution, and also, each scattering pattern is invariant under rotations of the frame of reference. The computation of expansion coefficients in the addition theorem is crucial for multiple sphere scattering, and multiple sphere scattering has been widely studied (e.g., Bruning and Lo, 1971a, b; Fuller and Kattawar, 1988a, b; Mackowski, 1994; Xu, 1995, 1997; Mackowski and Mishchenko, 1996; Fuller and Mackowski, 2000). The addition theorem of the scalar spherical wave functions is presented first, and then, the addition theorem of the vector spherical wave functions is given in terms of the addition theorem of the scalar functions.

In frame 1, the position vector $\vec{r}_1 = \vec{r}_1(r_1, \theta_1, \varphi_1)$ and the translation direction vector $\vec{r}_0 = \vec{r}_0(r_0, \theta_0, \varphi_0)$ and, in frame 2, the position vector $\vec{r}_2 = \vec{r}_2(r_2, \theta_2, \varphi_2)$ and their corresponding unit vectors are denoted as $\hat{r}_1(\theta_1, \varphi_1)$, $\hat{r}_0(\theta_0, \varphi_0)$, and $\hat{r}_2(\theta_2, \varphi_2)$. Note that the unit vectors are invariant under translations of a frame of reference. A plane wave propagating in the direction $\hat{r}$ with wave vector $k\hat{r}$ can be expanded using the spherical Bessel functions and the (associated) Legendre polynomials as follows (Morse and Feshbach, 1953; Friedman and Russek, 1954; Arfken and Weber, 1995):

$$\exp\left(ik\hat{r} \cdot \vec{r}_1\right) = \exp(ikr_1 \cos\Theta_1) = \sum_{n=0}^{\infty} i^n (2n+1) j_n(kr_1) P_n(\cos\Theta_1)$$
$$= \sum_{n=0}^{\infty} \sum_{m=-n}^{n} i^n (2n+1) j_n(kr_1) \frac{(n-m)!}{(n+m)!} P_n^m(\cos\theta) P_n^m(\cos\theta_1) \exp\left[im(\varphi_1 - \varphi)\right], \tag{3.2.84}$$

where $\hat{r} = \hat{r}(\theta, \varphi)$ and $\cos\Theta_1 = \hat{r} \cdot \hat{r}_1 = \cos\theta\cos\theta_1 + \sin\theta\sin\theta_1\cos(\varphi - \varphi_1)$. The addition theorem for the Legendre polynomial in Eq. (3.2.26) is used in

Eq. (3.2.84). Using the orthogonality of the spherical harmonics, the regular scalar wave function has the following integral representation (Stratton, 1941):

$$j_n(kr_1)P_n^m(\cos\theta_1)\exp(im\varphi_1) = \frac{i^{-n}}{4\pi}\int_0^{2\pi} d\varphi \int_0^{\pi} \sin\theta d\theta \exp(ikr_1\cos\Theta_1)P_n^m(\cos\theta)\exp(im\varphi). \tag{3.2.85}$$

Also, it is apparent that

$$r_1\cos\Theta_1 = r_2\cos\Theta_2 + r_0\cos\Theta_0, \tag{3.2.86}$$

where $\cos\Theta_2 = \hat{r}\cdot\hat{r}_2$ and $\cos\Theta_0 = \hat{r}\cdot\hat{r}_0$. Substituting Eq. (3.2.86) into Eq. (3.2.85) and using Eq. (3.2.84) for $\exp(ikr_2\cos\Theta_2)$, Eq. (3.2.85) can be rearranged as

$$j_n(kr_1)P_n^m(\cos\theta_1)\exp(im\varphi_1) = \frac{i^{-n}}{4\pi}\sum_{\nu=0}^{\infty}\sum_{\mu=-\nu}^{\nu}\Bigg[i^{\nu}(2\nu+1)j_\nu(kr_2)\frac{(\nu-\mu)!}{(\nu+\mu)!}P_\nu^\mu(\cos\theta_2)$$

$$\exp(-i\mu\varphi_2)\cdot\int_0^{2\pi} d\varphi\int_0^{\pi}\sin\theta d\theta\exp(ikr_0\cos\Theta_0)P_\nu^\mu(\cos\theta)P_n^m(\cos\theta)\exp[i(m+\mu)\varphi]\Bigg]. \tag{3.2.87}$$

Similar to Eq. (3.2.83a), the product of two associated Legendre polynomials is given as (Friedman and Russek, 1954)

$$P_n^m(\cos\theta)P_\nu^\mu(\cos\theta) = \sum_p a(m,n|\mu,\nu|p)P_p^{m+\mu}(\cos\theta), \tag{3.2.88}$$

where coefficient $a$ is (Mishchenko et al., 2002)

$$a(m,n|\mu,\nu|p) = (-1)^{m+\mu}(2p+1)\sqrt{\frac{(n+m)!(\nu+\mu)!(p-m-\mu)!}{(n-m)!(\nu-\mu)!(p+m+\mu)!}}\cdot\begin{pmatrix} n & \nu & p \\ m & \mu & -(m+\mu)\end{pmatrix}\begin{pmatrix} n & \nu & p \\ 0 & 0 & 0\end{pmatrix}. \tag{3.2.89}$$

Using the relationship between the Wigner 3j-symbols and the CG coefficients in Eq. (3.2.81) and the symmetric relations for the CG coefficients in Eq. (3.2.82), we have

$$\begin{pmatrix} n & \nu & p \\ 0 & 0 & 0\end{pmatrix} = 0, \quad \text{if } (-1)^{n+\nu+p} = -1. \tag{3.2.90}$$

Accordingly, the summation over $p$ in Eq. (3.2.88) only covers $|n-\nu|, |n-\nu|+2, \ldots, n+\nu$, a total of $1+\min(n,\nu)$ terms, for nonvanishing $a(m,n|\mu,\nu|p)$. Substituting Eq. (3.2.88) into Eq. (3.2.87) and using Eq. (3.2.85), Eq. (3.2.87) can be rewritten as (Friedman and Russek, 1954)

$$j_n(kr_1)P_n^m(\cos\theta_1)\exp(im\varphi_1) = \sum_{\nu=0}^{\infty}\sum_{\mu=-\nu}^{\nu}\sum_{p}\Big[(-1)^{\mu}i^{\nu+p-n}(2\nu+1)a(m,n|-\mu,\nu|p)$$

$$\cdot j_\nu(kr_2)P_\nu^\mu(\cos\theta_2)\exp(i\mu\varphi_2)j_p(kr_0)P_p^{m-\mu}(\cos\theta_0)\exp[i(m-\mu)\varphi_0]\Big]. \quad (3.2.91)$$

The parity relation of the associated Legendre polynomial in Eq. (3.2.21) is also used in Eq. (3.2.91). Using the symbol for the regular scalar wave function $Rg\psi_{mn}$ in Eq. (3.2.37), the addition theorem for the regular scalar spherical wave functions can be written in a compact form as

$$\begin{aligned} Rg\psi_{mn}\left(k\vec{r}_1\right) = &\sum_{\nu=0}^{\infty}\sum_{\mu=-\nu}^{\nu}\sum_{p}(-1)^{\mu}i^{\nu+p-n}(2\nu+1)a(m,n|-\mu,\nu|p) \\ &Rg\psi_{m-\mu p}\left(k\vec{r}_0\right)Rg\psi_{\mu\nu}\left(k\vec{r}_2\right). \end{aligned} \quad (3.2.92)$$

The derivation of the addition theorem for the outgoing scalar spherical wave function $\psi_{mn}$ involves contour integrals, which are discussed in detail by Friedman and Russek (1954). The addition theorem for the outgoing scalar wave functions is directly given here without proof as follows (Friedman and Russek, 1954):

$$h_n^{(1)}(kr_1)P_n^m(\cos\theta_1)\exp(im\varphi_1) = \sum_{\nu=0}^{\infty}\sum_{\mu=-\nu}^{\nu}\sum_{p}\Big[(-1)^{\mu}i^{\nu+p-n}(2\nu+1)a(m,n|-\mu,\nu|p)$$

$$\cdot j_\nu(kr_<)P_\nu^\mu(\cos\theta_<)\exp(i\mu\varphi_<)h_p^{(1)}(kr_>)P_p^{m-\mu}(\cos\theta_>)\exp[i(m-\mu)\varphi_>]\Big], \quad (3.2.93)$$

where

$$\begin{cases} \vec{r}_<(r_<,\theta_<,\varphi_<) = \vec{r}_2(r_2,\theta_2,\varphi_2) \\ \vec{r}_>(r_>,\theta_>,\varphi_>) = \vec{r}_0(r_0,\theta_0,\varphi_0) \end{cases}, \quad \text{if } r_2 < r_0, \quad (3.2.94a)$$

$$\begin{cases} \vec{r}_<(r_<,\theta_<,\varphi_<) = \vec{r}_0(r_0,\theta_0,\varphi_0) \\ \vec{r}_>(r_>,\theta_>,\varphi_>) = \vec{r}_2(r_2,\theta_2,\varphi_2) \end{cases}, \quad \text{if } r_0 \le r_2. \quad (3.2.94b)$$

Depending on the comparison between $r_2$ and $r_0$, the addition theorem for outgoing scalar spherical wave functions has a different representation. Using Eqs. (3.2.91) and (3.2.93), it is apparent that the addition theorem for the scalar spherical wave functions using the spherical Neumann function $y_n$ or the spherical Hankel function of the second kind $h_n^{(2)}$ can be expressed as follows:

$$z_n(kr_1)P_n^m(\cos\theta_1)\exp(im\varphi_1) = \sum_{\nu=0}^{\infty}\sum_{\mu=-\nu}^{\nu}\sum_{p}\Big[(-1)^{\mu}i^{\nu+p-n}(2\nu+1)a(m,n|-\mu,\nu|p)$$
$$\cdot j_\nu(kr_<)P_\nu^\mu(\cos\theta_<)\exp(i\mu\varphi_<)z_p(kr_>)P_p^{m-\mu}(\cos\theta_>)\exp[i(m-\mu)\varphi_>]\Big] \tag{3.2.95}$$

where function $z_n$ can be $y_n$ or $h_n^{(2)}$.

For succinctness, we temporarily use $\Psi_{mn}$ in this subsection to generally denote the scalar spherical wave function, that is, $\Psi_{mn}=z_n(kr)P_n^m(\cos\theta)\exp(im\varphi)$, where $z_n(kr)$ could be the spherical Bessel function $j_n$, the spherical Neumann function $y_n$, or the spherical Hankel function of the first kind $h_n^{(1)}$ or the second kind $h_n^{(2)}$. Eqs. (3.2.91)–(3.2.95) can consistently be given as

$$\Psi_{mn}\left(k\vec{r}_1\right) = \sum_{\nu=0}^{\infty}\sum_{\mu=-\nu}^{\nu}\sum_{p}(-1)^{\mu}i^{\nu+p-n}(2\nu+1)a(m,n|-\mu,\nu|p)\Psi_{m-\mu p}\left(k\vec{r}_>\right)\mathrm{Rg}\psi_{\mu\nu}\left(k\vec{r}_<\right). \tag{3.2.96}$$

Note that the addition theorem of the scalar spherical wave function using the spherical Bessel function does not depend on the comparison between $r_2$ and $r_0$, because the two representations of the addition theorem are the same.

The translation direction vector $\vec{r}_0$ is a constant vector as opposed to the position vector $\vec{r}_2$. The addition theorem can be written as the expansion of the scalar spherical wave functions in frame 1 using the scalar spherical wave functions in frame 2 and the expansion coefficients involving the translation vector $\vec{r}_0$. When $r_2<r_0$, the addition theorem in Eq. (3.2.96) for the scalar spherical wave function in a compact form is

$$\Psi_{mn}\left(k\vec{r}_1\right) = \sum_{\nu=0}^{\infty}\sum_{\mu=-\nu}^{\nu}u(m,n|\mu,\nu)\mathrm{Rg}\psi_{\mu\nu}\left(k\vec{r}_2\right), \tag{3.2.97a}$$

$$u(m,n|\mu,\nu) = (-1)^{\mu}(2\nu+1)i^{\nu-n}\sum_{p}i^{p}a(m,n|-\mu,\nu|p)\Psi_{m-\mu p}\left(k\vec{r}_0\right). \tag{3.2.97b}$$

The expansion of the scalar spherical wave function when condition $r_2<r_0$ only involves the regular scalar spherical wave function, and the dependence of the spherical Bessel type is in the expansion coefficient $u$. When $r_2\geq r_0$, however, the $k\vec{r}_2$ dependence is in the function $\Psi_{m-\mu p}$ as in Eq. (3.2.96). The summation index has to be interchanged, so the compact form in Eq. (3.2.97) can be realized for that condition. After some algebra, the addition theorem in Eq. (3.2.97) when $r_2\geq r_0$ can be given as (Stein, 1961)

$$\Psi_{mn}\left(k\vec{r}_1\right) = \sum_{\nu=0}^{\infty}\sum_{\mu=-\nu}^{\nu}v(m,n|\mu,\nu)\Psi_{\mu\nu}\left(k\vec{r}_2\right), \tag{3.2.98a}$$

$$v(m,n|\mu,\nu)=(-1)^{\mu}(2\nu+1)i^{\nu-n}\sum_{p}i^{p}a(m,n|-\mu,\nu|p)\mathrm{Rg}\psi_{m-\mu p}\left(k\vec{r}_0\right). \quad (3.2.98b)$$

The dependence of the spherical Bessel type is in the expanding scalar spherical wave function, and the expansion coefficient only involves the regular spherical wave function. Since the involvement of the spherical Bessel type for the expanding scalar spherical wave function or the expansion coefficient is the only difference between the condition $r_2 \geq r_0$ or $r_2 < r_0$, the addition theorem for the scalar spherical wave functions can be concisely written as

$$\Psi_{mn}\left(k\vec{r}_1\right)=\sum_{\nu=0}^{\infty}\sum_{\mu=-\nu}^{\nu}u(m,n|\mu,\nu)\Psi'_{\mu\nu}\left(k\vec{r}_2\right). \quad (3.2.99)$$

The spherical Bessel-type dependence for the expanded scalar spherical wave function $\Psi_{mn}\left(k\vec{r}_1\right)$, the expanding scalar spherical wave function $\Psi'_{\mu\nu}\left(k\vec{r}_2\right)$, and the expansion coefficient $u(m,n\,|\,\mu,\nu)$ can be found in Eqs. (3.2.97) or (3.2.98). The prime on $\Psi'_{\mu\nu}\left(k\vec{r}_2\right)$ is used to denote that it may have a different spherical Bessel type than $\Psi_{mn}\left(k\vec{r}_1\right)$.

For convenience, we use $\mathrm{M}_{mn}$ and $\mathrm{N}_{mn}$ to generally denote the vector spherical wave functions, where the spherical Bessel type could be $j_n$, $y_n$, $h_n^{(1)}$, or $h_n^{(2)}$. The vector spherical wave functions $\mathrm{M}_{mn}$ and $\mathrm{N}_{mn}$ are generated using the scalar wave functions in Eqs. (3.2.38b) and (3.2.38c) as

$$\mathrm{M}_{mn}=\gamma_{mn}\nabla\times\left(\vec{r}\,\Psi_{mn}\right)=\gamma_{mn}(\nabla\Psi_{mn})\times\vec{r}\,, \quad (3.2.100a)$$

$$\mathrm{N}_{mn}=\frac{1}{k}\nabla\times\mathrm{M}_{mn}. \quad (3.2.100b)$$

Note that the dependence on the spherical Bessel type for the vector spherical wave functions $\mathrm{M}_{mn}$ and $\mathrm{N}_{mn}$ is the same as for the scalar spherical wave functions $\Psi_{mn}$. For instance, if $\Psi_{mn}$ uses the spherical Bessel function $j_n$, $\mathrm{M}_{mn}$ and $\mathrm{N}_{mn}$ use it as well, that is, $\mathrm{M}_{mn}=\mathrm{Rg}\mathbf{M}_{mn}$ and $\mathrm{N}_{mn}=\mathrm{Rg}\mathbf{N}_{mn}$. It is evident that the gradient and curl operators in Eq. (3.2.100) are invariant under translations of a frame of reference, that is, $\nabla_1=\nabla_2$ and $\nabla_1\times=\nabla_2\times$. Substituting Eq. (3.2.99) into Eq. (3.2.100a), we have

$$\begin{aligned}\mathrm{M}_{mn}\left(k\vec{r}_1\right)&=\sum_{\nu=1}^{\infty}\sum_{\mu=-\nu}^{\nu}\frac{\gamma_{mn}}{\gamma_{\mu\nu}}u(m,n|\mu,\nu)\gamma_{\mu\nu}\left(\nabla_2\Psi'_{\mu\nu}\left(k\vec{r}_2\right)\right)\times\left(\vec{r}_2+\vec{r}_0\right)\\&=\sum_{\nu=1}^{\infty}\sum_{\mu=-\nu}^{\nu}\frac{\gamma_{mn}}{\gamma_{\mu\nu}}u(m,n|\mu,\nu)\left[\gamma_{\mu\nu}\left(\nabla_2\Psi'_{\mu\nu}\left(k\vec{r}_2\right)\right)\times\vec{r}_0\right]\\&\quad+\sum_{\nu=1}^{\infty}\sum_{\mu=-\nu}^{\nu}\frac{\gamma_{mn}}{\gamma_{\mu\nu}}u(m,n|\mu,\nu)\mathrm{M}'_{\mu\nu}\left(k\vec{r}_2\right),\end{aligned} \quad (3.2.101)$$

where

$$\mathbf{M}'_{\mu\nu}\left(k\vec{r}_2\right) = \gamma_{\mu\nu}\left(\nabla_2\Psi'_{\mu\nu}\left(k\vec{r}_2\right)\right) \times \vec{r}_2. \tag{3.2.102}$$

The prime over $\mathbf{M}'_{\mu\nu}\left(k\vec{r}_2\right)$ denotes a possibly different spherical Bessel type than $\mathbf{M}_{mn}\left(k\vec{r}_1\right)$. The second term has already been expressed in frame 2. The translation of the first term to the expression in frame 2 is tricky and was elegantly done by Cruzan (1962). The translational vector $\vec{r}_0$ can be decomposed in Cartesian coordinates, and the unit vectors of the Cartesian coordinates can be decomposed using the unit vectors of the spherical coordinates in frame 2 as follows:

$$\vec{r}_0 = r_0(\sin\theta_0\cos\varphi_0\hat{x} + \sin\theta_0\sin\varphi_0\hat{y} + \cos\theta_0\hat{z}), \tag{3.2.103}$$

$$\hat{x} = \sin\theta_2\cos\varphi_2\hat{r}_2 + \cos\theta_2\cos\varphi_2\hat{\theta}_2 - \sin\varphi_2\hat{\varphi}_2, \tag{3.2.104a}$$

$$\hat{y} = \sin\theta_2\sin\varphi_2\hat{r}_2 + \cos\theta_2\sin\varphi_2\hat{\theta}_2 + \cos\varphi_2\hat{\varphi}_2, \tag{3.2.104b}$$

$$\hat{z} = \cos\theta_2\hat{r}_2 - \sin\theta_2\hat{\theta}_2, \tag{3.2.104c}$$

where $\hat{x}$, $\hat{y}$, and $\hat{z}$ are the unit vectors in Cartesian coordinates. Part of the first term can be decomposed into three terms as

$$\begin{aligned}\left(\nabla_2\Psi'_{\mu\nu}\right) \times \vec{r}_0 = r_0\Big[&\sin\theta_0\cos\varphi_0\left(\nabla_2\Psi'_{\mu\nu}\right) \times \hat{x} \\ &+ \sin\theta_0\sin\varphi_0\left(\nabla_2\Psi'_{\mu\nu}\right) \times \hat{y} + \cos\theta_0\left(\nabla_2\Psi'_{\mu\nu}\right) \times \hat{z}\Big].\end{aligned} \tag{3.2.105}$$

Using Eq. (3.2.104) along with the recurrence relations and the orthogonality properties of the angular and radial functions in the vector spherical wave functions, the three terms on the right-hand side of Eq. (3.2.105) can be expanded using the vector spherical wave functions in frame 2, as described in detail by Cruzan (1962). The coefficients of the two terms on the right-hand side of Eq. (3.2.101) can be combined and organized as the expansion of the vector spherical wave functions in frame 2 (Cruzan, 1962) as follows:

$$\mathbf{M}_{mn}\left(k\vec{r}_1\right) = \sum_{\nu=1}^{\infty}\sum_{\mu=-\nu}^{\nu}\left[A_{\mu\nu}^{mn}\mathbf{M}'_{\mu\nu}\left(k\vec{r}_2\right) + B_{\mu\nu}^{mn}\mathbf{N}'_{\mu\nu}\left(k\vec{r}_2\right)\right], \tag{3.2.106}$$

where the prime still denotes a possibly different spherical Bessel type. Also (Mishchenko et al., 2002),

$$A_{\mu\nu}^{mn} = \frac{\gamma_{mn}}{\gamma_{\mu\nu}}(-1)^{\mu}\sum_{p} a(m,n|-\mu,\nu|p)a(n,\nu,p)\Psi_{m-\mu p}\left(k\vec{r}_0\right), \tag{3.2.107a}$$

$$B_{\mu\nu}^{mn} = \frac{\gamma_{mn}}{\gamma_{\mu\nu}}(-1)^{\mu+1}\sum_{p} a(m,n|-\mu,\nu|p,p-1)b(n,\nu,p)\Psi_{m-\mu p}\left(k\vec{r}_0\right), \tag{3.2.107b}$$

and

$$a(m,n|\mu,\nu|p,q) = (-1)^{m+\mu}(2p+1)\sqrt{\frac{(n+m)!(\nu+\mu)!(p-m-\mu)!}{(n-m)!(\nu-\mu)!(p+m+\mu)!}} \cdot \begin{pmatrix} n & \nu & p \\ m & \mu & -(m+\mu) \end{pmatrix}\begin{pmatrix} n & \nu & q \\ 0 & 0 & 0 \end{pmatrix}, \tag{3.2.108}$$

$$a(n,\nu,p) = \frac{i^{\nu+p-n}(2\nu+1)}{2\nu(\nu+1)}[\nu(\nu+1)+n(n+1)-p(p+1)], \tag{3.2.109a}$$

$$b(n,\nu,p) = \frac{i^{\nu+p-n}(2\nu+1)}{2\nu(\nu+1)}\sqrt{(n+\nu+p+1)(\nu-n+p)(n-\nu+p)(n+\nu-p+1)}. \tag{3.2.109b}$$

The dependence of the spherical Bessel type in the expansion coefficients $A$ and $B$ is the same as the expansion coefficients $u$ or $v$ in Eqs. (3.2.97) and (3.2.98). Using Eq. (3.2.100b), the expansion for $\mathrm{N}_{mn}$ is straightforward as (Cruzan, 1962)

$$\mathrm{N}_{mn}\left(k\vec{r}_1\right) = \sum_{\nu=1}^{\infty}\sum_{\mu=-\nu}^{\nu}\left[B_{\mu\nu}^{mn}\mathrm{M}'_{\mu\nu}\left(k\vec{r}_2\right) + A_{\mu\nu}^{mn}\mathrm{N}'_{\mu\nu}\left(k\vec{r}_2\right)\right]. \tag{3.2.110}$$

From Eqs. (3.2.99)–(3.2.110), the spherical Bessel-type dependence of the spherical wave functions is always suppressed, and the corresponding relationships can be found in Eqs. (3.2.97) and (3.2.98).

Explicitly adding the spherical Bessel-type dependence, the expansions of the regular VSWF $\begin{pmatrix}\mathrm{Rg}\mathbf{M}_{mn}\\ \mathrm{Rg}\mathbf{N}_{mn}\end{pmatrix}$ and outgoing VSWF $\begin{pmatrix}\mathbf{M}_{mn}\\ \mathbf{N}_{mn}\end{pmatrix}$ using the VSWF in another frame of reference are explicitly given for reference:

$$\begin{pmatrix}\mathrm{Rg}\mathbf{M}_{mn}\left(k\vec{r}_1\right)\\ \mathrm{Rg}\mathbf{N}_{mn}\left(k\vec{r}_1\right)\end{pmatrix} = \sum_{\nu=1}^{\infty}\sum_{\mu=-\nu}^{\nu}\begin{pmatrix}\mathrm{Rg}A_{\mu\nu}^{mn}\left(k\vec{r}_0\right) & \mathrm{Rg}B_{\mu\nu}^{mn}\left(k\vec{r}_0\right)\\ \mathrm{Rg}B_{\mu\nu}^{mn}\left(k\vec{r}_0\right) & \mathrm{Rg}A_{\mu\nu}^{mn}\left(k\vec{r}_0\right)\end{pmatrix}\begin{pmatrix}\mathrm{Rg}\mathbf{M}_{\mu\nu}\left(k\vec{r}_2\right)\\ \mathrm{Rg}\mathbf{N}_{\mu\nu}\left(k\vec{r}_2\right)\end{pmatrix}, \tag{3.2.111}$$

$$\begin{pmatrix}\mathbf{M}_{mn}\left(k\vec{r}_1\right)\\ \mathbf{N}_{mn}\left(k\vec{r}_1\right)\end{pmatrix} = \sum_{\nu=1}^{\infty}\sum_{\mu=-\nu}^{\nu}\begin{pmatrix}\mathrm{Rg}A_{\mu\nu}^{mn}\left(k\vec{r}_0\right) & \mathrm{Rg}B_{\mu\nu}^{mn}\left(k\vec{r}_0\right)\\ \mathrm{Rg}B_{\mu\nu}^{mn}\left(k\vec{r}_0\right) & \mathrm{Rg}A_{\mu\nu}^{mn}\left(k\vec{r}_0\right)\end{pmatrix}\begin{pmatrix}\mathbf{M}_{\mu\nu}\left(k\vec{r}_2\right)\\ \mathbf{N}_{\mu\nu}\left(k\vec{r}_2\right)\end{pmatrix}, \quad r_2 \geq r_0, \tag{3.2.112a}$$

$$\begin{pmatrix} \mathbf{M}_{mn}\left(k\vec{r}_1\right) \\ \mathbf{N}_{mn}\left(k\vec{r}_1\right) \end{pmatrix} = \sum_{\nu=1}^{\infty}\sum_{\mu=-\nu}^{\nu} \begin{pmatrix} A_{\mu\nu}^{mn}\left(k\vec{r}_0\right) & B_{\mu\nu}^{mn}\left(k\vec{r}_0\right) \\ B_{\mu\nu}^{mn}\left(k\vec{r}_0\right) & A_{\mu\nu}^{mn}\left(k\vec{r}_0\right) \end{pmatrix} \begin{pmatrix} \mathrm{Rg}\mathbf{M}_{\mu\nu}\left(k\vec{r}_2\right) \\ \mathrm{Rg}\mathbf{N}_{\mu\nu}\left(k\vec{r}_2\right) \end{pmatrix}, \quad r_2 < r_0, \tag{3.2.112b}$$

and

$$\begin{pmatrix} \mathrm{Rg}\mathbf{M}_{mn}\left(k\vec{r}_2\right) \\ \mathrm{Rg}\mathbf{N}_{mn}\left(k\vec{r}_2\right) \end{pmatrix} = \sum_{\nu=1}^{\infty}\sum_{\mu=-\nu}^{\nu} \begin{pmatrix} \mathrm{Rg}A_{\mu\nu}^{mn}\left(-k\vec{r}_0\right) & \mathrm{Rg}B_{\mu\nu}^{mn}\left(-k\vec{r}_0\right) \\ \mathrm{Rg}B_{\mu\nu}^{mn}\left(-k\vec{r}_0\right) & \mathrm{Rg}A_{\mu\nu}^{mn}\left(-k\vec{r}_0\right) \end{pmatrix} \begin{pmatrix} \mathrm{Rg}\mathbf{M}_{\mu\nu}\left(k\vec{r}_1\right) \\ \mathrm{Rg}\mathbf{N}_{\mu\nu}\left(k\vec{r}_1\right) \end{pmatrix}, \tag{3.2.113}$$

$$\begin{pmatrix} \mathbf{M}_{mn}\left(k\vec{r}_2\right) \\ \mathbf{N}_{mn}\left(k\vec{r}_2\right) \end{pmatrix} = \sum_{\nu=1}^{\infty}\sum_{\mu=-\nu}^{\nu} \begin{pmatrix} \mathrm{Rg}A_{\mu\nu}^{mn}\left(-k\vec{r}_0\right) & \mathrm{Rg}B_{\mu\nu}^{mn}\left(-k\vec{r}_0\right) \\ \mathrm{Rg}B_{\mu\nu}^{mn}\left(-k\vec{r}_0\right) & \mathrm{Rg}A_{\mu\nu}^{mn}\left(-k\vec{r}_0\right) \end{pmatrix} \begin{pmatrix} \mathbf{M}_{\mu\nu}\left(k\vec{r}_1\right) \\ \mathbf{N}_{\mu\nu}\left(k\vec{r}_1\right) \end{pmatrix}, \quad r_1 \geq r_0, \tag{3.2.114a}$$

$$\begin{pmatrix} \mathbf{M}_{mn}\left(k\vec{r}_2\right) \\ \mathbf{N}_{mn}\left(k\vec{r}_2\right) \end{pmatrix} = \sum_{\nu=1}^{\infty}\sum_{\mu=-\nu}^{\nu} \begin{pmatrix} A_{\mu\nu}^{mn}\left(-k\vec{r}_0\right) & B_{\mu\nu}^{mn}\left(-k\vec{r}_0\right) \\ B_{\mu\nu}^{mn}\left(-k\vec{r}_0\right) & A_{\mu\nu}^{mn}\left(-k\vec{r}_0\right) \end{pmatrix} \begin{pmatrix} \mathrm{Rg}\mathbf{M}_{\mu\nu}\left(k\vec{r}_1\right) \\ \mathrm{Rg}\mathbf{N}_{\mu\nu}\left(k\vec{r}_1\right) \end{pmatrix}, \quad r_1 < r_0, \tag{3.2.114b}$$

where $\vec{r}_2 = \vec{r}_1 - \vec{r}_0$ as in Fig. 3.7.

The T-matrix method is implemented in terms of the expansions of the incident and scattered fields using the VSWF. The transformation of the VSWF under the translation of a frame of reference can be used similarly to obtain the T-matrix transformation under the same translation of the frame of reference. For single-particle scattering, the T-matrix transformation under the translation of the frame of reference can be given using Eqs. (3.2.111) and (3.2.112a) or Eqs. (3.2.113) and (3.2.114a), since the condition $r_1$, $r_2 \geq r_0$ is easily satisfied for the light scattering in the far field. Using Eq. (2.2.50), the incident and scattered electric fields can be expanded according to frame 1 and frame 2 as follows:

$$\begin{aligned} \mathbf{E}^{\mathrm{inc}}\left(\vec{r}_1\right) &= \sum_{n=1}^{\infty}\sum_{m=-n}^{n} \left(\mathrm{Rg}\mathbf{M}_{mn}\left(k\vec{r}_1\right) \ \mathrm{Rg}\mathbf{N}_{mn}\left(k\vec{r}_1\right)\right) \begin{pmatrix} a_{mn}^1 \\ b_{mn}^1 \end{pmatrix} \\ &= \sum_{n=1}^{\infty}\sum_{m=-n}^{n} \left(\mathrm{Rg}\mathbf{M}_{mn}\left(k\vec{r}_2\right) \ \mathrm{Rg}\mathbf{N}_{mn}\left(k\vec{r}_2\right)\right) \\ &\quad \left[\sum_{\nu=1}^{\infty}\sum_{\mu=-\nu}^{\nu} \begin{pmatrix} \mathrm{Rg}A_{mn}^{\mu\nu}\left(k\vec{r}_0\right) & \mathrm{Rg}B_{mn}^{\mu\nu}\left(k\vec{r}_0\right) \\ \mathrm{Rg}B_{mn}^{\mu\nu}\left(k\vec{r}_0\right) & \mathrm{Rg}A_{mn}^{\mu\nu}\left(k\vec{r}_0\right) \end{pmatrix} \begin{pmatrix} a_{\mu\nu}^1 \\ b_{\mu\nu}^1 \end{pmatrix}\right], \end{aligned} \tag{3.2.115a}$$

$$\mathbf{E}^{\mathrm{sca}}\left(\vec{r}_1\right)=\sum_{n=1}^{\infty}\sum_{m=-n}^{n}\left(\mathbf{M}_{mn}\left(k\vec{r}_1\right)\ \mathbf{N}_{mn}\left(k\vec{r}_1\right)\right)\begin{pmatrix}p_{mn}^{1}\\ q_{mn}^{1}\end{pmatrix}$$
$$=\sum_{n=1}^{\infty}\sum_{m=-n}^{n}\left(\mathbf{M}_{mn}\left(k\vec{r}_2\right)\ \mathbf{N}_{mn}\left(k\vec{r}_2\right)\right)$$
$$\left[\sum_{\nu=1}^{\infty}\sum_{\mu=-\nu}^{\nu}\begin{pmatrix}\mathrm{Rg}A_{mn}^{\mu\nu}\left(k\vec{r}_0\right) & \mathrm{Rg}B_{mn}^{\mu\nu}\left(k\vec{r}_0\right)\\ \mathrm{Rg}B_{mn}^{\mu\nu}\left(k\vec{r}_0\right) & \mathrm{Rg}A_{mn}^{\mu\nu}\left(k\vec{r}_0\right)\end{pmatrix}\begin{pmatrix}p_{\mu\nu}^{1}\\ q_{\mu\nu}^{1}\end{pmatrix}\right], \tag{3.2.115b}$$

and

$$\mathbf{E}^{\mathrm{inc}}\left(\vec{r}_2\right)=\sum_{n=1}^{\infty}\sum_{m=-n}^{n}\left(\mathrm{Rg}\mathbf{M}_{mn}\left(k\vec{r}_2\right)\ \mathrm{Rg}\mathbf{N}_{mn}\left(k\vec{r}_2\right)\right)\begin{pmatrix}a_{mn}^{2}\\ b_{mn}^{2}\end{pmatrix}$$
$$=\sum_{n=1}^{\infty}\sum_{m=-n}^{n}\left(\mathrm{Rg}\mathbf{M}_{mn}\left(k\vec{r}_1\right)\ \mathrm{Rg}\mathbf{N}_{mn}\left(k\vec{r}_1\right)\right)$$
$$\left[\sum_{\nu=1}^{\infty}\sum_{\mu=-\nu}^{\nu}\begin{pmatrix}\mathrm{Rg}A_{mn}^{\mu\nu}\left(-k\vec{r}_0\right) & \mathrm{Rg}B_{mn}^{\mu\nu}\left(-k\vec{r}_0\right)\\ \mathrm{Rg}B_{mn}^{\mu\nu}\left(-k\vec{r}_0\right) & \mathrm{Rg}A_{mn}^{\mu\nu}\left(-k\vec{r}_0\right)\end{pmatrix}\begin{pmatrix}a_{\mu\nu}^{2}\\ b_{\mu\nu}^{2}\end{pmatrix}\right], \tag{3.2.116a}$$

$$\mathbf{E}^{\mathrm{sca}}\left(\vec{r}_2\right)=\sum_{n=1}^{\infty}\sum_{m=-n}^{n}\left(\mathbf{M}_{mn}\left(k\vec{r}_2\right)\ \mathbf{N}_{mn}\left(k\vec{r}_2\right)\right)\begin{pmatrix}p_{mn}^{2}\\ q_{mn}^{2}\end{pmatrix}$$
$$=\sum_{n=1}^{\infty}\sum_{m=-n}^{n}\left(\mathbf{M}_{mn}\left(k\vec{r}_1\right)\ \mathbf{N}_{mn}\left(k\vec{r}_1\right)\right)$$
$$\left[\sum_{\nu=1}^{\infty}\sum_{\mu=-\nu}^{\nu}\begin{pmatrix}\mathrm{Rg}A_{mn}^{\mu\nu}\left(-k\vec{r}_0\right) & \mathrm{Rg}B_{mn}^{\mu\nu}\left(-k\vec{r}_0\right)\\ \mathrm{Rg}B_{mn}^{\mu\nu}\left(-k\vec{r}_0\right) & \mathrm{Rg}A_{mn}^{\mu\nu}\left(-k\vec{r}_0\right)\end{pmatrix}\begin{pmatrix}p_{\mu\nu}^{2}\\ q_{\mu\nu}^{2}\end{pmatrix}\right], \tag{3.2.116b}$$

where superscripts "1" and "2" denote the expansion coefficients in frame 1 and frame 2, respectively, and conditions $r_2 \geq r_0$ for Eq. (3.2.115b) and $r_1 \geq r_0$ for Eq. (3.2.116b) are required. Accordingly, using the definition of the T-matrix in Eq. (3.2.53), the T-matrix in one frame can be expressed by the T-matrix in another frame as

$$\begin{pmatrix}\mathrm{T}_{mnm'n'}^{11}(2) & \mathrm{T}_{mnm'n'}^{12}(2)\\ \mathrm{T}_{mnm'n'}^{21}(2) & \mathrm{T}_{mnm'n'}^{22}(2)\end{pmatrix}=\sum_{\nu=1}^{\infty}\sum_{\mu=-\nu}^{\nu}\sum_{\nu'=1}^{\infty}\sum_{\mu'=-\nu'}^{\nu'}\left[\begin{pmatrix}\mathrm{Rg}A_{mn}^{\mu\nu}\left(k\vec{r}_0\right) & \mathrm{Rg}B_{mn}^{\mu\nu}\left(k\vec{r}_0\right)\\ \mathrm{Rg}B_{mn}^{\mu\nu}\left(k\vec{r}_0\right) & \mathrm{Rg}A_{mn}^{\mu\nu}\left(k\vec{r}_0\right)\end{pmatrix}\right.$$
$$\left.\cdot\begin{pmatrix}\mathrm{T}_{\mu\nu\mu'\nu'}^{11}(1) & \mathrm{T}_{\mu\nu\mu'\nu'}^{12}(1)\\ \mathrm{T}_{\mu\nu\mu'\nu'}^{21}(1) & \mathrm{T}_{\mu\nu\mu'\nu'}^{22}(1)\end{pmatrix}\cdot\begin{pmatrix}\mathrm{Rg}A_{\mu'\nu'}^{m'n'}\left(-k\vec{r}_0\right) & \mathrm{Rg}B_{\mu'\nu'}^{m'n'}\left(-k\vec{r}_0\right)\\ \mathrm{Rg}B_{\mu'\nu'}^{m'n'}\left(-k\vec{r}_0\right) & \mathrm{Rg}A_{\mu'\nu'}^{m'n'}\left(-k\vec{r}_0\right)\end{pmatrix}\right], \tag{3.2.117a}$$

$$\begin{pmatrix} \mathrm{T}^{11}_{mnm'n'}(1) & \mathrm{T}^{12}_{mnm'n'}(1) \\ \mathrm{T}^{21}_{mnm'n'}(1) & \mathrm{T}^{22}_{mnm'n'}(1) \end{pmatrix} = \sum_{\nu=1}^{\infty}\sum_{\mu=-\nu}^{\nu}\sum_{\nu'=1}^{\infty}\sum_{\mu'=-\nu'}^{\nu'} \left[ \begin{pmatrix} \mathrm{Rg}A^{\mu\nu}_{mn}\left(-k\vec{r}_0\right) & \mathrm{Rg}B^{\mu\nu}_{mn}\left(-k\vec{r}_0\right) \\ \mathrm{Rg}B^{\mu\nu}_{mn}\left(-k\vec{r}_0\right) & \mathrm{Rg}A^{\mu\nu}_{mn}\left(-k\vec{r}_0\right) \end{pmatrix} \right.$$
$$\left. \cdot \begin{pmatrix} \mathrm{T}^{11}_{\mu\nu\mu'\nu'}(2) & \mathrm{T}^{12}_{\mu\nu\mu'\nu'}(2) \\ \mathrm{T}^{21}_{\mu\nu\mu'\nu'}(2) & \mathrm{T}^{22}_{\mu\nu\mu'\nu'}(2) \end{pmatrix} \cdot \begin{pmatrix} \mathrm{Rg}A^{m'n'}_{\mu'\nu'}\left(k\vec{r}_0\right) & \mathrm{Rg}B^{m'n'}_{\mu'\nu'}\left(k\vec{r}_0\right) \\ \mathrm{Rg}B^{m'n'}_{\mu'\nu'}\left(k\vec{r}_0\right) & \mathrm{Rg}A^{m'n'}_{\mu'\nu'}\left(k\vec{r}_0\right) \end{pmatrix} \right], \tag{3.2.117b}$$

where (1) and (2) denote the T-matrices corresponding to frame 1 and frame 2, respectively.

For multiple-particle scattering involving $N$ particles, the total scattered field is composed of the scattered fields of all particles. The actual incident field for an individual particle is crucial to the computation of the scattered field of the particle because each particle has its own T-matrix in the local frame of reference attached to the particle to connect the incident field and the scattered field. The incident field for an individual particle of an ensemble of particles consists of the original incident field plus the sum of the scattered field from other particles (Mishchenko et al., 2002; Doicu et al., 2006):

$$\mathbf{E}^{\mathrm{inc}}_i\left(\vec{r}\right) = \mathbf{E}^{\mathrm{inc}}_0\left(\vec{r}\right) + \sum_{\substack{j=1 \\ j\neq i}}^{N} \mathbf{E}^{\mathrm{sca}}_j\left(\vec{r}\right), \quad i = 1,\ldots,N. \tag{3.2.118}$$

Representing all the variables in the corresponding local frame of reference and using the expansion symbols in Eq. (3.2.51), Eq. (3.2.118) can be replaced using the expansion coefficients of the incident and scattered fields:

$$\begin{pmatrix} a^{(i)}_{mn} \\ b^{(i)}_{mn} \end{pmatrix} = \begin{pmatrix} a^{(i,o)}_{mn} \\ b^{(i,o)}_{mn} \end{pmatrix} + \sum_{\substack{j=1 \\ j\neq i}}^{N} \sum_{\nu=1}^{\infty}\sum_{\mu=-\nu}^{\nu} \begin{pmatrix} A^{mn}_{\mu\nu}\left(k\vec{r}_{ji}\right) & B^{mn}_{\mu\nu}\left(k\vec{r}_{ji}\right) \\ B^{mn}_{\mu\nu}\left(k\vec{r}_{ji}\right) & A^{mn}_{\mu\nu}\left(k\vec{r}_{ji}\right) \end{pmatrix} \begin{pmatrix} p^{(j)}_{\mu\nu} \\ q^{(j)}_{\mu\nu} \end{pmatrix}, \quad i = 1,\ldots,N, \tag{3.2.119a}$$

or in a compact form

$$\begin{pmatrix} \mathbf{a}^{(i)} \\ \mathbf{b}^{(i)} \end{pmatrix} = \begin{pmatrix} \mathbf{a}^{(i,o)} \\ \mathbf{b}^{(i,o)} \end{pmatrix} + \sum_{\substack{j=1 \\ j\neq i}}^{N} \begin{pmatrix} \mathbf{A}\left(k\vec{r}_{ji}\right) & \mathbf{B}\left(k\vec{r}_{ji}\right) \\ \mathbf{B}\left(k\vec{r}_{ji}\right) & \mathbf{A}\left(k\vec{r}_{ji}\right) \end{pmatrix} \begin{pmatrix} \mathbf{p}^{(j)} \\ \mathbf{q}^{(j)} \end{pmatrix}, \quad i = 1,\ldots,N, \tag{3.2.119b}$$

where superscript $(i)$ is associated with the $i$th local frame of reference attached to the $i$th particle and superscript $(i,o)$ represents the incident expansion coefficients of the original incident field in the $i$th local frame of reference; $\vec{r}_{ji}$ is the vector from the origin of the $j$th local frame of reference to the origin of the $i$th local frame of reference. The corresponding scattering expansion coefficients are stated in the following compact form (Mishchenko et al., 2002):

$$\begin{pmatrix} \mathbf{p}^{(i)} \\ \mathbf{q}^{(i)} \end{pmatrix} = \mathbf{T}^{(i)} \begin{pmatrix} \mathbf{a}^{(i,o)} \\ \mathbf{b}^{(i,o)} \end{pmatrix} + \mathbf{T}^{(i)} \sum_{\substack{j=1 \\ j \neq i}}^{N} \begin{pmatrix} \mathbf{A}\left(k\vec{r}_{ji}\right) & \mathbf{B}\left(k\vec{r}_{ji}\right) \\ \mathbf{B}\left(k\vec{r}_{ji}\right) & \mathbf{A}\left(k\vec{r}_{ji}\right) \end{pmatrix} \begin{pmatrix} \mathbf{p}^{(j)} \\ \mathbf{q}^{(j)} \end{pmatrix}, \quad i = 1,\ldots,N, \tag{3.2.120}$$

where $\mathbf{T}^{(i)}$ is the T-matrix of the $i$th particle in the local frame of reference. The expansion coefficients of the scattered field from the $i$th particle in Eq. (3.2.120) can be obtained in terms of an efficient iterative algorithm. Once the expansion coefficients of the scattered field from each particle are given, the total scattered field and the multiple-scattering properties can be straightforwardly computed. It is evident from Eq. (3.2.120) that the scattering properties obtained by the iterative algorithm depend on the incident direction. Accordingly, the scattering from randomly oriented multiple particles can only be determined by numerically averaging the results from different incident directions. On the other hand, Eq. (3.2.120) is a set of linear equations associated with the scattering expansion coefficients and the original incident expansion coefficients, and a matrix inversion can be used to obtain an expression as follows:

$$\begin{pmatrix} \left(\mathbf{p}^{(1)}, \mathbf{q}^{(1)}\right)^T \\ \cdots \\ \left(\mathbf{p}^{(i)}, \mathbf{q}^{(i)}\right)^T \\ \cdots \\ \left(\mathbf{p}^{(N)}, \mathbf{q}^{(N)}\right)^T \end{pmatrix} = \begin{pmatrix} \mathbf{I} & \cdots & -\mathbf{T}^{(1)}\mathbf{A}^{(1i)} & \cdots & -\mathbf{T}^{(1)}\mathbf{A}^{(1N)} \\ \cdots & \cdots & \cdots & \cdots & \cdots \\ -\mathbf{T}^{(i)}\mathbf{A}^{(i1)} & \cdots & \mathbf{I} & \cdots & -\mathbf{T}^{(i)}\mathbf{A}^{(iN)} \\ \cdots & \cdots & \cdots & \cdots & \cdots \\ -\mathbf{T}^{(N)}\mathbf{A}^{(N1)} & \cdots & -\mathbf{T}^{(N)}\mathbf{A}^{(Ni)} & \cdots & \mathbf{I} \end{pmatrix}^{-1}$$

$$\operatorname{diag}\begin{pmatrix} \mathbf{T}^{(1)} \\ \cdots \\ \mathbf{T}^{(i)} \\ \cdots \\ \mathbf{T}^{(N)} \end{pmatrix} \begin{pmatrix} \left(\mathbf{a}^{(1,o)}, \mathbf{b}^{(1,o)}\right)^T \\ \cdots \\ \left(\mathbf{a}^{(i,o)}, \mathbf{b}^{(i,o)}\right)^T \\ \cdots \\ \left(\mathbf{a}^{(N,o)}, \mathbf{b}^{(N,o)}\right)^T \end{pmatrix}, \tag{3.2.121}$$

where

$$\mathbf{A}^{(ij)} = \begin{pmatrix} \mathbf{A}\left(k\vec{r}_{ji}\right) & \mathbf{B}\left(k\vec{r}_{ji}\right) \\ \mathbf{B}\left(k\vec{r}_{ji}\right) & \mathbf{A}\left(k\vec{r}_{ji}\right) \end{pmatrix}. \tag{3.2.122}$$

Accordingly, the expansion coefficients can be formally given as (Mackowski, 1994; Mishchenko et al., 2002)

$$\begin{pmatrix} \mathbf{p}^{(i)} \\ \mathbf{q}^{(i)} \end{pmatrix} = \sum_{j=1}^{N} \mathbf{T}^{(ij)} \begin{pmatrix} \mathbf{a}^{(j,o)} \\ \mathbf{b}^{(j,o)} \end{pmatrix}, \quad i = 1,\ldots,N. \tag{3.2.123}$$

Matrix $\mathbf{T}^{(ij)}$ is a block matrix and can be obtained from Eq. (3.2.121). The physical meaning of Eq. (3.2.123) is evident. The incident and scattered expansion coefficients in Eq. (3.2.123) are represented in the local frame of reference. Transforming all the expansion coefficients from the local frame of reference to the laboratory frame of reference and summing the scattered fields from all particles, the total T-matrix can be obtained as (Peterson and Ström, 1973; Mackowski, 1994; Mishchenko et al., 2002)

$$\mathbf{T} = \sum_{i,j=1}^{N} \begin{pmatrix} \mathrm{Rg}\mathbf{A}\left(k\vec{r}_{i0}\right) & \mathrm{Rg}\mathbf{B}\left(k\vec{r}_{i0}\right) \\ \mathrm{Rg}\mathbf{B}\left(k\vec{r}_{i0}\right) & \mathrm{Rg}\mathbf{A}\left(k\vec{r}_{i0}\right) \end{pmatrix} \mathbf{T}^{(ij)} \begin{pmatrix} \mathrm{Rg}\mathbf{A}\left(k\vec{r}_{0j}\right) & \mathrm{Rg}\mathbf{B}\left(k\vec{r}_{0j}\right) \\ \mathrm{Rg}\mathbf{B}\left(k\vec{r}_{0j}\right) & \mathrm{Rg}\mathbf{A}\left(k\vec{r}_{0j}\right) \end{pmatrix}, \tag{3.2.124}$$

where $\vec{r}_{i0}$ is the vector from the origin of the $i$th frame of reference to the origin of the laboratory frame of reference and $\vec{r}_{0j}$ is the vector from the origin of the laboratory frame of reference to the origin of the $j$th frame of reference. Although direct calculation of the T-matrix in Eq. (3.2.124) is time-consuming because it involves the inversion of a huge matrix in Eq. (3.2.121), it can also be iteratively obtained (Mackowski and Mishchenko, 2011).

## 3.3 A common approach to compute the T-matrix

### 3.3.1 Free space dyadic green function expansion using VSWF

A general Green function is defined as the solution at field point $\vec{r}$ of an equation associated with a point source at $\vec{r}\,'$. Physically, the Green function serves as an integral operator or a convolution transforming a volume or surface source to a field point. Consequently, the Green function of a scalar field equation should also be scalar, while the Green function of a vector field equation should be a tensor or a dyad. Conforming to the scalar and vector field equations given in Eq. (3.2.4), one can define the scalar and vector Green functions.

The scalar Green function satisfying the scalar field equation given in Eq. (3.2.4a) and its well-known solution are (Morse and Feshbach, 1953; Tsang et al., 2000)

$$\nabla^2 g\left(\vec{r}, \vec{r}\,'\right) + k^2 g\left(\vec{r}, \vec{r}\,'\right) = -\delta\left(\vec{r} - \vec{r}\,'\right), \tag{3.3.1a}$$

$$g\left(\vec{r}, \vec{r}\,'\right) = \frac{\exp\left(ik\left|\vec{r} - \vec{r}\,'\right|\right)}{4\pi\left|\vec{r} - \vec{r}\,'\right|}, \tag{3.3.1b}$$

where the left argument of the scalar Green function denotes the field point while the right one denotes the source point. It is evident that the scalar Green function is symmetric:

$$g\left(\vec{r}\,', \vec{r}\right) = g\left(\vec{r}, \vec{r}\,'\right). \tag{3.3.2}$$

Moreover, the gradient operator on the scalar Green function is

$$\nabla g\left(\vec{r}, \vec{r}\,'\right) = -\nabla' g\left(\vec{r}, \vec{r}\,'\right), \tag{3.3.3}$$

where the primed gradient operator indicates an operation expressed in the primed coordinates. A Green function corresponding to a vector field equation is a dyad and named as dyadic Green function. In this book, several vector field equations are involved such as the vector Helmholtz equation given in Eq. (3.2.4b), the vector wave equations with respect to the electric field or the magnetic field. The Green functions can be defined corresponding to different vector field equations, and the details can be obtained in books given by Morse and Feshbach (1953) and by Tai (1994). The dyadic Green function with respect to the magnetic field is discussed in Section 4.1.3. In this chapter, only the dyadic Green function with respect to the electric field is discussed, and without causing confusion, the dyadic Green function is usually used to refer to the electric field. The free space dyadic Green function satisfying the vector wave equation with respect to the electric field and the corresponding solution are (Morse and Feshbach, 1953; Tsang et al., 2000)

$$\nabla \times \nabla \times \overleftrightarrow{G}\left(\vec{r}, \vec{r}\,'\right) - k^2 \overleftrightarrow{G}\left(\vec{r}, \vec{r}\,'\right) = \overleftrightarrow{I}\delta\left(\vec{r} - \vec{r}\,'\right), \tag{3.3.4a}$$

$$\overleftrightarrow{G}\left(\vec{r}, \vec{r}\,'\right) = \left(\overleftrightarrow{I} + \frac{1}{k^2}\nabla\nabla\right) g\left(\vec{r}, \vec{r}\,'\right), \tag{3.3.4b}$$

where the differential operator in Eq. (3.3.4b) can be explicitly given as

$$\overleftrightarrow{G}\left(\vec{r}, \vec{r}\,'\right) = \left\{\left(1 + \frac{i}{kR} - \frac{1}{k^2R^2}\right)\overleftrightarrow{I} - \left(1 + \frac{3i}{kR} - \frac{3}{k^2R^2}\right)\hat{R}\hat{R}\right\} \frac{\exp(ikR)}{4\pi R}, \tag{3.3.4c}$$

where $R = \left|\vec{r} - \vec{r}\,'\right|$ and $\hat{R} = \frac{\vec{r} - \vec{r}\,'}{\left|\vec{r} - \vec{r}\,'\right|}$. The free space dyadic Green function has the symmetric property (Morse and Feshbach, 1953; Tai, 1994):

$$\overleftrightarrow{G}^T\left(\vec{r}\,', \vec{r}\right) = \overleftrightarrow{G}\left(\vec{r}, \vec{r}\,'\right). \tag{3.3.5}$$

At infinity, the distance between the source point and the field point can be approximated by ignoring the small second- and higher-order quantities as

$$\left|\vec{r}-\vec{r}\,'\right|\Big|_{r\to\infty} = \sqrt{r^2\left(1-\frac{2\vec{r}\cdot\vec{r}\,'}{r^2}+\frac{r'^2}{r^2}\right)}\Bigg|_{r\to\infty} = r-\hat{r}\cdot\vec{r}\,', \tag{3.3.6}$$

where $\hat{r}$ is the unit vector for vector $\vec{r}$, $r'$ is a finite number, and $r'/r$ is a small first-order quantity since $r \gg r'$. The scalar Green function in Eq. (3.3.1b) at infinity becomes

$$g\left(\vec{r},\vec{r}\,'\right)\Big|_{r\to\infty} = \frac{\exp(ikr)}{4\pi r}\exp\left(-ik\hat{r}\cdot\vec{r}\,'\right)+O\left(\frac{r'}{r}\right), \tag{3.3.7}$$

where $O\left(\frac{r'}{r}\right)$ denotes the infinitesimal of the second and higher orders. The free space dyadic Green function in Eq. (3.3.4c) at infinity becomes

$$\overleftrightarrow{G}\left(\vec{r},\vec{r}\,'\right)\Big|_{r\to\infty} = \left(\overleftrightarrow{\mathrm{I}}-\hat{r}\hat{r}\right)\frac{\exp(ikr)}{4\pi r}\exp\left(-ik\hat{r}\cdot\vec{r}\,'\right)+O\left(\frac{r'}{r}\right). \tag{3.3.8}$$

In the applications of Eqs. (3.3.7)–(3.3.8), $O\left(\frac{r'}{r}\right)$ is ignored.

Mathematically, the Dirac delta function can be expanded using a set of complete orthogonal functions $\exp\left(i\vec{p}\cdot\vec{r}\right)$ as follows (Tsang et al., 2000):

$$\delta\left(\vec{r}-\vec{r}\,'\right) = \frac{1}{(2\pi)^3}\int \exp\left[i\vec{p}\cdot\left(\vec{r}-\vec{r}\,'\right)\right]d^3\vec{p}\,. \tag{3.3.9}$$

One can multiply both sides of Eq. (3.2.50) by $\exp\left(-i\vec{p}\cdot\vec{r}\,'\right)/(2\pi)^3$ and integrate over $\int d^3\vec{p}$ in full $p$ space. For future usage, we seek to expand the dyadic Green function using the vector spherical wave functions. The method used here is called the Ohm-Rayleigh method, introduced by Sommerfeld (1949, p. 179). It is actually an expansion method using eigenfunctions (Tai, 1994). The key step in this method is to expand the delta function in terms of the corresponding eigenfunctions. The eigenfunctions of the vector Helmholtz equation are the vector spherical wave functions. After a series of algebraic steps, Eq. (3.2.50) is changed to (Tsang et al., 2000)

$$\begin{aligned}\overleftrightarrow{\mathrm{I}}\delta\left(\vec{r}-\vec{r}\,'\right) = \sum_{n=0}^{\infty}\sum_{m=-n}^{n}(-1)^m\frac{2}{\pi}\int_0^{\infty}p^2dp[&\mathrm{Rg}\mathbf{L}_{mn}(pr,\theta,\varphi)\mathrm{Rg}\mathbf{L}_{-mn}(pr',\theta',\varphi')\\ &+\mathrm{Rg}\mathbf{M}_{mn}(pr,\theta,\varphi)\mathrm{Rg}\mathbf{M}_{-mn}(pr',\theta',\varphi')\\ &+\mathrm{Rg}\mathbf{N}_{mn}(pr,\theta,\varphi)\mathrm{Rg}\mathbf{N}_{-mn}(pr',\theta',\varphi')].\end{aligned} \tag{3.3.10}$$

The regular vector spherical wave functions are the general solutions of the vector Helmholtz equation. The $\{\mathrm{Rg}\mathbf{M}_{mn}\}$ and $\{\mathrm{Rg}\mathbf{N}_{mn}\}$ are divergence free, while the $\{\mathrm{Rg}\mathbf{L}_{mn}\}$ is not. The regular vector spherical wave functions have the following relations:

$$\nabla\times\nabla\times\mathrm{Rg}\mathbf{L}_{mn}(pr,\theta,\varphi)=0, \tag{3.3.11}$$

$$\nabla\times\nabla\times\mathrm{Rg}\mathbf{M}_{mn}(pr,\theta,\varphi)=p^2\mathrm{Rg}\mathbf{M}_{mn}(pr,\theta,\varphi), \tag{3.3.12a}$$

$$\nabla\times\nabla\times\mathrm{Rg}\mathbf{N}_{mn}(pr,\theta,\varphi)=p^2\mathrm{Rg}\mathbf{N}_{mn}(pr,\theta,\varphi). \tag{3.3.12b}$$

Consequently, the dyadic Green function in the integral representation using Eqs. (3.3.4) and (3.3.10)–(3.3.12) is (Tsang et al., 2000)

$$\begin{aligned}\overleftrightarrow{G}\left(\vec{r},\vec{r}\,'\right)=\sum_{n=0}^{\infty}\sum_{m=-n}^{n}(-1)^m\frac{2}{\pi}\int_0^{\infty}p^2dp\Bigg[&-\frac{\mathrm{Rg}\mathbf{L}_{mn}(pr,\theta,\varphi)\mathrm{Rg}\mathbf{L}_{-mn}(pr',\theta',\varphi')}{k^2}\\&+\frac{\mathrm{Rg}\mathbf{M}_{mn}(pr,\theta,\varphi)\mathrm{Rg}\mathbf{M}_{-mn}(pr',\theta',\varphi')}{p^2-k^2}\\&+\frac{\mathrm{Rg}\mathbf{N}_{mn}(pr,\theta,\varphi)\mathrm{Rg}\mathbf{N}_{-mn}(pr',\theta',\varphi')}{p^2-k^2}\Bigg].\end{aligned} \tag{3.3.13}$$

The integral element $p$ appears only in the radial functions of the regular vector spherical wave functions. As shown in Eq. (3.3.13), there are singular poles at $p=\pm k$. The integration in the real plane is extended to integration in the complex plane and is solved using contour integration by choosing the path of integration to deal with the singular poles (Morse and Feshbach, 1953; Tsang and Kong, 1980). The dyadic Green function can be expanded using the vector spherical wave functions as follows (Tai, 1994; Tsang et al., 2000):

$$\begin{aligned}&\overleftrightarrow{G}\left(\vec{r},\vec{r}\,'\right)=-\hat{r}\hat{r}\frac{\delta\left(\vec{r}-\vec{r}\,'\right)}{k^2}+ik\sum_{n=1}^{\infty}\sum_{m=-n}^{n}(-1)^m\\&\begin{cases}\mathbf{M}_{-mn}(kr,\theta,\varphi)\mathrm{Rg}\mathbf{M}_{mn}(kr',\theta',\varphi')+\mathbf{N}_{-mn}(kr,\theta,\varphi)\mathrm{Rg}\mathbf{N}_{mn}(kr',\theta',\varphi'), & \text{for } r>r',\\ \mathrm{Rg}\mathbf{M}_{-mn}(kr,\theta,\varphi)\mathbf{M}_{mn}(kr',\theta',\varphi')+\mathrm{Rg}\mathbf{N}_{-mn}(kr,\theta,\varphi)\mathbf{N}_{mn}(kr',\theta',\varphi'), & \text{for } r<r'.\end{cases}\end{aligned} \tag{3.3.14}$$

It is evident in Eq. (3.3.14) that the dyadic Green function has the symmetric property shown in Eq. (3.3.5) (Tsang et al., 2000).

### 3.3.2 Extended boundary condition method

Calculation of the T-matrix of a scattering particle was first accomplished using the EBCM by Waterman (1965, 1971). To summarize, the EBCM expresses the incident field and scattered field as a surface integral over the dyadic Green function and the

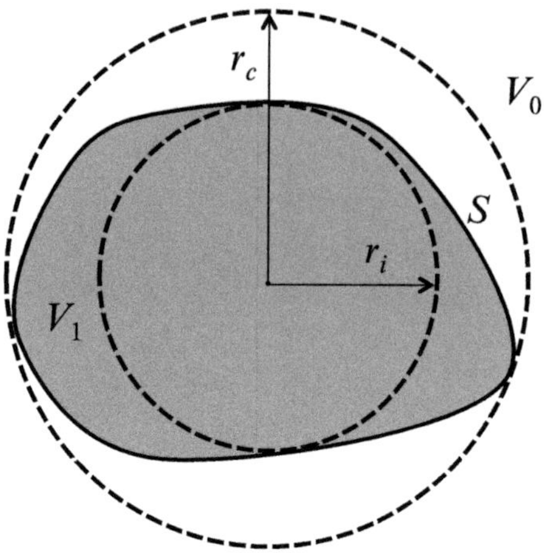


**Fig. 3.8** A homogeneous scattering particle. The shaded region $V_1$ represents the volume of the scattering particle, enclosed by surface area S. The exterior region $V_0$ indicates the entire volume outside the scattering particle including the volume within the circumscribed sphere and outside $V_1$, and $r_c$ and $r_i$ are the radii of the smallest circumscribed sphere and the largest inscribed sphere of the particle.

surface electromagnetic field. The electromagnetic field and the dyadic Green function can be expanded using VSWF. Using the vector Green's theorem, the scattered and the incident electromagnetic fields are represented as surface integrals in the exterior and interior regions of the scattering particle, respectively. According to the boundary conditions, the tangential component of the exterior surface electromagnetic field can be matched to that of the interior surface electromagnetic field. Moreover, the electromagnetic field in the interior region can also be expanded using the VSWF. Accordingly, the incident expansion coefficients and the scattered expansion coefficients both have a linear relation with the expansion coefficients of the interior field. Finally, the T-matrix can be obtained by the relation matrix between the scattered coefficients and the interior coefficients multiplied by the inversion of the relation matrix between the incident coefficients and the interior coefficients.

Fig. 3.8 shows an arbitrary homogeneous dielectric particle enclosed by a surface S, centered at the origin. The interior region of the scattering particle is denoted as $V_1$, and the exterior region outside the scattering particle is denoted as $V_0$. The radii $r_c$ and $r_i$ are radii of the smallest circumscribed sphere and the largest inscribed sphere of the particle. The wave numbers inside and outside the scattering particle are denoted as $k_1$ and $k$, respectively. According to Maxwell's equations given in Chapter 1, the electric fields in regions $V_0$ and $V_1$ satisfy the following equations:

$$\nabla \times \nabla \times \mathbf{E}\left(\vec{r}\right) - k^2 \mathbf{E}\left(\vec{r}\right) = 0, \quad \vec{r} \in \mathrm{V}_0, \tag{3.3.15a}$$

$$\nabla \times \nabla \times \mathbf{E}\left(\vec{r}\right) - k_1^2 \mathbf{E}\left(\vec{r}\right) = 0, \quad \vec{r} \in \mathrm{V}_1. \tag{3.3.15b}$$

Here, $\mathbf{E}\left(\vec{r}\right)$ is the total electric field, that is,

$$\mathbf{E}\left(\vec{r}\right) = \mathbf{E}^{\mathrm{inc}}\left(\vec{r}\right) + \mathbf{E}^{\mathrm{sca}}\left(\vec{r}\right), \tag{3.3.16}$$

where the incident field satisfies the same expression as Eq. (3.3.15a). The surface integrals for the incident field and the scattered field can actually be obtained in terms of the vector Green's theorem (Morse and Feshbach, 1953; Tsang et al., 2000; Mishchenko et al., 2002). The vector Green's theorem is expressed as

$$\int_{V} dV[\mathbf{A}\cdot(\nabla\times\nabla\times\mathbf{B}) - \mathbf{B}\cdot(\nabla\times\nabla\times\mathbf{A})] = \oint_{S} dS\hat{n}_s\cdot(\mathbf{B}\times\nabla\times\mathbf{A} - \mathbf{A}\times\nabla\times\mathbf{B}), \tag{3.3.17}$$

where volume V is enclosed by the area S, $\hat{n}_s$ is the outward unit vector normal to the local surface and is a function of the position vector, and **A** and **B** are arbitrary vectors.

The first step is to insert the following expressions into the earlier equation and to apply volume integration to the full space, including regions $V_0$ and $V_1$:

$$\mathbf{A} = \mathbf{E}^{\mathrm{inc}}, \quad \mathbf{B} = \overleftrightarrow{G}\left(\vec{r}, \vec{r}\,'\right)\cdot\vec{a}, \tag{3.3.18}$$

where $\overleftrightarrow{G}\left(\vec{r}, \vec{r}\,'\right)$ is the free space dyadic Green function and $\vec{a}$ is an arbitrary vector. The vector Green's theorem then gives the identity of the incident field at infinity:

$$\begin{aligned}\mathbf{E}^{\mathrm{inc}}\left(\vec{r}\,'\right)\cdot\vec{a} = -\oint_{S_\infty} dS\hat{n}_s\cdot\Big\{\left[\nabla\times\mathbf{E}^{\mathrm{inc}}\left(\vec{r}\right)\right]\times\left(\overleftrightarrow{G}\left(\vec{r}, \vec{r}\,'\right)\cdot\vec{a}\right)\\ +\mathbf{E}^{\mathrm{inc}}\left(\vec{r}\right)\times\left[\nabla\times\left(\overleftrightarrow{G}\left(\vec{r}, \vec{r}\,'\right)\cdot\vec{a}\right)\right]\Big\},\end{aligned} \tag{3.3.19}$$

where $S_\infty$ is the spherical surface at infinity.

Next, insert the following expression for the vector **A** into the vector Green's theorem and apply volume integration to the exterior region $V_0$:

$$\mathbf{A} = \mathbf{E}, \quad \mathbf{B} = \overleftrightarrow{G}\left(\vec{r}, \vec{r}\,'\right)\cdot\vec{a}, \tag{3.3.20}$$

The surfaces enclosing the exterior region $V_0$ are composed of two parts: the surface at infinity $S_\infty$ and the particle surface S. Since the normal vectors on both surfaces point in an outward direction, the surface integral over the particle surface should use the inward normal or the negative of the outward normal vectors. The vector Green's theorem gives the following expressions for the two different field regions $V_1$ (inside the particle) and $V_0$ (outside the particle):

$$\left.\begin{array}{r}\mathbf{E}\left(\vec{r}\,'\right)\cdot\vec{a},\ \vec{r}\,'\in \mathrm{V}_0\\ 0,\ \vec{r}\,'\in \mathrm{V}_1\end{array}\right\}=\left[-\oint\limits_{S_\infty}+\oint\limits_{S}\right]dS\hat{n}_s\cdot\left\{\left[\nabla\times\mathbf{E}^{\mathrm{ext}}\left(\vec{r}\right)\right]\times\left(\overleftrightarrow{G}\left(\vec{r},\vec{r}\,'\right)\cdot\vec{a}\right)\right.$$

$$\left.+\mathbf{E}^{\mathrm{ext}}\left(\vec{r}\right)\times\left[\nabla\times\left(\overleftrightarrow{G}\left(\vec{r},\vec{r}\,'\right)\cdot\vec{a}\right)\right]\right\}. \tag{3.3.21}$$

where $\hat{n}_s$ still represents the outward normal to the local surface and the superscript "ext" represents the exterior surface. The total electric field in the earlier equation is composed of the incident field and the scattered field, as was shown in Eq. (3.3.16). For the first surface integral on the right-hand side of the earlier equation, the incident electric field representation in the integrand gives the incident field as in Eq. (3.3.19). The Sommerfeld radiation condition with respect to the dyadic Green function can be written as (Tai, 1994)

$$\lim_{r\to\infty} r\left[\nabla\times\overleftrightarrow{G}-ik\hat{r}\times\overleftrightarrow{G}\right]=0. \tag{3.3.22}$$

Using the asymptotic behavior of the scattered field given in Eq. (3.2.56) and the earlier Sommerfeld radiation condition, the surface integral at infinity of the scattered electric field representation goes to zero. Since $\vec{a}$ is an arbitrary vector, it can be canceled from both sides of Eq. (3.3.21). Finally, Eq. (3.3.21) can be simplified as follows:

$$\left.\begin{array}{r}\mathbf{E}\left(\vec{r}\,'\right),\ \vec{r}\,'\in \mathrm{V}_0\\ 0,\ \vec{r}\,'\in \mathrm{V}_1\end{array}\right\}=\mathbf{E}^{\mathrm{inc}}\left(\vec{r}\,'\right)+\oint\limits_{S}dS\left\{i\omega\mu_0\left[\hat{n}_s\times\mathbf{H}^{\mathrm{ext}}\left(\vec{r}\right)\right]\cdot\overleftrightarrow{G}\left(\vec{r},\vec{r}\,'\right)\right.$$

$$\left.+\left[\hat{n}_s\times\mathbf{E}^{\mathrm{ext}}\left(\vec{r}\right)\right]\cdot\left[\nabla\times\overleftrightarrow{G}\left(\vec{r},\vec{r}\,'\right)\right]\right\}. \tag{3.3.23}$$

The curl of the electric field is expressed in the magnetic field using Maxwell's equations and $\mu=\mu_0$ associated with nonferromagnetic dielectric particles. For further derivations in the succeeding text, the scattered field in the exterior region and the incident field in the interior region are written separately as (Waterman, 1971; Tsang et al., 2000; Mishchenko et al., 2002; Doicu et al., 2006)

$$\mathbf{E}^{\mathrm{sca}}\left(\vec{r}\,'\right)=\oint\limits_{S}dS\left\{i\omega\mu_0\left[\hat{n}_s\times\mathbf{H}^{\mathrm{int}}\left(\vec{r}\right)\right]\cdot\overleftrightarrow{G}\left(\vec{r},\vec{r}\,'\right)\right.$$

$$\left.+\left[\hat{n}_s\times\mathbf{E}^{\mathrm{int}}\left(\vec{r}\right)\right]\cdot\left[\nabla\times\overleftrightarrow{G}\left(\vec{r},\vec{r}\,'\right)\right]\right\},\quad \vec{r}\,'\in\mathrm{V}_0, \tag{3.3.24a}$$

$$\mathbf{E}^{\mathrm{inc}}\left(\vec{r}\,'\right)=-\oint\limits_{S}dS\left\{i\omega\mu_0\left[\hat{n}_s\times\mathbf{H}^{\mathrm{int}}\left(\vec{r}\right)\right]\cdot\overleftrightarrow{G}\left(\vec{r},\vec{r}\,'\right)\right.$$

$$\left.+\left[\hat{n}_s\times\mathbf{E}^{\mathrm{int}}\left(\vec{r}\right)\right]\cdot\left[\nabla\times\overleftrightarrow{G}\left(\vec{r},\vec{r}\,'\right)\right]\right\},\quad \vec{r}\,'\in\mathrm{V}_1, \tag{3.3.24b}$$

where the superscript "int" represents the interior surface and boundary conditions of tangential continuity have been used between the exterior and the interior electromagnetic field. Actually, Eq. (3.3.24a) is the so-called surface integral equation for the scattered electric field. The electromagnetic field exterior to the particle surface at any point can be obtained by mapping the surface tangential electromagnetic field in terms of the dyadic Green function. Accordingly, Eqs. (3.3.24a) and (3.3.24b) are also called the Huygens' principle and the extinction theorem (Tsang et al., 2000). The surface integral in Eq. (3.3.24b) is called the extended boundary condition because the field point extends from the exterior to the interior region of the scattering particle (Mishchenko et al., 2002). The surface tangential electromagnetic field is mapped to the incident field in the interior region in terms of the dyadic Green function. Moreover, the EBCM is also known as the null field method, due to Eq. (3.3.24b).

Eq. (3.3.24) gives the surface integral equations of the incident and scattered fields associated with the free space dyadic Green function and the internal electromagnetic field. The incident field, the scattered field, and the free space dyadic Green function can be expanded using the VSWF as shown in Eqs. (3.2.52), (3.3.14), respectively. As shown in Eq. (3.3.14), the unprimed and primed quantities in the free space dyadic Green function are decoupled using the VSWF. Using Eqs. (3.3.24), (3.2.52), the expansion coefficients associated with the incident and scattered fields can be explicitly expressed as surface integrals with respect to the internal electromagnetic fields and the VSWF. If one can continue to expand the internal electromagnetic fields using the VSWF, the relation between the expansion coefficients associated with the scattered field and the expansion coefficients associated with the internal field can be explicitly given as a surface integral using Eq. (3.3.24a), and the relation between the expansion coefficients associated with the incident field and the expansion coefficients associated with the internal field can be explicitly given as a surface integral using Eq. (3.3.24b). The expansion coefficients associated with the incident and scattered fields can be connected in terms of the expansion coefficients associated with the internal field, so the T-matrix formally given in Eq. (3.2.53) can be explicitly obtained using matrix inversion.

The conditions $r' > r_c$ in Eq. (3.3.24a) and $r' < r_i$ in Eq. (3.3.24b) are required to avoid singularities of the Green function when $r = r'$. Using Eqs. (3.2.52), (3.3.14), the incident and scattered expansion coefficients can be obtained by comparing the coefficients of regular and outgoing VSWF (Rg)$\mathbf{M}_{mn}$ and (Rg)$\mathbf{N}_{mn}$:

$$\begin{aligned}\begin{pmatrix} a_{mn} \\ b_{mn} \end{pmatrix} = k(-1)^m \oint_S dS \Big\{ & \omega\mu_0 \left[\hat{n}_s \times \mathbf{H}^{\text{int}}\left(\vec{r}\right)\right] \cdot \begin{pmatrix} \mathbf{M}_{-mn}(kr,\theta,\varphi) \\ \mathbf{N}_{-mn}(kr,\theta,\varphi) \end{pmatrix} \\ & -ik\left[\hat{n}_s \times \mathbf{E}^{\text{int}}\left(\vec{r}\right)\right] \cdot \begin{pmatrix} \mathbf{N}_{-mn}(kr,\theta,\varphi) \\ \mathbf{M}_{-mn}(kr,\theta,\varphi) \end{pmatrix} \Big\},\end{aligned} \tag{3.3.25a}$$

$$\begin{aligned}\begin{pmatrix} p_{mn} \\ q_{mn} \end{pmatrix} = -k(-1)^m \oint_S dS \Big\{ & \omega\mu_0 \left[\hat{n}_s \times \mathbf{H}^{\text{int}}\left(\vec{r}\right)\right] \cdot \begin{pmatrix} \text{Rg}\mathbf{M}_{-mn}(kr,\theta,\varphi) \\ \text{Rg}\mathbf{N}_{-mn}(kr,\theta,\varphi) \end{pmatrix} \\ & -ik\left[\hat{n}_s \times \mathbf{E}^{\text{int}}\left(\vec{r}\right)\right] \cdot \begin{pmatrix} \text{Rg}\mathbf{N}_{-mn}(kr,\theta,\varphi) \\ \text{Rg}\mathbf{M}_{-mn}(kr,\theta,\varphi) \end{pmatrix} \Big\},\end{aligned} \tag{3.3.25b}$$

where the representations of the incident and scattered expansion coefficients have been described in a compact way. The internal surface electromagnetic field is expanded using the regular VSWF associated with wave number $k_1$ inside the particle as follows:

$$\mathbf{E}^{\text{int}}\left(\vec{r}\right)=\sum_{n=1}^{\infty}\sum_{m=-n}^{n}\left[c_{mn}\text{Rg}\mathbf{M}_{mn}(k_1r,\theta,\varphi)+d_{mn}\text{Rg}\mathbf{N}_{mn}(k_1r,\theta,\varphi)\right], \tag{3.3.26a}$$

$$\begin{aligned}\mathbf{H}^{\text{int}}\left(\vec{r}\right)&=\frac{\nabla\times\mathbf{E}^{\text{int}}\left(\vec{r}\right)}{i\omega\mu_0}\\&=\frac{k_1}{i\omega\mu_0}\sum_{n=1}^{\infty}\sum_{m=-n}^{n}\left[d_{mn}\text{Rg}\mathbf{M}_{mn}(k_1r,\theta,\varphi)+c_{mn}\text{Rg}\mathbf{N}_{mn}(k_1r,\theta,\varphi)\right],\end{aligned} \tag{3.3.26b}$$

Substituting Eq. (3.3.26) into Eq. (3.3.25), the incident and scattered expansion coefficients have linear relations to the interior expansion coefficients. The relations of the expansion coefficients can be organized using matrices $\mathbf{Q}$ and Rg$\mathbf{Q}$ as follows:

$$\begin{aligned}\begin{pmatrix}\mathbf{a}\\\mathbf{b}\end{pmatrix}&=\begin{pmatrix}\mathbf{Q}^{11}\,\mathbf{Q}^{12}\\\mathbf{Q}^{21}\,\mathbf{Q}^{22}\end{pmatrix}\begin{pmatrix}\mathbf{c}\\\mathbf{d}\end{pmatrix},\\\begin{pmatrix}a_{mn}\\b_{mn}\end{pmatrix}&=\sum_{n'=1}^{\infty}\sum_{m'=-n'}^{n'}\begin{pmatrix}\text{Q}_{mnm'n'}^{11}\,\text{Q}_{mnm'n'}^{12}\\\text{Q}_{mnm'n'}^{21}\,\text{Q}_{mnm'n'}^{22}\end{pmatrix}\begin{pmatrix}c_{m'n'}\\d_{m'n'}\end{pmatrix},\end{aligned} \tag{3.3.27a}$$

$$\begin{aligned}\begin{pmatrix}\mathbf{p}\\\mathbf{q}\end{pmatrix}&=-\begin{pmatrix}\text{Rg}\mathbf{Q}^{11}\,\text{Rg}\mathbf{Q}^{12}\\\text{Rg}\mathbf{Q}^{21}\,\text{Rg}\mathbf{Q}^{22}\end{pmatrix}\begin{pmatrix}\mathbf{c}\\\mathbf{d}\end{pmatrix},\\\begin{pmatrix}p_{mn}\\q_{mn}\end{pmatrix}&=-\sum_{n'=1}^{\infty}\sum_{m'=-n'}^{n'}\begin{pmatrix}\text{RgQ}_{mnm'n'}^{11}\,\text{RgQ}_{mnm'n'}^{12}\\\text{RgQ}_{mnm'n'}^{21}\,\text{RgQ}_{mnm'n'}^{22}\end{pmatrix}\begin{pmatrix}c_{m'n'}\\d_{m'n'}\end{pmatrix},\end{aligned} \tag{3.3.27b}$$

where the matrix elements of matrix (Rg)$\mathbf{Q}$ or $\mathbf{Q}$ are (Barber and Hill, 1990; Tsang et al., 2000; Mishchenko et al., 2002; Doicu et al., 2006)

$$\begin{pmatrix}(\text{Rg})\text{Q}_{mnm'n'}^{11}\\(\text{Rg})\text{Q}_{mnm'n'}^{12}\\(\text{Rg})\text{Q}_{mnm'n'}^{21}\\(\text{Rg})\text{Q}_{mnm'n'}^{22}\end{pmatrix}=-ikk_1\begin{pmatrix}(\text{Rg})\text{J}_{mnm'n'}^{21}\\(\text{Rg})\text{J}_{mnm'n'}^{11}\\(\text{Rg})\text{J}_{mnm'n'}^{22}\\(\text{Rg})\text{J}_{mnm'n'}^{12}\end{pmatrix}-ik^2\begin{pmatrix}(\text{Rg})\text{J}_{mnm'n'}^{12}\\(\text{Rg})\text{J}_{mnm'n'}^{22}\\(\text{Rg})\text{J}_{mnm'n'}^{11}\\(\text{Rg})\text{J}_{mnm'n'}^{21}\end{pmatrix}, \tag{3.3.28}$$

and

$$\begin{pmatrix}(\text{Rg})\text{J}_{mnm'n'}^{11}\\(\text{Rg})\text{J}_{mnm'n'}^{12}\\(\text{Rg})\text{J}_{mnm'n'}^{21}\\(\text{Rg})\text{J}_{mnm'n'}^{22}\end{pmatrix}=(-1)^m\oint_S dS\hat{n}_s\left(\vec{r}\right)\cdot\begin{pmatrix}\text{Rg}\mathbf{M}_{m'n'}(k_1r,\theta,\varphi)\times(\text{Rg})\mathbf{M}_{-mn}(kr,\theta,\varphi)\\\text{Rg}\mathbf{M}_{m'n'}(k_1r,\theta,\varphi)\times(\text{Rg})\mathbf{N}_{-mn}(kr,\theta,\varphi)\\\text{Rg}\mathbf{N}_{m'n'}(k_1r,\theta,\varphi)\times(\text{Rg})\mathbf{M}_{-mn}(kr,\theta,\varphi)\\\text{Rg}\mathbf{N}_{m'n'}(k_1r,\theta,\varphi)\times(\text{Rg})\mathbf{N}_{-mn}(kr,\theta,\varphi)\end{pmatrix}, \tag{3.3.29}$$

where "(Rg)" represents with or without "Rg" cases.

Using Eqs. (3.2.53), (3.3.27), The T-matrix can be formally written as

$$\mathbf{T} = -(\mathrm{Rg}\mathbf{Q})\mathbf{Q}^{-1}. \tag{3.3.30}$$

Based on the analysis of the matrix size for computations, the number of T-matrix elements is $4N^2(N+2)^2$, where $N$ is the truncation order associated with the index $n$ of the VSWF expansion of the electromagnetic field.

For computation of the T-matrix of a homogeneous dielectric particle, the variables are the permittivity and the surface shape based on Eq. (3.3.29). The permittivity stays in the argument of the regular VSWF, which is in the integrand of the surface integrals. The surface shape S in the calculation of the surface integrals is critical and determines the complexity of the computation of the T-matrix. In spherical coordinates, a surface is defined as

$$r - r(\theta, \varphi) = 0. \tag{3.3.31}$$

The outward normal to the local surface is the gradient of the left-hand side of Eq. (3.3.31) as follows:

$$\hat{n}_s\left(\vec{r}\right) = \nabla[r - r(\theta, \varphi)] = \hat{r} - \frac{1}{r}\frac{\partial r}{\partial \theta}\hat{\theta} - \frac{1}{r\sin\theta}\frac{\partial r}{\partial \varphi}\hat{\varphi}. \tag{3.3.32}$$

The vector surface integral element is evident as (Tsang et al., 2000; Mishchenko et al., 2002; Doicu et al., 2006)

$$d\vec{S} = dS\hat{n}_s\left(\vec{r}\right) = r^2\sin\theta\left(\hat{r} - \frac{1}{r}\frac{\partial r}{\partial \theta}\hat{\theta} - \frac{1}{r\sin\theta}\frac{\partial r}{\partial \varphi}\hat{\varphi}\right)d\theta d\varphi. \tag{3.3.33}$$

For axially symmetric particles, when the $z$-axis is along the rotational axis, the surface equation and the corresponding vector surface integral element are formally written as

$$r - r(\theta) = 0, \tag{3.3.34a}$$

$$dS\hat{n}_s\left(\vec{r}\right) = r^2\sin\theta\left(\hat{r} - \frac{1}{r}\frac{dr}{d\theta}\hat{\theta}\right)d\theta d\varphi. \tag{3.3.34b}$$

For homogeneous spherical particles, the surface equation and the corresponding vector surface integral element are simplified to

$$r - r_0 = 0, \tag{3.3.35a}$$

$$dS\hat{n}_s\left(\vec{r}\right) = \hat{r}r^2\sin\theta d\theta d\varphi. \tag{3.3.35b}$$

### 3.3.3 T-matrix for particles with axial symmetries and homogeneous spherical particles

When particles have axial symmetry, the surface integral element is azimuthally independent. The integral over azimuthal element $d\varphi$ decouples the coefficient matrix (Rg)**Q** associated with the azimuthal component $m$ as follows:

$$\int_0^{2\pi} d\varphi \exp\left[i(m'-m)\varphi\right] = 2\pi\delta_{mm'}. \tag{3.3.36}$$

The integration in Eq. (3.3.29) can be simplified into the following expressions (Tsang et al., 2000):

$$\begin{aligned}\mathrm{J}^{11}_{mnm'n'} &= 2\pi\delta_{mm'}\gamma_{mn'}\gamma_{-mn}(-1)^m \int_0^{\pi} d\theta \sin\theta \Big\{ r^2 j_{n'}(k_1 r) h_n^{(1)}(kr)\frac{m}{i\sin\theta} \\ &\left[P_{n'}^m(\cos\theta)\frac{dP_n^{-m}(\cos\theta)}{d\theta} + \frac{dP_{n'}^m(\cos\theta)}{d\theta}P_n^{-m}(\cos\theta)\right]\Big\},\end{aligned} \tag{3.3.37a}$$

$$\begin{aligned}\mathrm{J}^{12}_{mnm'n'} &= 2\pi\delta_{mm'}\gamma_{mn'}\gamma_{-mn}(-1)^m \int_0^{\pi} d\theta \sin\theta \Bigg\{ r^2 j_{n'}(k_1 r)\frac{1}{kr}\frac{d\left[krh_n^{(1)}(kr)\right]}{d(kr)} \\ &\cdot\left[\frac{m^2}{\sin^2\theta}P_{n'}^m(\cos\theta)P_n^{-m}(\cos\theta) + \frac{dP_{n'}^m(\cos\theta)}{d\theta}\frac{dP_n^{-m}(\cos\theta)}{d\theta}\right] \\ &+ r\frac{dr}{d\theta}j_{n'}(k_1 r)\frac{n(n+1)}{kr}h_n^{(1)}(kr)\frac{dP_{n'}^m(\cos\theta)}{d\theta}P_n^{-m}(\cos\theta)\Bigg\},\end{aligned} \tag{3.3.37b}$$

$$\begin{aligned}\mathrm{J}^{21}_{mnm'n'} &= 2\pi\delta_{mm'}\gamma_{mn'}\gamma_{-mn}(-1)^{m+1} \int_0^{\pi} d\theta \sin\theta \Bigg\{ r^2 \frac{1}{k_1 r}\frac{d[k_1 r j_{n'}(k_1 r)]}{d(k_1 r)} h_n^{(1)}(kr) \\ &\cdot\left[\frac{m^2}{\sin^2\theta}P_{n'}^m(\cos\theta)P_n^{-m}(\cos\theta) + \frac{dP_{n'}^m(\cos\theta)}{d\theta}\frac{dP_n^{-m}(\cos\theta)}{d\theta}\right] \\ &+ r\frac{dr}{d\theta}\frac{n'(n'+1)}{k_1 r}j_{n'}(k_1 r)h_n^{(1)}(kr)P_{n'}^m(\cos\theta)\frac{dP_n^{-m}(\cos\theta)}{d\theta}\Bigg\},\end{aligned} \tag{3.3.37c}$$

$$\begin{aligned}\mathrm{J}^{22}_{mnm'n'} &= 2\pi\delta_{mm'}\gamma_{mn'}\gamma_{-mn}(-1)^{m+1} \int_0^{\pi} d\theta \sin\theta \Bigg\{ r^2 \frac{1}{k_1 r}\frac{d[k_1 r j_{n'}(k_1 r)]}{d(k_1 r)}\frac{1}{kr}\frac{d\left[krh_n^{(1)}(kr)\right]}{d(kr)} \\ &\cdot\frac{im}{\sin\theta}\left[\frac{dP_{n'}^m(\cos\theta)}{d\theta}P_n^{-m}(\cos\theta) + P_{n'}^m(\cos\theta)\frac{dP_n^{-m}(\cos\theta)}{d\theta}\right] \\ &+ r\frac{dr}{d\theta}\frac{im}{\sin\theta}P_{n'}^m(\cos\theta)P_n^{-m}(\cos\theta)\cdot\Bigg[\frac{n'(n'+1)}{k_1 r}j_{n'}(k_1 r)\frac{1}{kr}\frac{d\left[krh_n^{(1)}(kr)\right]}{d(kr)} \\ &+ \frac{1}{k_1 r}\frac{d[k_1 r j_{n'}(k_1 r)]}{d(k_1 r)}\frac{n(n+1)}{kr}h_n^{(1)}(kr)\Bigg]\Bigg\},\end{aligned} \tag{3.3.37d}$$

where $r = r(\theta)$. The integrations for RgJ in Eq. (3.3.29) are performed to substitute $j_n(kr)$ for $h_n^{(1)}(kr)$ in Eq. (3.3.37). When the coefficient matrices associated with the azimuthal component $m$ are decoupled, the T-matrix is also decoupled. Integration over zenith angles can be numerically implemented, such as by using Gaussian quadrature.

For a homogeneous spherical particle, the following orthogonality is used (Mishchenko et al., 2002):

$$\int_0^\pi d\theta \left[ P_{n'}^m(\cos\theta) \frac{dP_n^{-m}(\cos\theta)}{d\theta} + \frac{dP_{n'}^m(\cos\theta)}{d\theta} P_n^{-m}(\cos\theta) \right] = 0, \tag{3.3.38a}$$

$$\int_0^\pi d\theta \sin\theta \left[ \frac{dP_{n'}^m(\cos\theta)}{d\theta} \frac{dP_n^{-m}(\cos\theta)}{d\theta} + \frac{m^2}{\sin^2\theta} P_{n'}^m(\cos\theta) P_n^{-m}(\cos\theta) \right] = (-1)^m \delta_{nn'} \frac{2n(n+1)}{2n+1}. \tag{3.3.38b}$$

The orthogonality and recurrence relations for associated Legendre polynomials are used in Eq. (3.3.38). Using $\frac{dr}{d\theta} = 0$, the integrations in Eq. (3.3.37) are further simplified to the analytical solutions (Tsang et al., 2000; Mishchenko et al., 2002; Doicu et al., 2006):

$$(\mathrm{Rg})\mathrm{J}_{mnm'n'}^{11} = 0, \quad (\mathrm{Rg})\mathrm{J}_{mnm'n'}^{22} = 0, \tag{3.3.39a}$$

$$\mathrm{J}_{mnm'n'}^{12} = \delta_{mm'} \delta_{nn'} r^2 j_n(k_1 r) \frac{1}{kr} \frac{d\left[ kr h_n^{(1)}(kr) \right]}{d(kr)}, \quad \mathrm{RgJ}_{mnm'n'}^{12} = \delta_{mm'} \delta_{nn'} r^2 j_n(k_1 r) \frac{1}{kr} \frac{d[kr j_n(kr)]}{d(kr)}, \tag{3.3.39b}$$

$$\mathrm{J}_{mnm'n'}^{21} = -\delta_{mm'} \delta_{nn'} r^2 \frac{1}{k_1 r} \frac{d[k_1 r j_n(k_1 r)]}{d(k_1 r)} h_n^{(1)}(kr), \ \mathrm{RgJ}_{mnm'n'}^{21} = -\delta_{mm'} \delta_{nn'} r^2 \frac{1}{k_1 r} \frac{d[k_1 r j_n(k_1 r)]}{d(k_1 r)} j_n(kr). \tag{3.3.39c}$$

The T-matrix for spherical particles is fully decoupled and is only dependent on $n$. The following Riccati–Bessel function notations are used to reproduce the Lorenz–Mie coefficients (van de Hulst, 1957; Bohren and Huffman, 1983):

$$\zeta_n(x) = x j_n(x), \ \xi_n(x) = x h_n^{(1)}(x), \tag{3.3.40a}$$

$$x = kr, m = k_1/k. \tag{3.3.40b}$$

Here, $m$ is the refractive index of the scattering particle relative to the surrounding media, not the azimuthal component index. The T-matrix is reduced to the Lorenz–Mie coefficients (Bohren and Huffman, 1983):

$$\mathrm{T}_{mnm'n'}^{12} = \mathrm{T}_{mnm'n'}^{21} = 0, \tag{3.3.41a}$$

$$\mathrm{T}_{mnm'n'}^{11} = -\delta_{mm'} \delta_{nn'} b_n, \quad b_n = \frac{\zeta_n(mx)\zeta_n'(x) - m\zeta_n(x)\zeta_n'(mx)}{\zeta_n(mx)\xi_n'(x) - m\xi_n(x)\zeta_n'(mx)}, \tag{3.3.41b}$$

$$T^{22}_{mnm'n'} = -\delta_{mm'}\delta_{nn'}a_n, \quad a_n = \frac{m\zeta_n(mx)\zeta_n'(x) - \zeta_n(x)\zeta_n'(mx)}{m\zeta_n(mx)\xi_n'(x) - \xi_n(x)\zeta_n'(mx)}. \tag{3.3.41c}$$

where the prime denotes the derivative over the argument in the parenthesis.

## 3.4 Unitarity, symmetries, and random orientations

The EBCM in Section 3.3 is briefly introduced to compute the T-matrix of a scattering particle. EBCM is only one of the methods to compute the T-matrix. In this section, we discuss general T-matrix properties: unitarity and symmetry. The unitarity is derived in terms of energy conservation. The symmetry includes temporal and spatial symmetries, in which the time reversal symmetry is intrinsic and named as reciprocity. Scattering of a particle in random orientation can be analytically computed using the T-matrix of the particle, and the computation can be simplified using symmetry, especially the spatial symmetry.

### 3.4.1 Unitarity

The total electric field at infinity can be decomposed into outgoing and incoming components. If the scattering particle is nonabsorptive, energy associated with the incoming and outgoing components should be conserved. Unitarity is related to energy conservation. Moreover, the total electric field is composed of the incident field and the scattered field, which are related using the T-matrix. Therefore, one can specifically express the unitarity with respect to the T-matrix. Using the VSWF, the total electric field can be described as

$$\begin{aligned}\mathbf{E}\left(\vec{r}\right) = &\sum_{n=1}^{\infty}\sum_{m=-n}^{n}\sum_{n'=1}^{\infty}\sum_{m'=-n'}^{n'}\left[\left(T^{11}_{mnm'n'}a_{m'n'} + T^{12}_{mnm'n'}b_{m'n'}\right)\mathbf{M}_{mn}\left(k\vec{r}\right)\right.\\ &\left.+\left(T^{21}_{mnm'n'}a_{m'n'} + T^{22}_{mnm'n'}b_{m'n'}\right)\mathbf{N}_{mn}\left(k\vec{r}\right)\right]\\ &+\sum_{n=1}^{\infty}\sum_{m=-n}^{n}\left[a_{mn}\mathrm{Rg}\mathbf{M}_{mn}\left(k\vec{r}\right) + b_{mn}\mathrm{Rg}\mathbf{N}_{mn}\left(k\vec{r}\right)\right].\end{aligned} \tag{3.4.1}$$

Using Eqs. (3.2.46), (3.2.51), the total electric field in the far field can be described in terms of the outgoing and incoming waves. After a series of algebraic steps, Eq. (3.4.1) can be rewritten as follows (Tsang et al., 2000; Mishchenko et al., 2002):

$$\mathbf{E}\left(\vec{r}\right)\Big|_{kr\to\infty} = \frac{\exp(ikr)}{2kr}\mathbf{E}_1(\theta,\varphi) + \frac{\exp(-ikr)}{2kr}\mathbf{E}_2(\theta,\varphi), \tag{3.4.2}$$

where

$$\mathbf{E}_1(\theta,\varphi)=\sum_{n=1}^{\infty}\sum_{m=-n}^{n}\sum_{n'=1}^{\infty}\sum_{m'=-n}^{n}\Big\{\Big[\Big(2\mathrm{T}^{11}_{mnm'n'}+\delta_{mm'}\delta_{nn'}\Big)a_{m'n'}+2\mathrm{T}^{12}_{mnm'n'}b_{m'n'}\Big](-i)^{n+1}\gamma_{mn}\mathbf{C}_{mn}(\theta,\varphi)$$
$$+\Big[2\mathrm{T}^{21}_{mnm'n'}a_{m'n'}+\Big(2\mathrm{T}^{22}_{mnm'n'}+\delta_{mm'}\delta_{nn'}\Big)b_{m'n'}\Big](-i)^{n}\gamma_{mn}\mathbf{B}_{mn}(\theta,\varphi)\Big\}, \tag{3.4.3a}$$

$$\mathbf{E}_2(\theta,\varphi)=\sum_{n=1}^{\infty}\sum_{m=-n}^{n}\left[a_{mn}i^{n+1}\gamma_{mn}\mathbf{C}_{mn}(\theta,\varphi)+b_{mn}i^{n}\gamma_{mn}\mathbf{B}_{mn}(\theta,\varphi)\right]. \tag{3.4.3b}$$

The electromagnetic field at infinity only has the transverse component.

The scattering matrix S is defined as follows:

$$\mathrm{S}=\mathbf{I}+2\mathbf{T}, \tag{3.4.4a}$$

where

$$\mathrm{S}^{11}_{mnm'n'}=2\mathrm{T}^{11}_{mnm'n'}+\delta_{mm'}\delta_{nn'},\mathrm{S}^{12}_{mnm'n'}=2\mathrm{T}^{12}_{mnm'n'}, \tag{3.4.4b}$$

$$\mathrm{S}^{21}_{mnm'n'}=2\mathrm{T}^{21}_{mnm'n'},\mathrm{S}^{22}_{mnm'n'}=2\mathrm{T}^{22}_{mnm'n'}+\delta_{mm'}\delta_{nn'}. \tag{3.4.4c}$$

Note that the scattering matrix S has a different physical meaning from the scattering matrix introduced in Eq. (2.2.11). We just follow the convention to name Eq. (3.4.4) as the scattering matrix. Eq. (3.4.3) can be written in compact forms as follows:

$$\mathbf{E}_1(\theta,\varphi)=\sum_{n=1}^{\infty}\sum_{m=-n}^{n}\sum_{n'=1}^{\infty}\sum_{m'=-n'}^{n'}\gamma_{mn}(-i)^{n+1}\left(\mathbf{C}_{mn}(\theta,\varphi)\ \ i\mathbf{B}_{mn}(\theta,\varphi)\right)\begin{pmatrix}\mathrm{S}^{11}_{mnm'n'} & \mathrm{S}^{12}_{mnm'n'}\\ \mathrm{S}^{21}_{mnm'n'} & \mathrm{S}^{22}_{mnm'n'}\end{pmatrix}\begin{pmatrix}a_{m'n'}\\ b_{m'n'}\end{pmatrix}$$
$$=\left(\mathbf{C}_1\ \ \mathbf{B}_1\right)\begin{pmatrix}\mathrm{S}^{11} & \mathrm{S}^{12}\\ \mathrm{S}^{21} & \mathrm{S}^{22}\end{pmatrix}\begin{pmatrix}\mathbf{a}\\ \mathbf{b}\end{pmatrix}, \tag{3.4.5a}$$

$$\mathbf{E}_2(\theta,\varphi)=\sum_{n=1}^{\infty}\sum_{m=-n}^{n}\gamma_{mn}i^{n}\left(i\mathbf{C}_{mn}(\theta,\varphi)\ \ \mathbf{B}_{mn}(\theta,\varphi)\right)\begin{pmatrix}a_{mn}\\ b_{mn}\end{pmatrix}$$
$$=\left(\mathbf{C}_2\ \ \mathbf{B}_2\right)\begin{pmatrix}\mathbf{a}\\ \mathbf{b}\end{pmatrix}, \tag{3.4.5b}$$

where

$$(\mathbf{C}_1)_{mn}=\gamma_{mn}(-i)^{n+1}\mathbf{C}_{mn}(\theta,\varphi),\ \ (\mathbf{C}_2)_{mn}=\gamma_{mn}i^{n+1}\mathbf{C}_{mn}(\theta,\varphi), \tag{3.4.6a}$$

$$(\mathbf{B}_1)_{mn}=\gamma_{mn}(-i)^{n}\mathbf{B}_{mn}(\theta,\varphi),\ \ (\mathbf{B}_2)_{mn}=\gamma_{mn}i^{n}\mathbf{B}_{mn}(\theta,\varphi), \tag{3.4.6b}$$

Applying the orthogonality of vector spherical harmonics described in Eq. (3.2.42), the following orthogonality is straightforward:

$$\int_{4\pi} d\Omega (\mathbf{C}_i)^*_{mn} \cdot (\mathbf{B}_i)_{m'n'} = 0, \quad i = 1,2, \tag{3.4.7a}$$

$$\int_{4\pi} d\Omega (\mathbf{C}_i)^*_{mn} \cdot (\mathbf{C}_i)_{m'n'} = \int_{4\pi} d\Omega (\mathbf{B}_i)^*_{mn} \cdot (\mathbf{B}_i)_{m'n'} = \delta_{mm'}\delta_{nn'}, \quad i = 1,2. \tag{3.4.7b}$$

The earlier expression can be written in the following compact form:

$$\int_{4\pi} d\Omega (\mathbf{C}_i \;\; \mathbf{B}_i)^\dagger (\mathbf{C}_i \;\; \mathbf{B}_i) = \mathbf{I}, \quad i = 1,2, \tag{3.4.8}$$

where superscript "†" refers to the complex conjugate and transpose and **I** is the unit matrix.

The directional energy flux of the electromagnetic field is represented by the Poynting vector. The electromagnetic field at infinity is transverse. The time-averaged Poynting vector is described as (Tsang et al., 2000; Mishchenko et al., 2002)

$$\begin{aligned} \left\langle \mathbf{S}\left(\vec{r}\right)\right\rangle &= \frac{1}{2} Re\left[\mathbf{E}\left(\vec{r}\right) \times \mathbf{H}^*\left(\vec{r}\right)\right]\Big|_{kr\to\infty} \\ &= \hat{r}\frac{1}{8(kr)^2}\sqrt{\frac{\varepsilon}{\mu_0}}\left\{|\mathbf{E}_1(\theta,\varphi)|^2 - |\mathbf{E}_2(\theta,\varphi)|^2\right\}, \end{aligned} \tag{3.4.9}$$

where $\varepsilon$ is the permittivity of surrounding media. If the scattering particle is nonabsorptive, energy should be conserved. Accordingly, one can have

$$\int_{4\pi} d\Omega |\mathbf{E}_1(\theta,\varphi)|^2 = \int_{4\pi} d\Omega |\mathbf{E}_2(\theta,\varphi)|^2. \tag{3.4.10}$$

The physical meaning of Eq. (3.4.10) shows that the energy coming out of the area is equal to the energy going into the area. Using the orthogonality of vector spherical harmonics described in Eq. (3.4.7), the spherical surface integrals in Eq. (3.4.10) can be explicitly given as

$$\begin{aligned} \int_{4\pi} d\Omega |\mathbf{E}_1(\theta,\varphi)|^2 &= \int_{4\pi} d\Omega \mathbf{E}_1^\dagger\left(\vec{r}\right) \cdot \mathbf{E}_1\left(\vec{r}\right) \\ &= \begin{pmatrix} \mathbf{a} \\ \mathbf{b} \end{pmatrix}^\dagger \begin{pmatrix} \mathrm{S}^{11} & \mathrm{S}^{12} \\ \mathrm{S}^{21} & \mathrm{S}^{22} \end{pmatrix}^\dagger \begin{pmatrix} \mathrm{S}^{11} & \mathrm{S}^{12} \\ \mathrm{S}^{21} & \mathrm{S}^{22} \end{pmatrix} \begin{pmatrix} \mathbf{a} \\ \mathbf{b} \end{pmatrix}, \end{aligned} \tag{3.4.11a}$$

$$\int_{4\pi} d\Omega |\mathbf{E}_2(\theta,\varphi)|^2 = \int_{4\pi} d\Omega \mathbf{E}_2^\dagger(\vec{r}) \cdot \mathbf{E}_2(\vec{r}) = \begin{pmatrix} \mathbf{a} \\ \mathbf{b} \end{pmatrix}^\dagger \begin{pmatrix} \mathbf{a} \\ \mathbf{b} \end{pmatrix}. \tag{3.4.11b}$$

Using Eq. (3.4.11), Eq. (3.4.10) gives

$$\begin{pmatrix} \mathbf{a} \\ \mathbf{b} \end{pmatrix}^\dagger \left[ \begin{pmatrix} S^{11} & S^{12} \\ S^{21} & S^{22} \end{pmatrix}^\dagger \begin{pmatrix} S^{11} & S^{12} \\ S^{21} & S^{22} \end{pmatrix} - \mathbf{I} \right] \begin{pmatrix} \mathbf{a} \\ \mathbf{b} \end{pmatrix} = 0. \tag{3.4.12}$$

The incident expansion coefficients are arbitrary, so an identity for the scattering matrix can be obtained as follows (Waterman, 1971; Tsang et al., 2000; Mishchenko et al., 2002):

$$\begin{pmatrix} S^{11} & S^{12} \\ S^{21} & S^{22} \end{pmatrix}^\dagger \begin{pmatrix} S^{11} & S^{12} \\ S^{21} & S^{22} \end{pmatrix} = S^\dagger S = \mathbf{I}. \tag{3.4.13}$$

Scattering matrix S is a unitary matrix if the scattering particle is nonabsorptive. Equivalently, the unitarity can also be described by the T-matrix of the particle:

$$\mathbf{T}^\dagger \mathbf{T} = -\frac{1}{2}\left(\mathbf{T}^\dagger + \mathbf{T}\right). \tag{3.4.14}$$

The unitarity turns into an inequality for an absorptive scattering particle:

$$S^\dagger S \le \mathbf{I}, \tag{3.4.15a}$$

$$\mathbf{T}^\dagger \mathbf{T} \le -\frac{1}{2}\left(\mathbf{T}^\dagger + \mathbf{T}\right). \tag{3.4.15b}$$

### 3.4.2 Symmetry

Intrinsic symmetry is related to the reciprocity relation. The amplitude scattering matrix associated with meridional planes has the following symmetric relation from Eq. (2.2.71) (van de Hulst, 1957; Mishchenko et al., 2002; Hovenier et al., 2004):

$$\mathbf{S}(-\hat{e}_i, -\hat{e}_s) = \begin{pmatrix} S_{11}(\hat{e}_s, \hat{e}_i) & -S_{21}(\hat{e}_s, \hat{e}_i) \\ -S_{12}(\hat{e}_s, \hat{e}_i) & S_{22}(\hat{e}_s, \hat{e}_i) \end{pmatrix}, \tag{3.4.16}$$

where the left and right arguments denote the scattered and incident directions, and the arguments for the incident field and scattered directions are

$$\hat{e}_i = \hat{e}_i(\theta_i, \varphi_i), \quad \hat{e}_s = \hat{e}_s(\theta_s, \varphi_s), \tag{3.4.17}$$

and

$$-\hat{e}_i(\theta_i,\varphi_i)=\hat{e}_i(\pi-\theta_i,\pi+\varphi_i),\quad -\hat{e}_s(\theta_s,\varphi_s)=\hat{e}_s(\pi-\theta_s,\pi+\varphi_s). \tag{3.4.18}$$

In terms of the relation between the amplitude scattering matrix and the T-matrix in Eqs. (3.2.58) and (3.2.59), an intrinsic symmetric relation for the T-matrix is given using Eq. (3.4.16) (Waterman, 1971; Tsang et al., 2000; Mishchenko et al., 2002):

$$\mathrm{T}^{ij}_{-mn-m'n'}=(-1)^{m+m'}\mathrm{T}^{ji}_{mnm'n'},\quad i,j=1,2. \tag{3.4.19}$$

The symmetry shown in Eq. (3.4.19) occurs due to reciprocity and applies to an arbitrary particle.

On the other hand, the scattering particle itself may have spatial symmetry, which can produce additional symmetries for its T-matrix. For example, when the particle has axial symmetry as in Section 3.3.3, the T-matrix is fully decoupled from the azimuthal component $m$. Accordingly, the computation burden, including time and memory usage, is significantly reduced. This subsection discusses mirror symmetry and rotational symmetry. Since the electromagnetic field in the T-matrix method is expanded using the VSWF, the properties of the T-matrix associated with mirror symmetry or rotational symmetry can be obtained by applying a mirror or rotational transformation of VSWF. In Section 3.2.4, the rotations of the spherical harmonics and then the scalar and vector spherical wave functions using the Wigner functions are given, and the corresponding T-matrix rotation is expressed in terms of the Wigner functions. For mirror symmetry, the properties of the VSWF are exploited here. In Cartesian coordinates, the mirror symmetry of an arbitrary vector $(x,y,z)$ is straightforward. For instance, the mirror symmetries over the $x$-$y$ plane $\sigma_{xy}$, the $x$-$z$ plane $\sigma_{xz}$, or the $y$-$z$ plane $\sigma_{yz}$ can be represented as $\sigma_{xy}[(x,y,z)]=(x,y,-z)$, $\sigma_{xz}[(x,y,z)]=(x,-y,z)$, and $\sigma_{yz}[(x,y,z)]=(-x,y,z)$, respectively. For an arbitrary vector $\mathbf{A}$ in spherical coordinates, that is, $\mathbf{A}=\mathrm{A}_r\hat{r}+\mathrm{A}_\theta\hat{\theta}+\mathrm{A}_\varphi\hat{\varphi}$, the strategy is to find the change of the unit vectors under the corresponding mirror symmetry. Formally, it can be written as

$$\sigma[\mathbf{A}]=\mathrm{A}_r\sigma[\hat{r}]+\mathrm{A}_\theta\sigma\left[\hat{\theta}\right]+\mathrm{A}_\varphi\sigma[\hat{\varphi}], \tag{3.4.20}$$

where $\sigma$ indicates one of the earlier mirror symmetries. The unit vectors in spherical coordinates are given in Eq. (3.2.1) and the corresponding mirror symmetries can be represented as follows:

$$\sigma_{xy}\left[\left(\hat{r}(\theta,\varphi),\hat{\theta}(\theta,\varphi),\hat{\varphi}(\theta,\varphi)\right)\right]=\left(\hat{r}(\pi-\theta,\varphi),-\hat{\theta}(\pi-\theta,\varphi),\hat{\varphi}(\pi-\theta,\varphi)\right), \tag{3.4.21a}$$

$$\sigma_{xz}\left[\left(\hat{r}(\theta,\varphi),\hat{\theta}(\theta,\varphi),\hat{\varphi}(\theta,\varphi)\right)\right]=\left(\hat{r}(\theta,-\varphi),\hat{\theta}(\theta,-\varphi),-\hat{\varphi}(\theta,-\varphi)\right), \tag{3.4.21b}$$

$$\sigma_{yz}\left[\left(\hat{r}(\theta,\varphi),\hat{\theta}(\theta,\varphi),\hat{\varphi}(\theta,\varphi)\right)\right]=\left(\hat{r}(\theta,\pi-\varphi),\hat{\theta}(\theta,\pi-\varphi),-\hat{\varphi}(\theta,\pi-\varphi)\right). \tag{3.4.21c}$$

The mirror symmetry only affects vector spherical harmonics, so the regular and outgoing vector spherical wave functions have the same transformation. After some algebraic steps, the operations of mirror symmetry on the VSWF can be written as (Schulz et al., 1999):

$$\sigma_{xy}\left[\begin{pmatrix}(\mathrm{Rg})\mathbf{M}_{mn}\left(k\vec{r}\right)\\(\mathrm{Rg})\mathbf{N}_{mn}\left(k\vec{r}\right)\end{pmatrix}\right]=\begin{pmatrix}(-1)^{n+m+1}(\mathrm{Rg})\mathbf{M}_{mn}\left(k\sigma_{xy}\left[\vec{r}\right]\right)\\(-1)^{n+m}(\mathrm{Rg})\mathbf{N}_{mn}\left(k\sigma_{xy}\left[\vec{r}\right]\right)\end{pmatrix}, \tag{3.4.22a}$$

$$\sigma_{xz}\left[\begin{pmatrix}(\mathrm{Rg})\mathbf{M}_{mn}\left(k\vec{r}\right)\\(\mathrm{Rg})\mathbf{N}_{mn}\left(k\vec{r}\right)\end{pmatrix}\right]=\begin{pmatrix}(-1)^{m+1}(\mathrm{Rg})\mathbf{M}_{-mn}\left(k\sigma_{xz}\left[\vec{r}\right]\right)\\(-1)^{m}(\mathrm{Rg})\mathbf{N}_{-mn}\left(k\sigma_{xz}\left[\vec{r}\right]\right)\end{pmatrix}, \tag{3.4.22b}$$

$$\sigma_{yz}\left[\begin{pmatrix}(\mathrm{Rg})\mathbf{M}_{mn}\left(k\vec{r}\right)\\(\mathrm{Rg})\mathbf{N}_{mn}\left(k\vec{r}\right)\end{pmatrix}\right]=\begin{pmatrix}-(\mathrm{Rg})\mathbf{M}_{-mn}\left(k\sigma_{yz}\left[\vec{r}\right]\right)\\(\mathrm{Rg})\mathbf{N}_{-mn}\left(k\sigma_{yz}\left[\vec{r}\right]\right)\end{pmatrix}, \tag{3.4.22c}$$

where Eq. (3.4.22) holds both for regular and outgoing vector spherical wave functions. Accordingly, the T-matrix under these mirror symmetries can be described as

$$\sigma_{xy}\left[\begin{pmatrix}\mathrm{T}^{11}_{mnm'n'} & \mathrm{T}^{12}_{mnm'n'}\\\mathrm{T}^{21}_{mnm'n'} & \mathrm{T}^{22}_{mnm'n'}\end{pmatrix}\right]=(-1)^{n+m+n'+m'}\begin{pmatrix}\mathrm{T}^{11}_{mnm'n'} & -\mathrm{T}^{12}_{mnm'n'}\\-\mathrm{T}^{21}_{mnm'n'} & \mathrm{T}^{22}_{mnm'n'}\end{pmatrix}, \tag{3.4.23a}$$

$$\sigma_{xz}\left[\begin{pmatrix}\mathrm{T}^{11}_{mnm'n'} & \mathrm{T}^{12}_{mnm'n'}\\\mathrm{T}^{21}_{mnm'n'} & \mathrm{T}^{22}_{mnm'n'}\end{pmatrix}\right]=(-1)^{m+m'}\begin{pmatrix}\mathrm{T}^{11}_{-mn-m'n'} & -\mathrm{T}^{12}_{-mn-m'n'}\\-\mathrm{T}^{21}_{-mn-m'n'} & \mathrm{T}^{22}_{-mn-m'n'}\end{pmatrix}, \tag{3.4.23b}$$

$$\sigma_{yz}\left[\begin{pmatrix}\mathrm{T}^{11}_{mnm'n'} & \mathrm{T}^{12}_{mnm'n'}\\\mathrm{T}^{21}_{mnm'n'} & \mathrm{T}^{22}_{mnm'n'}\end{pmatrix}\right]=\begin{pmatrix}\mathrm{T}^{11}_{-mn-m'n'} & -\mathrm{T}^{12}_{-mn-m'n'}\\-\mathrm{T}^{21}_{-mn-m'n'} & \mathrm{T}^{22}_{-mn-m'n'}\end{pmatrix}. \tag{3.4.23c}$$

If a particle has the aforementioned mirror symmetries, the T-matrix fulfills the following symmetric relations (Schulz et al., 1999):

$$\sigma_{xy}:\mathrm{T}^{ij}_{mnm'n'}=0,\text{if }(-1)^{n+m+n'+m'+i+j}\neq 1, i,j=1,2, \tag{3.4.24a}$$

$$\sigma_{xz}:\mathrm{T}^{ij}_{-mn-m'-n'}=(-1)^{m+m'+i+j}\mathrm{T}^{ij}_{mnm'n'}, i,j=1,2, \tag{3.4.24b}$$

$$\sigma_{yz}:\mathrm{T}^{ij}_{-mn-m'-n'}=(-1)^{i+j}\mathrm{T}^{ij}_{mnm'n'}, i,j=1,2. \tag{3.4.24c}$$

In the T-matrix method, the $z$-axis of the frame of reference for the T-matrix is directed along the rotational axis to use the particle symmetry. The T-matrix associated with a rotation of Euler angles $(\alpha,\beta,\gamma)$ to the original frame of reference can be described using the Wigner D-function in Eq. (3.2.75a) and the Wigner D-function is related to the Wigner d-function given in Eq. (3.2.70). For a particle with $N$-fold rotational symmetry, the Euler angles can be chosen as

$$\alpha = \frac{2\pi}{N}l, \;\; l = 1,2,\ldots,N-1; \; \beta = \gamma = 0. \tag{3.4.25}$$

The rotations in Eq. (3.4.25) produce the identical particle relative to the frame of reference. Substituting Eq. (3.4.25) into Eq. (3.2.75a), the following relation can be obtained (Schulz et al., 1999):

$$\exp\left[-i2\pi(m-m')l/N\right] = 1, \;\; l = 1,2,\ldots,N-1. \tag{3.4.26}$$

To satisfy Eq. (3.4.26), the azimuthal components $m$ and $m'$ are related and satisfy the relationship $m-m'=N$k, where k is an integer. Accordingly, the T-matrix is partially decoupled into $N$ submatrices according to the remainder of the azimuthal component over the fold number $N$, that is, $j=\text{mod}(m \text{ or } m',N)$. Using a hexagonal column with a sixfold rotational symmetry as an example, $m$ can take the following values:

$$\begin{aligned}
j &= \text{mod}(m,6) = 0, \;\; m = \ldots, -18, -12, -6, 0, 6, 12, 18, \ldots, \\
j &= \text{mod}(m,6) = 1, \;\; m = \ldots, -17, -11, -5, 1, 7, 13, 19, \ldots, \\
j &= \text{mod}(m,6) = 2, \;\; m = \ldots, -16, -10, -4, 2, 8, 14, 20, \ldots, \\
j &= \text{mod}(m,6) = 3, \;\; m = \ldots, -15, -9, -3, 3, 9, 15, 21, \ldots, \\
j &= \text{mod}(m,6) = 4, \;\; m = \ldots, -14, -8, -2, 4, 10, 16, 22, \ldots, \\
j &= \text{mod}(m,6) = 5, \;\; m = \ldots, -13, -7, -1, 5, 11, 17, 23, \ldots
\end{aligned} \tag{3.4.27}$$

It is evident that the $m$ values with remainder 1 are the negative of the $m$ values with remainder 5 and the $m$ values with remainder 2 are negative of the $m$ values with remainder 4. Accordingly, the computation of the T-matrix is furthermore reduced because only the elements with $m$ having remainder 0–3 are necessary and elements with $m$ having remainder 4–5 can be obtained using Eq. (3.4.19) or using Eq. (3.4.24b) or (3.4.24c) if the particle has corresponding mirror symmetry. The matrix index shown in Eq. (3.2.54) must be changed corresponding to different remainder. Table 3.1 is one way to arrange the elements of the partially decoupled T-matrix.

For a particle with $N$-fold rotational symmetry, the T-matrix is decoupled into $N$ submatrices according to the remainder of the azimuthal component, that is, mod($m$,$N$), where the remainder is from 0 to ($N$-1). The azimuthal component $m$ of a submatrix with remainder $j$ is the negative of the azimuthal component $m$ of another

**Table 3.1** The unique correspondence between matrix index $l$ and index $(n,m)$ for a particle with sixfold symmetry.

(a)

| **(n,m)** | **(1,0)** | **(2,0)** | **(3,0)** | **(4,0)** | **(5,0)** | **(6,−6)** | **(6,0)** | **(6,6)** | **(7,−6)** | **(7,0)** | **(7,6)** | ... |
|---|---|---|---|---|---|---|---|---|---|---|---|---|
| $l$ | 1 | 2 | 3 | 4 | 5 | 6 | 7 | 8 | 9 | 10 | 11 | ... |

(b)

| **(n,m)** | **(1,1)** | **(2,1)** | **(3,1)** | **(4,1)** | **(5,−5)** | **(5,1)** | **(6,−5)** | **(6,1)** | **(7,−5)** | **(7,1)** | **(7,7)** | ... |
|---|---|---|---|---|---|---|---|---|---|---|---|---|
| $l$ | 1 | 2 | 3 | 4 | 5 | 6 | 7 | 8 | 9 | 10 | 11 | ... |

(c)

| **(n,m)** | **(2,2)** | **(3,2)** | **(4,−4)** | **(4,2)** | **(5,−4)** | **(5,2)** | **(6,−4)** | **(6,2)** | **(7,−4)** | **(7,2)** | **(8,−4)** | ... |
|---|---|---|---|---|---|---|---|---|---|---|---|---|
| $l$ | 1 | 2 | 3 | 4 | 5 | 6 | 7 | 8 | 9 | 10 | 11 | ... |

(d)

| **(n,m)** | **(3,−3)** | **(3,3)** | **(4,−3)** | **(4,3)** | **(5,−3)** | **(5,3)** | **(6,−3)** | **(6,3)** | **(7,−3)** | **(7,3)** | **(8,−3)** | ... |
|---|---|---|---|---|---|---|---|---|---|---|---|---|
| $l$ | 1 | 2 | 3 | 4 | 5 | 6 | 7 | 8 | 9 | 10 | 11 | ... |

Tables (a), (b), (c), and (d) correspond to remainders 0, 1, 2, and 3, respectively.

submatrix with remainder $(N-j)$, where $j \neq 0$ and $j \neq (N-j)$. The pair with remainders $j$ and $(N-j)$ is not independent and is related using the reciprocity relation or mirror symmetry. Accordingly, if $N$ is odd, the submatrices with remainders from 0 to $(N-1)/2$ are independent and must be calculated, while if $N$ is even, only submatrices with remainders from 0 to $N/2$ are independent.

If $N$ is infinity, that is, the particle has complete rotational symmetry or axial symmetry, the condition for Eq. (3.4.26) is $m=m'$. A particle with rotational symmetry always has vertical mirror symmetry. Using Eq. (3.4.24b) or (3.4.24c), the T-matrix of an axially symmetric particle has the following symmetry:

$$\mathrm{T}^{ij}_{-mn-m'-n'} = (-1)^{i+j}\delta_{mm'}\mathrm{T}^{ij}_{mnm'n'}, \quad i,j = 1,2. \tag{3.4.28}$$

The T-matrix is decoupled according to the azimuthal component $m$. The matrix index can be arranged solely according to the azimuthal component $m$, that is, the index is $m$-dominant. Moreover, using Eq. (3.4.19) or Eq. (3.4.28), only the components with $m \geq 0$ of the T-matrix are sufficient to describe all light scattering properties.

For a particle with spherical symmetry (a homogeneous sphere or a multilayered sphere), the T-matrix after arbitrary rotations is invariant:

$$\begin{aligned} \mathrm{T}^{ij}_{mnmn'}(0,0,0) = & \sum_{m_1=-\min(n,n')}^{\min(n,n')} D^{n}_{mm_1}(\alpha,\beta,\gamma)\mathrm{T}^{ij}_{m_1nm_1n'}(0,0,0) \\ & D^{n'}_{m_1m}(-\gamma,-\beta,-\alpha), \quad i,j = 1,2, \end{aligned} \tag{3.4.29}$$

Since Eq. (3.4.29) must be satisfied for arbitrary Euler angles $(\alpha,\beta,\gamma)$, the T-matrix of the particle with spherical symmetry is diagonal as follows:

$$\mathrm{T}^{ij}_{mnm'n'} = \delta_{ij}\delta_{mm'}\delta_{nn'}\mathrm{T}^{ii}_{mnmn}, \quad i,j = 1,2. \tag{3.4.30}$$

Moreover, using Eqs. (3.4.29)–(3.4.30) and the unitary condition of the Wigner-D function in Eq. (3.2.77a), the T-matrix is independent of index $m$. In summary, the T-matrix of a particle with spherical symmetry can be given by

$$\mathrm{T}^{11}_{mnm'n'} = -\delta_{mm'}\delta_{nn'}b_n, \tag{3.4.31}$$

$$\mathrm{T}^{22}_{mnm'n'} = -\delta_{mm'}\delta_{nn'}a_n, \tag{3.4.32}$$

$$\mathrm{T}^{12}_{mnm'n'} = \mathrm{T}^{21}_{mnm'n'} = 0, \tag{3.4.33}$$

where the Lorenz-Mie coefficients $a_n$ and $b_n$ are given in Eq.(3.3.41).

A more systematic treatment of symmetries in the T-matrix formulation can be done in the framework of the theory of point groups and irreducible representations. This treatment allows us to construct a change of basis functions adapted to a particle's symmetry group. In this so-called irreducible basis, the matrices **T**, **Q**, and **RgQ** become block diagonal. In this block-diagonal form, the inversion of the Q-matrix becomes numerically both faster and more stable. The details can be found in Kahnert (2005).

### 3.4.3 Completely random orientation for asymmetric particles

Atmospheric science computations usually (but not always) assume that particles are randomly oriented. With random orientations, the scattering phase matrix is only related to the scattering angle. The scattering phase matrix expansion using generalized spherical functions is broadly used in radiative transfer calculations (e.g., Kuščer and Ribarič, 1959; Domke, 1974; Siewert, 1981; de Haan et al., 1987; Stamnes et al., 2000). For numerically accurate methods, such as FDTD, DDA, PSTD, and MFMA mentioned in Chapter 2, the random orientation assumption can only be numerically implemented by summing a finite number of orientations. The number of needed orientations for accurate results significantly increases when the particle size increases, causing the computational burden to be extremely time-consuming for large particles. The T-matrix, however, describes the intrinsic properties of a particle, such as the morphology, refractive index, and the location of the frame of reference. It not only contains all scattering information but also is independent of the incident state. Consequently, it has been expected to be possible to derive an analytical formula for the T-matrix of a particle with random orientation. An analytical method for computations associated with randomly oriented and axially symmetric particles was developed by Mishchenko (1991). Later, the process was generalized to random orientation associated with any particle morphology (Khlebtsov, 1992; Mackowski and Mishchenko, 1996; Bi and Yang, 2014).

The computational formulas for random orientation associated with an asymmetric particle are derived in the following section, and the simplifications associated with symmetric particles are applied where possible. The amplitude scattering matrix and scattering phase matrix are derived with the circular representation in the succeeding text because the circular representation formulas are simpler than with the linear representation in the derivation of the T-matrix for randomly oriented particles (e.g., Kuščer and Ribarič, 1959; Domke, 1974; Hovenier and van der Mee, 1983; Mishchenko et al., 2002). Random orientation is accounted for by the rotation of the frame of reference in terms of the Euler angles $(\alpha,\beta,\gamma)$. The result is independent of the original incident direction, so the incident angles can be set to zero, that is, $\theta_i = 0$ and $\varphi_i = 0$. Euler angle rotation $\alpha$ is equivalent to selecting the scattering azimuthal angle. Accordingly, the scattering azimuthal angle is also set to be zero, that is, $\varphi_s = 0$. With this alignment of coordinate axes, the scattering zenith angle $\theta_s$ is the scattering angle $\Theta$ in the scattering plane. The following orthogonality relations can then be used in the integration over Euler angles:

$$\frac{1}{2\pi}\int_0^{2\pi} \exp\left[i(m_1 - m_2)\alpha\right] d\alpha = \delta_{m_1 m_2}, \tag{3.4.34}$$

$$\frac{1}{2\pi}\int_0^{2\pi} \exp\left[i(\kappa_1-\kappa_2)\gamma\right]d\gamma = \delta_{\kappa_1\kappa_2}, \tag{3.4.35}$$

$$\frac{1}{2}\int_0^{\pi} d^n_{m_1m_2}(\beta)d^{n'}_{m_1m_2}(\beta)\sin\beta d\beta = \frac{\delta_{nn'}}{2n+1}. \tag{3.4.36}$$

The integrated scattering properties, that is, the extinction and scattering cross sections, for an oriented particle are only related to the incident direction or the orientation after rotations using the Euler angles with incident angles set to $(\theta_i=0,\varphi_i=0)$. Using Eqs. (3.2.60) and (3.2.61), the extinction and scattering cross sections can be described as follows:

$$\langle C_{\text{ext}}\rangle = \frac{1}{8\pi^2}\int_0^{2\pi} d\alpha\int_0^{\pi}\sin\beta d\beta\int_0^{2\pi} d\gamma\left\{-\frac{1}{k^2}\frac{\sum\limits_{n=1}^{\infty}\sum\limits_{m=-n}^{n} Re\left[a_{mn}p^*_{mn}+b_{mn}q^*_{mn}\right]}{|\mathbf{E}_0|^2}\right\}, \tag{3.4.37a}$$

$$\langle C_{\text{sca}}\rangle = \frac{1}{8\pi^2}\int_0^{2\pi} d\alpha\int_0^{\pi}\sin\beta d\beta\int_0^{2\pi} d\gamma\left\{\frac{1}{k^2}\frac{\sum\limits_{n=1}^{\infty}\sum\limits_{m=-n}^{n}\left[|p_{mn}|^2+|q_{mn}|^2\right]}{|\mathbf{E}_0|^2}\right\}, \tag{3.4.37b}$$

where the arguments for $a_{mn}$ and $b_{mn}$ are $(\theta_i=0,\varphi_i=0)$ and the arguments for $p_{mn}$ and $q_{mn}$ are $(\alpha,\beta,\gamma;\theta_i=0,\varphi_i=0)$. The expansion coefficients in these arguments can be simplified as

$$a_{mn}=0,\quad b_{mn}=0,\quad \text{if } m\neq\pm 1, \tag{3.4.38a}$$

$$b_{1n}=a_{1n}=-\sqrt{\pi(2n+1)}\,i^{n-1}\left(\mathbf{E}_0\cdot\hat{\theta}_i-i\mathbf{E}_0\cdot\hat{\varphi}_i\right), \tag{3.4.38b}$$

$$b_{-1n}=-a_{-1n}=\sqrt{\pi(2n+1)}\,i^{n-1}\left(\mathbf{E}_0\cdot\hat{\theta}_i+i\mathbf{E}_0\cdot\hat{\varphi}_i\right), \tag{3.4.38c}$$

$$p_{mn}=\sum_{n'=1}^{\infty}\left\{\left[\mathrm{T}^{11}_{mn1n'}(\alpha,\beta,\gamma)+\mathrm{T}^{12}_{mn1n'}(\alpha,\beta,\gamma)\right]a_{1n'}+\left[\mathrm{T}^{11}_{mn-1n'}(\alpha,\beta,\gamma)-\mathrm{T}^{12}_{mn-1n'}(\alpha,\beta,\gamma)\right]a_{-1n'}\right\}, \tag{3.4.38d}$$

$$q_{mn}=\sum_{n'=1}^{\infty}\left\{\left[\mathrm{T}^{21}_{mn1n'}(\alpha,\beta,\gamma)+\mathrm{T}^{22}_{mn1n'}(\alpha,\beta,\gamma)\right]a_{1n'}+\left[\mathrm{T}^{21}_{mn-1n'}(\alpha,\beta,\gamma)-\mathrm{T}^{22}_{mn-1n'}(\alpha,\beta,\gamma)\right]a_{-1n'}\right\}. \tag{3.4.38e}$$

The cross sections in Eq. (3.4.37) are independent of the magnitude of the incident field. Consequently, they can be obtained by averaging the cases with two orthogonal polarizations, that is,

$$\langle C_{\text{ext,sca}}\rangle = \frac{1}{2}\left[\langle C_{\text{ext,sca}}\rangle_{\hat{\theta}_i} + \langle C_{\text{ext,sca}}\rangle_{\hat{\varphi}_i}\right], \tag{3.4.39}$$

where subscripts $\hat{\theta}_i$ and $\hat{\varphi}_i$ represent the cases when $\mathbf{E}_0 = \hat{\theta}_i$ and $\mathbf{E}_0 = \hat{\varphi}_i$, respectively. Using Eq. (3.2.75a), the T-matrix after rotations can be described using the T-matrix in the particle frame of reference $\mathbf{T}$(P) as follows:

$$\mathrm{T}^{ij}_{mnm'n'}(\alpha,\beta,\gamma) = \exp\left[i(m'-m)\alpha\right]\sum_{m_1=-n}^{n}\sum_{m_2=-n'}^{n'}\exp\left[i(m_2-m_1)\gamma\right]d^{n}_{mm_1}(\beta)d^{n'}_{m'm_2}(\beta)\mathrm{T}^{ij}_{m_1nm_2n'}(\mathrm{P}), \tag{3.4.40}$$

where the rotations described by Euler angles $(\alpha,\beta,\gamma)$ are defined in Section 3.2.4.1.

Using the orthogonality of Euler angles in Eqs. (3.4.34)–(3.4.36), the extinction and scattering cross sections in random orientation are as follows (Mishchenko et al., 2002):

$$\langle C_{\text{ext}}\rangle = -\frac{2\pi}{k^2}\sum_{n=1}^{\infty}\sum_{m=-n}^{n} Re\left[\mathrm{T}^{11}_{mnmn} + \mathrm{T}^{22}_{mnmn}\right], \tag{3.4.41a}$$

$$\langle C_{\text{sca}}\rangle = \frac{2\pi}{k^2}\sum_{n=1}^{\infty}\sum_{m=-n}^{n}\sum_{n'=1}^{\infty}\sum_{m'=-n'}^{n'}\left[\left|\mathrm{T}^{11}_{mnm'n'}\right|^2 + \left|\mathrm{T}^{12}_{mnm'n'}\right|^2 + \left|\mathrm{T}^{21}_{mnm'n'}\right|^2 + \left|\mathrm{T}^{22}_{mnm'n'}\right|^2\right]. \tag{3.4.41b}$$

The cross sections in random orientation are independent of the particle orientation. Eq. (3.4.41) states that the real part of the trace of the T-matrix and the sum of the squares of the absolute values of all T-matrix elements are independent of the particle orientation. That is, regardless of the orientation of the particle, the trace and the sum of the absolute squares of all elements of the T-matrix are invariant.

From Eqs. (3.2.39), (3.2.58), the elements of the amplitude scattering matrix in the circular representation with Euler angles rotation $(\alpha,\beta,\gamma)$ can be stated as follows:

$$\mathbb{C}_{--}(\Theta;\alpha,\beta,\gamma) = \sum_{n=1}^{\infty}\sum_{m=-n}^{n}\sum_{n'=1}^{\infty} t_{mnn'}\left[\pi_{mn}(\Theta) - \tau_{mn}(\Theta)\right]$$
$$\left[\mathrm{T}^{11}_{mn-1n'}(\alpha,\beta,\gamma) - \mathrm{T}^{21}_{mn-1n'}(\alpha,\beta,\gamma) - \mathrm{T}^{12}_{mn-1n'}(\alpha,\beta,\gamma) + \mathrm{T}^{22}_{mn-1n'}(\alpha,\beta,\gamma)\right], \tag{3.4.42a}$$

$$\mathbb{C}_{-+}(\Theta;\alpha,\beta,\gamma) = \sum_{n=1}^{\infty}\sum_{m=-n}^{n}\sum_{n'=1}^{\infty} t_{mnn'}\left[\pi_{mn}(\Theta) - \tau_{mn}(\Theta)\right]$$
$$\left[\mathrm{T}^{11}_{mn1n'}(\alpha,\beta,\gamma) - \mathrm{T}^{21}_{mn1n'}(\alpha,\beta,\gamma) + \mathrm{T}^{12}_{mn1n'}(\alpha,\beta,\gamma) - \mathrm{T}^{22}_{mn1n'}(\alpha,\beta,\gamma)\right], \tag{3.4.42b}$$

$$\mathbb{C}_{+-}(\Theta;\alpha,\beta,\gamma)=\sum_{n=1}^{\infty}\sum_{m=-n}^{n}\sum_{n'=1}^{\infty}t_{mnn'}[\pi_{mn}(\Theta)+\tau_{mn}(\Theta)]$$
$$\left[\mathrm{T}^{11}_{mn-1n'}(\alpha,\beta,\gamma)+\mathrm{T}^{21}_{mn-1n'}(\alpha,\beta,\gamma)-\mathrm{T}^{12}_{mn-1n'}(\alpha,\beta,\gamma)-\mathrm{T}^{22}_{mn-1n'}(\alpha,\beta,\gamma)\right], \tag{3.4.42c}$$

$$\mathbb{C}_{++}(\Theta;\alpha,\beta,\gamma)=\sum_{n=1}^{\infty}\sum_{m=-n}^{n}\sum_{n'=1}^{\infty}t_{mnn'}[\pi_{mn}(\Theta)+\tau_{mn}(\Theta)]$$
$$\left[\mathrm{T}^{11}_{mn1n'}(\alpha,\beta,\gamma)+\mathrm{T}^{21}_{mn1n'}(\alpha,\beta,\gamma)+\mathrm{T}^{12}_{mn1n'}(\alpha,\beta,\gamma)+\mathrm{T}^{22}_{mn1n'}(\alpha,\beta,\gamma)\right], \tag{3.4.42d}$$

$$t_{mnn'}=\frac{1}{2}i^{n'-n}(-1)^m\sqrt{\frac{(2n+1)(2n'+1)}{n(n+1)}}, \tag{3.4.42e}$$

where the following relations for the incident zenith angle have been used in Eq. (3.4.42):

$$\pi_{m'n'}(0)=\delta_{m'\pm1}\frac{1}{2}\sqrt{n'(n'+1)}, \tag{3.4.43a}$$

$$\tau_{m'n'}(0)=m'\delta_{m'\pm1}\frac{1}{2}\sqrt{n'(n'+1)}. \tag{3.4.43b}$$

The functions $\pi$ and $\tau$ using the Wigner-d function have the following relations:

$$\pi_{mn}+\tau_{mn}=\sqrt{n(n+1)}d^{n}_{1m}, \tag{3.4.44a}$$

$$\pi_{mn}-\tau_{mn}=\sqrt{n(n+1)}d^{n}_{-1m}. \tag{3.4.44b}$$

Using the product relation for the Wigner-d functions in Eq. (3.2.83), the T-matrix after Euler angle rotations given in Eq. (3.4.40) can be further compacted by

$$\mathrm{T}^{ij}_{mnm'n'}(\alpha,\beta,\gamma)=\exp[i(m'-m)\alpha]\sum_{m_1=-n}^{n}\sum_{m_2=-n'}^{n'}\left\{\exp[i(m_2-m_1)\gamma]\mathrm{T}^{ij}_{m_1nm_2n'}(\mathrm{P})\right.$$
$$\left.\cdot\sum_{n_1=|n-n'|}^{n+n'}C^{n_1,m+m'}_{nmn'm'}d^{n_1}_{m+m',m_1+m_2}(\beta)C^{n_1,m_1+m_2}_{nm_1n'm_2}\right\}, \tag{3.4.45}$$

where $C$ is the CG coefficient introduced in Section 3.2.4.1. Consequently, the T-matrix after rotations can be given by

$$\mathrm{T}^{ij}_{mn1n'}(\alpha,\beta,\gamma)=\exp[-i(m-1)\alpha]\sum_{m_1=-n}^{n}\sum_{m_2=-n'}^{n'}\left\{(-1)^{m+1}(-1)^{\kappa}\exp(i\kappa\gamma)\mathrm{T}^{ij}_{m_1nm_1+\kappa n'}(\mathrm{P})\right.$$
$$\left.\cdot\sum_{n_1=|n-n'|}^{n+n'}\frac{2n_1+1}{2n'+1}C^{n',1}_{nmn_11-m}d^{n_1}_{m-1,-\kappa}(\beta)C^{n',m_1+\kappa}_{nm_1n_1\kappa}\right\}, \tag{3.4.46a}$$

$$\mathrm{T}^{ij}_{-mn-1n'}(\alpha,\beta,\gamma) = \exp[i(m-1)\alpha]\sum_{m_1=-n}^{n}\sum_{m_2=-n'}^{n'}\left\{(-1)^{m+1}(-1)^{\kappa}\exp(-i\kappa\gamma)\mathrm{T}^{ij}_{-m_1 n(-m_1-\kappa)n'}(\mathrm{P})\right.$$
$$\left.\cdot\sum_{n_1=|n-n'|}^{n+n'}\frac{2n_1+1}{2n'+1}C^{n',1}_{nmn_1 1-m}d^{n_1}_{-m+1,\kappa}(\beta)C^{n',m_1+\kappa}_{nm_1n_1\kappa}\right\}, \tag{3.4.46b}$$

where $\kappa = m_2 - m_1$ and the symmetric relations of the CG coefficients and the Wigner-d functions introduced in Section 3.2.4.1 have been used. After rearranging the summations, the amplitude scattering matrix in the circular representation in Eq. (3.4.42) can be rewritten as follows:

$$\mathbb{C}_{--}(\Theta;\alpha,\beta,\gamma) = \sum_{n=1}^{\infty}\sum_{m=-n}^{n} d^{n}_{-1-m}(\Theta)\exp[i(m-1)\alpha]$$
$$\sum_{n_1=|m-1|}^{\infty} f_{nn_1}\sum_{\kappa=-n_1}^{n_1}\exp(-i\kappa\gamma)(-1)^{\kappa}d^{n_1}_{-m+1,\kappa}(\beta)B^{3}_{\kappa mnn_1}, \tag{3.4.47a}$$

$$\mathbb{C}_{-+}(\Theta;\alpha,\beta,\gamma) = \sum_{n=1}^{\infty}\sum_{m=-n}^{n} d^{n}_{-1m}(\Theta)\exp[-i(m-1)\alpha]$$
$$\sum_{n_1=|m-1|}^{\infty} f_{nn_1}\sum_{\kappa=-n_1}^{n_1}\exp(i\kappa\gamma)(-1)^{\kappa}d^{n_1}_{m-1,-\kappa}(\beta)B^{2}_{\kappa mnn_1}, \tag{3.4.47b}$$

$$\mathbb{C}_{+-}(\Theta;\alpha,\beta,\gamma) = \sum_{n=1}^{\infty}\sum_{m=-n}^{n} d^{n}_{1-m}(\Theta)\exp[i(m-1)\alpha]$$
$$\sum_{n_1=|m-1|}^{\infty} f_{nn_1}\sum_{\kappa=-n_1}^{n_1}\exp(-i\kappa\gamma)(-1)^{\kappa}d^{n_1}_{-m+1,\kappa}(\beta)B^{4}_{\kappa mnn_1}, \tag{3.4.47c}$$

$$\mathbb{C}_{++}(\Theta;\alpha,\beta,\gamma) = \sum_{n=1}^{\infty}\sum_{m=-n}^{n} d^{n}_{1m}(\Theta)\exp[-i(m-1)\alpha]$$
$$\sum_{n_1=|m-1|}^{\infty} f_{nn_1}\sum_{\kappa=-n_1}^{n_1}\exp(i\kappa\gamma)(-1)^{\kappa}d^{n_1}_{m-1,-\kappa}(\beta)B^{1}_{\kappa mnn_1}, \tag{3.4.47d}$$

where

$$f_{nn_1} = -\frac{1}{2}(2n_1+1)\sqrt{2n+1}, \tag{3.4.48a}$$

$$B^{j}_{\kappa mnn_1} = \sum_{n'=\max(1,|n-n_1|)}^{n+n_1} C^{n',1}_{nmn_1 1-m}A^{j}_{\kappa nn'n_1},\ j=1,2,3,4, \tag{3.4.48b}$$

$$A^{\mathrm{j}}_{\kappa nn'n_1} = \frac{i^{n'-n}}{\sqrt{2n'+1}} \sum_{m_1=-n}^{n} C^{n',m_1+\kappa}_{nm_1n_1\kappa} \mathrm{T}^{\mathrm{j}}_{m_1n(m_1+\kappa)n'}, \tag{3.4.48c}$$

$$\begin{aligned}\mathrm{T}^{3}_{m_1n(m_1+\kappa)n'} &= \mathrm{T}^{11}_{-m_1n(-m_1-\kappa)n'}(\mathrm{P}) - \mathrm{T}^{21}_{-m_1n(-m_1-\kappa)n'}(\mathrm{P}) \\ &\quad - \mathrm{T}^{12}_{-m_1n(-m_1-\kappa)n'}(\mathrm{P}) + \mathrm{T}^{22}_{-m_1n(-m_1-\kappa)n'}(\mathrm{P}),\end{aligned} \tag{3.4.48d}$$

$$\begin{aligned}\mathrm{T}^{2}_{m_1n(m_1+\kappa)n'} &= \mathrm{T}^{11}_{m_1n(m_1+\kappa)n'}(\mathrm{P}) - \mathrm{T}^{21}_{m_1n(m_1+\kappa)n'}(\mathrm{P}) + \mathrm{T}^{12}_{m_1n(m_1+\kappa)n'}(\mathrm{P}) \\ &\quad - \mathrm{T}^{22}_{m_1n(m_1+\kappa)n'}(\mathrm{P}),\end{aligned} \tag{3.4.48e}$$

$$\begin{aligned}\mathrm{T}^{4}_{m_1n(m_1+\kappa)n'} &= \mathrm{T}^{11}_{-m_1n(-m_1-\kappa)n'}(\mathrm{P}) + \mathrm{T}^{21}_{-m_1n(-m_1-\kappa)n'}(\mathrm{P}) \\ &\quad - \mathrm{T}^{12}_{-m_1n(-m_1-\kappa)n'}(\mathrm{P}) - \mathrm{T}^{22}_{-m_1n(-m_1-\kappa)n'}(\mathrm{P}),\end{aligned} \tag{3.4.48f}$$

$$\mathrm{T}^{1}_{m_1n(m_1+\kappa)n'} = \mathrm{T}^{11}_{m_1n(m_1+\kappa)n'}(\mathrm{P}) + \mathrm{T}^{21}_{m_1n(m_1+\kappa)n'}(\mathrm{P}) + \mathrm{T}^{12}_{m_1n(m_1+\kappa)n'}(\mathrm{P}) + \mathrm{T}^{22}_{m_1n(m_1+\kappa)n'}(\mathrm{P}). \tag{3.4.48g}$$

The arrangement of the Euler angles in Eq. (3.4.47) is convenient to take advantage of the orthogonality of the Wigner-d function and the exponential functions.

The analytical formula with random orientation can be implemented in terms of integration over the Euler angles as follows:

$$\mathbf{P}^{\mathrm{CP}}(\Theta) = \frac{4\pi}{k^2\langle C_{\mathrm{sca}}\rangle} \frac{1}{8\pi^2} \int_0^{2\pi} d\alpha \int_0^{\pi} \sin\beta d\beta \int_0^{2\pi} d\gamma \mathbb{F}^{\mathrm{CP}}(\Theta; \alpha, \beta, \gamma), \tag{3.4.49}$$

where the scattering phase matrix in circular polarization with respect to the scattering plane is the same as the one with respect to the meridional planes when $\theta_i = 0$ as shown in Section 2.2.2. The factor $4\pi/k^2\langle C_{\mathrm{sca}}\rangle$ ensures the normalization of phase function $\mathrm{P}_{11}$ as shown in Eq. (2.2.119), that is, $1/2\int_0^{\pi} \mathrm{P}_{11}(\Theta)\sin\Theta d\Theta = 1$. The scattering matrix $\mathbb{F}^{\mathrm{CP}}(\Theta; \alpha, \beta, \gamma)$ in the circular polarization representation is given in Eq. (2.2.44) in terms of the amplitude scattering matrix in circular polarization representation and is restated here as follows:

$$\mathbb{F}^{\mathrm{CP}} = \begin{pmatrix} \mathbb{C}_{++}\mathbb{C}^*_{--} & \mathbb{C}_{+-}\mathbb{C}^*_{--} & \mathbb{C}_{++}\mathbb{C}^*_{-+} & \mathbb{C}_{+-}\mathbb{C}^*_{-+} \\ \mathbb{C}_{-+}\mathbb{C}^*_{--} & \mathbb{C}_{--}\mathbb{C}^*_{--} & \mathbb{C}_{-+}\mathbb{C}^*_{-+} & \mathbb{C}_{--}\mathbb{C}^*_{-+} \\ \mathbb{C}_{++}\mathbb{C}^*_{+-} & \mathbb{C}_{+-}\mathbb{C}^*_{+-} & \mathbb{C}_{++}\mathbb{C}^*_{++} & \mathbb{C}_{+-}\mathbb{C}^*_{++} \\ \mathbb{C}_{-+}\mathbb{C}^*_{+-} & \mathbb{C}_{--}\mathbb{C}^*_{+-} & \mathbb{C}_{-+}\mathbb{C}^*_{++} & \mathbb{C}_{--}\mathbb{C}^*_{++} \end{pmatrix}, \tag{3.4.50}$$

where the arguments $(\Theta;\alpha,\beta,\gamma)$ are suppressed. To be consistent with the Wigner-d function symbols, the index order (2,0,−0,−2) is used to enumerate the elements (Mishchenko et al., 2002). For example, $\mathbb{F}^{\mathrm{CP}}_{-02}(\Theta)$ is the element in the third row and the fourth column. For a particle without mirror symmetry, the scattering phase matrix associated with the linear representation is given in Eqs. (2.2.73), (2.2.119) by

$$\mathbf{P} = \begin{pmatrix} P_{11} & P_{12} & P_{13} & P_{14} \\ P_{12} & P_{22} & P_{23} & P_{24} \\ -P_{13} & -P_{23} & P_{33} & P_{34} \\ P_{14} & P_{24} & -P_{34} & P_{44} \end{pmatrix}, \tag{3.4.51}$$

where only upper triangular elements are independent. Accordingly, the scattering matrix in the circular representation can be given using Eq. (2.2.33) in terms of the scattering phase matrix in the linear representation as follows (Bi and Yang, 2014, and references cited therein):

$$\mathbf{P}^{CP} = \frac{1}{2}\begin{pmatrix} (P_{22}+P_{33})-i2P_{23} & (P_{12}+P_{24})+i(P_{34}-P_{13}) & (P_{12}-P_{24})-i(P_{34}+P_{13}) & P_{22}-P_{33} \\ (P_{12}+P_{24})+i(P_{34}-P_{13}) & P_{11}+P_{44}+2P_{14} & P_{11}-P_{44} & (P_{12}+P_{24})-i(P_{34}-P_{13}) \\ (P_{12}-P_{24})-i(P_{34}+P_{13}) & P_{11}-P_{44} & P_{11}+P_{44}-2P_{14} & (P_{12}-P_{24})+i(P_{34}+P_{13}) \\ P_{22}-P_{33} & (P_{12}+P_{24})-i(P_{34}-P_{13}) & (P_{12}-P_{24})+i(P_{34}+P_{13}) & (P_{22}+P_{33})+i2P_{23} \end{pmatrix}, \tag{3.4.52}$$

The following symmetric relations are evident from Eq. (3.4.52):

$$P_{pq}^{CP} = P_{qp}^{CP}, \quad p,q = 2,0,-0,-2, \tag{3.4.53a}$$

$$P_{22}^{CP} = \left(P_{-2-2}^{CP}\right)^*, \quad P_{20}^{CP} = \left(P_{0-2}^{CP}\right)^*, \quad P_{2-0}^{CP} = \left(P_{-0-2}^{CP}\right)^*. \tag{3.4.53b}$$

Generally, the elements $P_{00}^{CP}$, $P_{-0-0}^{CP}$, $P_{0-0}^{CP}$, $P_{-00}^{CP}$, $P_{2-2}^{CP}$, and $P_{-22}^{CP}$ are real, and the others are complex. Substituting Eq. (3.4.47) into Eqs. (3.4.49) and (3.4.50) and performing many algebraic steps, the scattering phase matrix in the circular polarization representation in random orientations can be written in an elegant way (Mishchenko et al., 2002; Bi and Yang, 2014):

$$P_{pq}^{CP}(\Theta) = \sum_{s=0}^{\infty} g_{pq}^{s} d_{pq}^{s}(\Theta), \quad p,q = 2,0,-0,-2. \tag{3.4.54}$$

Eq. (3.4.54) states that the scattering phase matrix in the circular polarization representation can be expanded using the Wigner-d functions, and $\{g_{pq}^{s}\}$ is the corresponding expansion coefficients. The Wigner-d functions are real and have the following symmetric properties:

$$d_{pq}^{s} = d_{qp}^{s}, \; d_{pq}^{s} = d_{-p-q}^{s}, \; p,q = 2,0,-0,-2. \tag{3.4.55}$$

Using Eqs. (3.4.53)–(3.4.55), the coefficients $g_{pq}^{s}$ have the same symmetric properties as the scattering phase matrix in circular representation:

$$g_{pq}^{s} = g_{qp}^{s}, \quad p,q = 2,0,-0,-2, \tag{3.4.56a}$$

$$g^s_{22} = \left(g^s_{-2-2}\right)^*, \quad g^s_{20} = \left(g^s_{0-2}\right)^*, \quad g^s_{2-0} = \left(g^s_{-0-2}\right)^*. \tag{3.4.56b}$$

Accordingly, only the coefficients $g^s_{22}$, $g^s_{00}$, $g^s_{-0-0}$, $g^s_{02}$, $g^s_{-02}$, $g^s_{2-2}$, and $g^s_{0-0}$ are independent, where $g^s_{00}$, $g^s_{-0-0}$, $g^s_{0-0}$, and $g^s_{2-2}$ are real numbers and $g^s_{22}$, $g^s_{02}$, and $g^s_{-02}$ are generally complex numbers. The independent coefficients have the following expressions:

$$g^s_{00} = \sum_{n=1}^{\infty} \sum_{j=\max(1,|n-s|)}^{n+s} h^+_{snj} C^{j1}_{n1s0} \sum_{m=-\min(n,j)}^{\min(n,j)} C^{jm}_{nms0} D^{00}_{mnj}, \tag{3.4.57a}$$

$$g^s_{-0-0} = \sum_{n=1}^{\infty} \sum_{j=\max(1,|n-s|)}^{n+s} h^+_{snj} C^{j1}_{n1s0} \sum_{m=-\min(n,j)}^{\min(n,j)} C^{jm}_{nms0} D^{-0-0}_{mnj}, \tag{3.4.57b}$$

$$g^s_{0-0} = \sum_{n=1}^{\infty} \sum_{j=\max(1,|n-s|)}^{n+s} h^-_{snj} C^{j1}_{n1s0} \sum_{m=-\min(n,j)}^{\min(n,j)} C^{jm}_{nms0} D^{0-0}_{mnj}, \tag{3.4.57c}$$

$$g^s_{2-2} = \sum_{n=1}^{\infty} \sum_{j=\max(1,|n-s|)}^{n+s} h^-_{snj} C^{j1}_{n-1s2} \sum_{m=-\min(n,j-2)}^{\min(n,j+2)} C^{j2-m}_{n-ms2} D^{2-2}_{mnj}, \tag{3.4.57d}$$

$$g^s_{22} = \sum_{n=1}^{\infty} \sum_{j=\max(1,|n-s|)}^{n+s} h^+_{snj} C^{j1}_{n-1s2} \sum_{m=-\min(n,j-2)}^{\min(n,j+2)} C^{j2-m}_{n-ms2} D^{22}_{mnj}, \tag{3.4.57e}$$

$$g^s_{02} = \sum_{n=1}^{\infty} \sum_{j=\max(1,|n-s|)}^{n+s} h^+_{snj} C^{j1}_{n1s0} \sum_{m=-\min(n,j-2)}^{\min(n,j+2)} C^{j2-m}_{n-ms2} D^{02}_{mnj}, \tag{3.4.57f}$$

$$g^s_{-02} = \sum_{n=1}^{\infty} \sum_{j=\max(1,|n-s|)}^{n+s} h^-_{snj} C^{j1}_{n1s0} \sum_{m=-\min(n,j-2)}^{\min(n,j+2)} C^{j2-m}_{n-ms2} D^{-02}_{mnj}, \tag{3.4.57g}$$

where

$$h^+_{snj} = \frac{(2s+1)\pi}{k^2\langle C_{\text{sca}}\rangle} \sqrt{\frac{2n+1}{2j+1}}, \tag{3.4.58a}$$

$$h^-_{snj} = (-1)^{s+n+j} h^+_{snj}, \tag{3.4.58b}$$

and

$$D^{00}_{mnj} = \sum_{n_1=|m-1|}^{\infty} (2n_1+1) \sum_{\kappa=-n_1}^{n_1} B^3_{\kappa mnn_1} \left(B^3_{\kappa mjn_1}\right)^*, \tag{3.4.59a}$$

$$D^{-0-0}_{mnj} = \sum_{n_1=|m-1|}^{\infty} (2n_1+1) \sum_{\kappa=-n_1}^{n_1} B^{1}_{\kappa mnn_1}\left(B^{1}_{\kappa mjn_1}\right)^{*}, \tag{3.4.59b}$$

$$D^{0-0}_{mnj} = \sum_{n_1=|m-1|}^{\infty} (2n_1+1) \sum_{\kappa=-n_1}^{n_1} B^{2}_{\kappa mnn_1}\left(B^{2}_{\kappa mjn_1}\right)^{*}, \tag{3.4.59c}$$

$$D^{2-2}_{mnj} = \sum_{n_1=|m-1|}^{\infty} (2n_1+1) \sum_{\kappa=-n_1}^{n_1} B^{4}_{\kappa mnn_1}\left(B^{2}_{-\kappa 2-mjn_1}\right)^{*}, \tag{3.4.59d}$$

$$D^{22}_{mnj} = \sum_{n_1=|m-1|}^{\infty} (2n_1+1) \sum_{\kappa=-n_1}^{n_1} B^{1}_{\kappa mnn_1}\left(B^{3}_{-\kappa 2-mjn_1}\right)^{*}, \tag{3.4.59e}$$

$$D^{02}_{mnj} = \sum_{n_1=|m-1|}^{\infty} (2n_1+1) \sum_{\kappa=-n_1}^{n_1} B^{2}_{\kappa mnn_1}\left(B^{3}_{-\kappa 2-mjn_1}\right)^{*}, \tag{3.4.59f}$$

$$D^{-02}_{mnj} = \sum_{n_1=|m-1|}^{\infty} (2n_1+1) \sum_{\kappa=-n_1}^{n_1} B^{1}_{\kappa mnn_1}\left(B^{4}_{-\kappa 2-mjn_1}\right)^{*}. \tag{3.4.59g}$$

Using Eq. (3.4.52) and Eqs. (3.4.54)–(3.4.56), the independent scattering phase matrix elements in the linear representation have the following expressions:

$$\mathrm{P}_{11}(\Theta) = \sum_{s=0}^{\infty} \alpha_1^s d^s_{00}(\Theta), \tag{3.4.60a}$$

$$\mathrm{P}_{22}(\Theta) + \mathrm{P}_{33}(\Theta) = \sum_{s=2}^{\infty} \left(\alpha_2^s + \alpha_3^s\right) d^s_{22}(\Theta), \tag{3.4.60b}$$

$$\mathrm{P}_{22}(\Theta) - \mathrm{P}_{33}(\Theta) = \sum_{s=2}^{\infty} \left(\alpha_2^s - \alpha_3^s\right) d^s_{2-2}(\Theta), \tag{3.4.60c}$$

$$\mathrm{P}_{44}(\Theta) = \sum_{s=0}^{\infty} \alpha_4^s d^s_{00}(\Theta), \tag{3.4.60d}$$

$$\mathrm{P}_{12}(\Theta) = \sum_{s=2}^{\infty} \beta_1^s d^s_{02}(\Theta), \tag{3.4.60e}$$

$$\mathrm{P}_{34}(\Theta) = \sum_{s=2}^{\infty} \beta_2^s d^s_{02}(\Theta), \tag{3.4.60f}$$

$$\mathrm{P}_{13}(\Theta) = \sum_{s=2}^{\infty} \beta_3^s d^s_{02}(\Theta), \tag{3.4.60g}$$

$$P_{23}(\Theta) = \sum_{s=2}^{\infty} \beta_4^s d_{22}^s(\Theta), \tag{3.4.60h}$$

$$P_{14}(\Theta) = \sum_{s=0}^{\infty} \beta_5^s d_{00}^s(\Theta), \tag{3.4.60i}$$

$$P_{24}(\Theta) = \sum_{s=2}^{\infty} \beta_6^s d_{02}^s(\Theta), \tag{3.4.60j}$$

where $\alpha$ and $\beta$ should be distinguished from the Euler angles and are conventionally used to denote the expansion coefficients of the diagonal and the upper off-diagonal elements of the scattering phase matrix, respectively, and

$$\alpha_1^s = \frac{1}{2}\left(g_{00}^s + g_{-0-0}^s\right) + g_{0-0}^s, \tag{3.4.61a}$$

$$\alpha_2^s = Re\left[g_{22}^s\right] + g_{2-2}^s, \tag{3.4.61b}$$

$$\alpha_3^s = Re\left[g_{22}^s\right] - g_{2-2}^s, \tag{3.4.61c}$$

$$\alpha_4^s = \frac{1}{2}\left(g_{00}^s + g_{-0-0}^s\right) - g_{0-0}^s, \tag{3.4.61d}$$

$$\beta_1^s = Re\left[g_{02}^s\right] + Re\left[g_{-02}^s\right], \tag{3.4.61e}$$

$$\beta_2^s = Im\left[g_{02}^s\right] - Im\left[g_{-02}^s\right], \tag{3.4.61f}$$

$$\beta_3^s = -Im\left[g_{02}^s\right] - Im\left[g_{-02}^s\right], \tag{3.4.61g}$$

$$\beta_4^s = -Im\left[g_{22}^s\right], \tag{3.4.61h}$$

$$\beta_5^s = \frac{1}{2}\left(g_{00}^s - g_{-0-0}^s\right), \tag{3.4.61i}$$

$$\beta_6^s = Re\left[g_{02}^s\right] - Re\left[g_{-02}^s\right]. \tag{3.4.61j}$$

It is evident that the asymmetry factor $g$ in the random orientation is $\langle g \rangle = \alpha_1^1/3$.

The computation process of the random orientation of a particle is described by a flowchart in Fig. 3.9.

There are several notes as follows:

1. In radiative transfer, the generalized spherical functions are generally used to expand the elements of the scattering phase matrix. The generalized spherical function and the corresponding Wigner-d function have a factor difference as follows (Hovenier and van der Mee, 1983; Mishchenko et al., 2002):

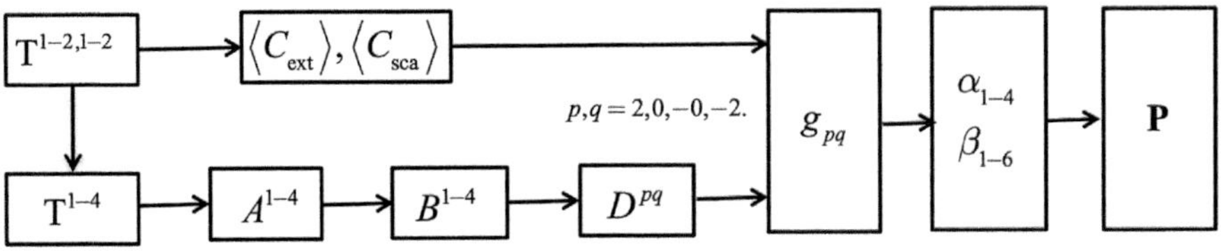


**Fig. 3.9** Random orientation computation.

$$P^s_{mm_1}(\cos\Theta) = i^{m-m_1} d^s_{mm_1}(\Theta), \tag{3.4.62a}$$

$$P^s_{00} = d^s_{00},\ P^s_{22} = d^s_{22},\ P^s_{2-2} = d^s_{2-2}, \tag{3.4.62b}$$

$$P^s_{02} = -d^s_{02}. \tag{3.4.62c}$$

The expansion coefficients using the generalized spherical functions for $\alpha^s_{1-4}$, $\beta^s_{1-3}$, and $\beta^s_6$ are the same as in Eq. (3.4.61), and the expansion coefficients for the others have a minus sign compared with those in Eq. (3.4.61).

2. For particles without any symmetry, the original T-matrix of a particle is arranged in the $n$-dominant order as given in Eq. (3.2.54). For a particle with finite-fold symmetry, the original T-matrix is still arranged in the $n$-dominant order (e.g., for a hexagonal column, the order is given in Table 3.1). For a particle with axial symmetry, the T-matrix is fully decoupled associated with the azimuthal component $m$ and is arranged in the $m$-dominant order. For the convenience of random orientation computation, the T-matrix $T^{1-4}$ for any particle is arranged in the $m$-dominant order.

### 3.4.4 Completely random orientations for symmetric particles

For symmetric particles, the $z$-axis in the particle frame of reference is along the rotational axis, so the T-matrix can be sufficiently decoupled. For particles with $j$-fold rotational symmetry, the azimuthal components of the T-matrix $T^{ij}_{mnm'n'}$ are correlated as follows:

$$\mathrm{mod}(m' - m, j) = 0, \tag{3.4.63}$$

where mod is a remainder function. The computation for random orientation is partially decoupled into several components according to the remainder of the azimuthal component divided by the fold number $j$, that is, mod($m$ and $m'$,$j$). Using the symmetry described in Section 3.4.2, the computation time can be significantly reduced. The equations to compute the scattering phase matrix of a particle with finite-fold rotational symmetry in random orientation are the same with an exception that $\kappa$ values in equations given in Section 3.4.3 can only be multiples of $j$; for instance, $\kappa$ can only be a multiple of 6 for a hexagonal column.

Furthermore, if the particle has axial symmetry, the computation is completely decoupled and associated with the azimuthal component $m$. All computations are changed to be $m$-dominant. For particles with axial symmetry, T-matrix has the following symmetric property:

$$\mathrm{T}^{ij}_{-mn-m'n'} = \delta_{mm'}(-1)^{i+j}\mathrm{T}^{ij}_{mnmn'}. \tag{3.4.64}$$

For the integrated scattering properties, the extinction cross section in random orientation has the same expression as Eq. (3.4.41a), while the scattering cross section can be rewritten as (Mishchenko et al., 2002)

$$\langle C_{\mathrm{sca}}\rangle = \frac{2\pi}{k^2}\sum_{n=1}^{\infty}\sum_{n'=1}^{\infty}\sum_{m=0}^{\min(n,n')}(2-\delta_{m0})\left[\left|\mathrm{T}^{11}_{mnmn'}\right|^2+\left|\mathrm{T}^{12}_{mnmn'}\right|^2+\left|\mathrm{T}^{21}_{mnmn'}\right|^2+\left|\mathrm{T}^{22}_{mnmn'}\right|^2\right]. \tag{3.4.65}$$

For computation of the scattering phase matrix, the representations in Eq. (3.4.48) are simplified to

$$\kappa = 0, \tag{3.4.66a}$$

$$\mathrm{T}^1_{m_1nn'} = \mathrm{T}^1_{m_1nm_1n'} = \mathrm{T}^3_{m_1nm_1n'},\quad \mathrm{T}^2_{m_1nn'} = \mathrm{T}^2_{m_1nm_1n'} = \mathrm{T}^4_{m_1nm_1n'}, \tag{3.4.66b}$$

$$B^1_{mnn_1} = B^1_{0mnn_1} = B^3_{0mnn_1},\quad B^2_{mnn_1} = B^2_{0mnn_1} = B^4_{0mnn_1}, \tag{3.4.66c}$$

$$A^1_{nn'n_1} = A^1_{0nn'n_1} = A^3_{0nn'n_1},\quad A^2_{nn'n_1} = A^2_{0nn'n_1} = A^4_{0nn'n_1}. \tag{3.4.66d}$$

All summations over $\kappa$ disappear in the computation of particles with axial symmetry. The elements of the amplitude scattering matrix in circular polarization representation are simplified as follows:

$$\mathbb{C}_{--}(\Theta;\alpha,\beta,\gamma) = \sum_{n=1}^{\infty}\sum_{m=-n}^{n} d^n_{-1-m}(\Theta)\exp[i(m-1)\alpha]\sum_{n_1=|m-1|}^{\infty} f_{nn_1}d^{n_1}_{-m+1,0}(\beta)B^1_{mnn_1}, \tag{3.4.67a}$$

$$\mathbb{C}_{-+}(\Theta;\alpha,\beta,\gamma) = \sum_{n=1}^{\infty}\sum_{m=-n}^{n} d^n_{-1m}(\Theta)\exp[-i(m-1)\alpha]\sum_{n_1=|m-1|}^{\infty} f_{nn_1}d^{n_1}_{m-1,0}(\beta)B^2_{mnn_1}, \tag{3.4.67b}$$

$$\mathbb{C}_{+-}(\Theta;\alpha,\beta,\gamma) = \sum_{n=1}^{\infty}\sum_{m=-n}^{n} d^n_{1-m}(\Theta)\exp[i(m-1)\alpha]\sum_{n_1=|m-1|}^{\infty} f_{nn_1}d^{n_1}_{-m+1,0}(\beta)B^2_{mnn_1}, \tag{3.4.67c}$$

$$\mathbb{C}_{++}(\Theta;\alpha,\beta,\gamma) = \sum_{n=1}^{\infty}\sum_{m=-n}^{n} d^n_{1m}(\Theta)\exp[-i(m-1)\alpha]\sum_{n_1=|m-1|}^{\infty} f_{nn_1}d^{n_1}_{m-1,0}(\beta)B^1_{mnn_1}, \tag{3.4.67d}$$

where $f_{nn_1}$ is given by Eq. (3.4.48a) and

$$B^{\mathrm{j}}_{mnn_1} = \sum_{n'=\max(1,|n-n_1|)}^{n+n_1} C^{n',1}_{nmn_1 1-m} A^{\mathrm{j}}_{nn'n_1}, \quad j=1,2, \tag{3.4.68a}$$

$$A^{\mathrm{j}}_{nn'n_1} = \frac{i^{n'-n}}{\sqrt{2n'+1}} \sum_{m_1=-\min(n,n')}^{\min(n,n')} C^{n',m_1}_{nm_1n_10} \mathrm{T}^{\mathrm{j}}_{m_1nn'}, \tag{3.4.68b}$$

$$\mathrm{T}^{2}_{m_1nn'} = \mathrm{T}^{11}_{m_1nm_1n'}(\mathrm{P}) - \mathrm{T}^{21}_{m_1nm_1n'}(\mathrm{P}) + \mathrm{T}^{12}_{m_1nm_1n'}(\mathrm{P}) - \mathrm{T}^{22}_{m_1nm_1n'}(\mathrm{P}), \tag{3.4.68c}$$

$$\mathrm{T}^{1}_{m_1nn'} = \mathrm{T}^{11}_{m_1nm_1n'}(\mathrm{P}) + \mathrm{T}^{21}_{m_1nm_1n'}(\mathrm{P}) + \mathrm{T}^{12}_{m_1nm_1n'}(\mathrm{P}) + \mathrm{T}^{22}_{m_1nm_1n'}(\mathrm{P}). \tag{3.4.68d}$$

A particle with axial symmetry also has mirror symmetry. The scattering phase matrix associated with random particle orientations in the linear representation is given in Eqs. (2.2.78), (2.2.119), which is block diagonal and only has six independent parameters, and is restated here as:

$$\mathbf{P} = \begin{pmatrix} \mathrm{P}_{11} & \mathrm{P}_{12} & 0 & 0 \\ \mathrm{P}_{12} & \mathrm{P}_{22} & 0 & 0 \\ 0 & 0 & \mathrm{P}_{33} & \mathrm{P}_{34} \\ 0 & 0 & -\mathrm{P}_{34} & \mathrm{P}_{44} \end{pmatrix}. \tag{3.4.69}$$

The scattering phase matrix in the circular polarization representation can be obtained using Eq. (2.2.33) in the following form (Hovenier and van der Mee, 1983; Mishchenko et al., 2002):

$$\mathbf{P}^{\mathrm{CP}} = \frac{1}{2} \begin{pmatrix} \mathrm{P}_{22}+\mathrm{P}_{33} & \mathrm{P}_{12}+i\mathrm{P}_{34} & \mathrm{P}_{12}-i\mathrm{P}_{34} & \mathrm{P}_{22}-\mathrm{P}_{33} \\ \mathrm{P}_{12}+i\mathrm{P}_{34} & \mathrm{P}_{11}+\mathrm{P}_{44} & \mathrm{P}_{11}-\mathrm{P}_{44} & \mathrm{P}_{12}-i\mathrm{P}_{34} \\ \mathrm{P}_{12}-i\mathrm{P}_{34} & \mathrm{P}_{11}-\mathrm{P}_{44} & \mathrm{P}_{11}+\mathrm{P}_{44} & \mathrm{P}_{12}+i\mathrm{P}_{34} \\ \mathrm{P}_{22}-\mathrm{P}_{33} & \mathrm{P}_{12}-i\mathrm{P}_{34} & \mathrm{P}_{12}+i\mathrm{P}_{34} & \mathrm{P}_{22}+\mathrm{P}_{33} \end{pmatrix}. \tag{3.4.70}$$

The scattering phase matrix in the circular polarization representation has the following symmetry:

$$\mathrm{P}^{\mathrm{CP}}_{pq} = \mathrm{P}^{\mathrm{CP}}_{qp}, \quad \mathrm{P}^{\mathrm{CP}}_{pq} = \mathrm{P}^{\mathrm{CP}}_{-q-p}, \quad p,q = 2,0,-0,-2, \tag{3.4.71a}$$

$$\mathrm{P}^{\mathrm{CP}}_{20} = \left(\mathrm{P}^{\mathrm{CP}}_{2-0}\right)^*. \tag{3.4.71b}$$

The scattering phase matrix is given by

$$\mathrm{P}^{\mathrm{CP}}_{pq}(\Theta) = \sum_{s=0}^{\infty} g^s_{pq} d^s_{pq}(\Theta), \quad p,q = 2,0,-0,-2, \tag{3.4.72}$$

where only the expansion coefficients $g^s_{22}$, $g^s_{02}$, $g^s_{00}$, $g^s_{0-0}$, and $g^s_{2-2}$ are independent; $g^s_{22}$, $g^s_{00}$, $g^s_{0-0}$, and $g^s_{2-2}$ are real; and $g^s_{02}$ is generally complex. The independent expansion coefficients are

$$g^s_{22} = \sum_{n=1}^{\infty} \sum_{j=\max(1,|n-s|)}^{n+s} h^+_{snj} C^{j1}_{n-1s2} \sum_{m=-\min(n,j-2)}^{\min(n,j+2)} C^{j2-m}_{n-ms2} D^{22}_{mnj}, \tag{3.4.73a}$$

$$g^s_{00} = \sum_{n=1}^{\infty} \sum_{j=\max(1,|n-s|)}^{n+s} h^+_{snj} C^{j1}_{n1s0} \sum_{m=-\min(n,j)}^{\min(n,j)} C^{jm}_{nms0} D^{00}_{mnj}, \tag{3.4.73b}$$

$$g^s_{0-0} = \sum_{n=1}^{\infty} \sum_{j=\max(1,|n-s|)}^{n+s} h^-_{snj} C^{j1}_{n1s0} \sum_{m=-\min(n,j)}^{\min(n,j)} C^{jm}_{nms0} D^{0-0}_{mnj}, \tag{3.4.73c}$$

$$g^s_{2-2} = \sum_{n=1}^{\infty} \sum_{j=\max(1,|n-s|)}^{n+s} h^-_{snj} C^{j1}_{n-1s2} \sum_{m=-\min(n,j-2)}^{\min(n,j+2)} C^{j2-m}_{n-ms2} D^{2-2}_{mnj}, \tag{3.4.73d}$$

$$g^s_{02} = \sum_{n=1}^{\infty} \sum_{j=\max(1,|n-s|)}^{n+s} h^+_{snj} C^{j1}_{n1s0} \sum_{m=-\min(n,j-2)}^{\min(n,j+2)} C^{j2-m}_{n-ms2} D^{02}_{mnj}, \tag{3.4.73e}$$

where $h^{\pm}_{snj}$ expressions are given in Eq. (3.4.58) and

$$D^{22}_{mnj} = \sum_{n_1=|m-1|}^{\infty} (2n_1+1) B^1_{mnn_1} \left(B^1_{2-mjn_1}\right)^*, \tag{3.4.74a}$$

$$D^{00}_{mnj} = \sum_{n_1=|m-1|}^{\infty} (2n_1+1) B^1_{mnn_1} \left(B^1_{mjn_1}\right)^*, \tag{3.4.74b}$$

$$D^{0-0}_{mnj} = \sum_{n_1=|m-1|}^{\infty} (2n_1+1) B^2_{mnn_1} \left(B^2_{mjn_1}\right)^*, \tag{3.4.74c}$$

$$D^{2-2}_{mnj} = \sum_{n_1=|m-1|}^{\infty} (2n_1+1) B^2_{mnn_1} \left(B^2_{2-mjn_1}\right)^*, \tag{3.4.74d}$$

$$D^{02}_{mnj} = \sum_{n_1=|m-1|}^{\infty} (2n_1+1) B^2_{mnn_1} \left(B^1_{2-mjn_1}\right)^*. \tag{3.4.74e}$$

The scattering phase matrix elements in linear representation can be stated as follows:

$$\mathrm{P}_{11}(\Theta) = \sum_{s=0}^{\infty} \alpha^s_1 d^s_{00}(\Theta), \tag{3.4.75a}$$

$$\mathrm{P}_{22}(\Theta)+\mathrm{P}_{33}(\Theta)=\sum_{s=2}^{\infty}\left(\alpha_2^s+\alpha_3^s\right)d_{22}^s(\Theta), \tag{3.4.75b}$$

$$\mathrm{P}_{22}(\Theta)-\mathrm{P}_{33}(\Theta)=\sum_{s=0}^{\infty}\left(\alpha_2^s-\alpha_3^s\right)d_{2-2}^s(\Theta), \tag{3.4.75c}$$

$$\mathrm{P}_{44}(\Theta)=\sum_{s=0}^{\infty}\alpha_4^s d_{00}^s(\Theta), \tag{3.4.75d}$$

$$\mathrm{P}_{12}(\Theta)=\sum_{s=2}^{\infty}\beta_1^s d_{02}^s(\Theta), \tag{3.4.75e}$$

$$\mathrm{P}_{34}(\Theta)=\sum_{s=2}^{\infty}\beta_2^s d_{02}^s(\Theta), \tag{3.4.75f}$$

where

$$\alpha_1^s=g_{00}^s+g_{0-0}^s, \tag{3.4.76a}$$

$$\alpha_2^s=g_{22}^s+g_{2-2}^s, \tag{3.4.76b}$$

$$\alpha_3^s=g_{22}^s-g_{2-2}^s, \tag{3.4.76c}$$

$$\alpha_4^s=g_{00}^s-g_{0-0}^s, \tag{3.4.76d}$$

$$\beta_1^s=2\,\mathrm{Re}\left[g_{02}^s\right], \tag{3.4.76e}$$

$$\beta_2^s=2\,\mathrm{Im}\left[g_{02}^s\right]. \tag{3.4.76f}$$

The random orientation computation for a particle with axial symmetry is also given by Mishchenko et al. (2002).

Finally, if a particle has spherical symmetry, the T-matrix becomes diagonal and is independent of index $m$ as shown in Eq. (3.3.41). The T-matrix is invariant under arbitrary rotations. The T-matrix elements are described using the Lorenz-Mie coefficients $a_n$ and $b_n$ in Eq. (3.3.41). The extinction and scattering cross sections are simplified to (van de Hulst, 1957; Bohren and Huffman, 1983; Mishchenko et al., 2002)

$$C_{\mathrm{ext}}=\frac{2\pi}{k^2}\sum_{n=1}^{\infty}(2n+1)Re\left[a_n+b_n\right], \tag{3.4.77}$$

$$C_{\mathrm{sca}}=\frac{2\pi}{k^2}\sum_{n=1}^{\infty}(2n+1)\left(|a_n|^2+|b_n|^2\right). \tag{3.4.77}$$

The amplitude scattering matrix in the circular polarization representation in Eq. (3.4.42) can be simplified to

$$\mathrm{C}_{++}(\Theta) = \mathrm{C}_{--}(\Theta) = \sum_{n=1}^{\infty} \frac{2n+1}{2} d_{11}^{n}(\Theta)(a_n + b_n), \tag{3.4.78a}$$

$$\mathrm{C}_{+-}(\Theta) = \mathrm{C}_{-+}(\Theta) = \sum_{n=1}^{\infty} \frac{2n+1}{2} d_{-11}^{n}(\Theta)(a_n - b_n). \tag{3.4.78b}$$

The T-matrix of a particle with spherical symmetry is diagonal, so the scattering phase matrix in the linear polarization representation is reduced to

$$\mathbf{P} = \begin{pmatrix} \mathrm{P}_{11} & \mathrm{P}_{12} & 0 & 0 \\ \mathrm{P}_{12} & \mathrm{P}_{11} & 0 & 0 \\ 0 & 0 & \mathrm{P}_{44} & \mathrm{P}_{34} \\ 0 & 0 & -\mathrm{P}_{34} & \mathrm{P}_{44} \end{pmatrix}. \tag{3.4.79}$$

Consequently, the scattering phase matrix in the circular polarization representation is simplified to

$$\mathbf{P}^{\mathrm{CP}} = \frac{1}{2} \begin{pmatrix} \mathrm{P}_{11} + \mathrm{P}_{44} & \mathrm{P}_{12} + i\mathrm{P}_{34} & \mathrm{P}_{12} - i\mathrm{P}_{34} & \mathrm{P}_{11} - \mathrm{P}_{44} \\ \mathrm{P}_{12} + i\mathrm{P}_{34} & \mathrm{P}_{11} + \mathrm{P}_{44} & \mathrm{P}_{11} - \mathrm{P}_{44} & \mathrm{P}_{12} - i\mathrm{P}_{34} \\ \mathrm{P}_{12} - i\mathrm{P}_{34} & \mathrm{P}_{11} - \mathrm{P}_{44} & \mathrm{P}_{11} + \mathrm{P}_{44} & \mathrm{P}_{12} + i\mathrm{P}_{34} \\ \mathrm{P}_{11} - \mathrm{P}_{44} & \mathrm{P}_{12} - i\mathrm{P}_{34} & \mathrm{P}_{12} + i\mathrm{P}_{34} & \mathrm{P}_{11} + \mathrm{P}_{44} \end{pmatrix}. \tag{3.4.80}$$

Then, the scattering phase matrix can be organized as

$$\mathrm{P}_{pq}^{\mathrm{CP}}(\Theta) = \sum_{s=0}^{\infty} g_{pq}^{s} d_{pq}^{s}(\Theta), \quad p,q = 2, 0, -0, -2, \tag{3.4.81}$$

where only the expansion coefficients $g_{00}^{s}$, $g_{0-0}^{s}$, and $g_{02}^{s}$ are independent; $g_{00}^{s}$ and $g_{0-0}{}^{s}$ are real; and $g_{02}^{s}$ is usually complex. The independent expansion coefficients are

$$g_{00}^{s} = \sum_{n=1}^{\infty} h_{sn} \sum_{n'=\max(1,|n-s|)}^{n+s} \left(C_{n1s0}^{n'1}\right)^2 (a_n + b_n)\left(a_{n'}^{*} + b_{n'}^{*}\right), \tag{3.4.82a}$$

$$g_{0-0}^{s} = \sum_{n=1}^{\infty} h_{sn} \sum_{n'=\max(1,|n-s|)}^{n+s} (-1)^{n+s+n'} \left(C_{n1s0}^{n'1}\right)^2 (a_n - b_n)\left(a_{n'}^{*} - b_{n'}^{*}\right), \tag{3.4.82b}$$

$$g_{02}^{s} = \sum_{n=1}^{\infty} h_{sn} \sum_{n'=\max(1,|n-s|)}^{n+s} C_{n1s0}^{n'1} C_{n-1s2}^{n'1} (a_n - b_n)\left(a_{n'}^{*} + b_{n'}^{*}\right), \tag{3.4.82c}$$

where

$$h_{sn} = \frac{\pi(2s+1)(2n+1)}{k^2 C_{\text{sca}}}. \tag{3.4.82d}$$

Since expansion coefficients $g_{00}^{s}$ and $g_{0-0}^{s}$ are real, Eqs. (3.4.82a)–(3.4.82b) can be simplified to

$$g_{00}^{s} = \sum_{n=1}^{\infty} h_{sn} \sum_{n'=\max(1,|n-s|)}^{n+s} \left(C_{n1s0}^{n'1}\right)^2 Re\left[(a_n + b_n)\left(a_{n'}^{*} + b_{n'}^{*}\right)\right], \tag{3.4.83a}$$

$$g_{0-0}^{s} = \sum_{n=1}^{\infty} h_{sn} \sum_{n'=\max(1,|n-s|)}^{n+s} (-1)^{n+s+n'} \left(C_{n1s0}^{n'1}\right)^2 Re\left[(a_n - b_n)\left(a_{n'}^{*} - b_{n'}^{*}\right)\right]. \tag{3.4.83b}$$

The independent scattering phase matrix elements are organized as follows:

$$P_{11}(\Theta) = \sum_{s=0}^{\infty} \alpha_1^s d_{00}^s(\Theta), \tag{3.4.84a}$$

$$P_{44}(\Theta) = \sum_{s=0}^{\infty} \alpha_4^s d_{00}^s(\Theta), \tag{3.4.84b}$$

$$P_{12}(\Theta) = \sum_{s=2}^{\infty} \beta_1^s d_{02}^s(\Theta), \tag{3.4.84c}$$

$$P_{34}(\Theta) = \sum_{s=2}^{\infty} \beta_2^s d_{02}^s(\Theta), \tag{3.4.84d}$$

where

$$\alpha_1^s = g_{00}^s + g_{0-0}^s, \tag{3.4.85a}$$

$$\alpha_4^s = g_{00}^s - g_{0-0}^s, \tag{3.4.85b}$$

$$\beta_1^s = 2Re\left[g_{02}^s\right], \tag{3.4.85c}$$

$$\beta_2^s = 2Im\left[g_{02}^s\right]. \tag{3.4.85d}$$

The coefficients $\alpha_1^s$ and $\alpha_4^s$ can be organized as follows:

$$\alpha_1^s = \sum_{n=1}^{\infty} h_{sn} \sum_{n'=\max(1,|n-s|)}^{n+s} \left\{ \left[1 + (-1)^{n+s+n'}\right] \left(C_{n1s0}^{n'1}\right)^2 Re\left[a_n a_{n'}^{*} + b_n b_{n'}^{*}\right] + \left[1 - (-1)^{n+s+n'}\right] \left(C_{n1s0}^{n'1}\right)^2 Re\left[a_n b_{n'}^{*} + b_n a_{n'}^{*}\right] \right\}, \tag{3.4.86}$$

$$\alpha_4^s = \sum_{n=1}^{\infty} h_{sn} \sum_{n'=\max(1,|n-s|)}^{n+s} \left\{ \left[1-(-1)^{n+s+n'}\right] \left(C_{n1s0}^{n'1}\right)^2 Re\left[a_n a_{n'}^* + b_n b_{n'}^*\right] + \left[1+(-1)^{n+s+n'}\right] \left(C_{n1s0}^{n'1}\right)^2 Re\left[a_n b_{n'}^* + b_n a_{n'}^*\right] \right\}. \tag{3.4.87}$$

For $s=1$, the CG coefficients are

$$C_{n110}^{n-11} = -\sqrt{\frac{(n+1)(n-1)}{n(2n+1)}},\ C_{n110}^{n1} = \sqrt{\frac{1}{n(n+1)}},\ C_{n110}^{n+11} = \sqrt{\frac{n(n+2)}{(n+1)(2n+1)}}. \tag{3.4.88}$$

Substituting Eq. (3.4.88) into Eq. (3.4.87), after some algebraic steps, the asymmetry factor can be given by (van de Hulst, 1957; Mishchenko et al., 2002)

$$\begin{aligned} g &= \frac{\alpha_1^1}{3} \\ &= \frac{4\pi}{k^2 C_{\text{sca}}} \sum_{n=1}^{\infty} \left\{ \frac{n(n+2)}{n+1} Re\left[a_n a_{n+1}^* + b_n b_{n+1}^*\right] + \frac{2n+1}{n(n+1)} Re\left[a_n b_n^*\right] \right\}. \end{aligned} \tag{3.4.89}$$

### 3.4.5 Partially random orientations

Euler angle $\alpha$ denotes the azimuthal rotation of the laboratory frame of reference, and Euler angle $\gamma$ denotes the azimuthal rotation of the particle frame of reference. Particles in the atmosphere or ocean in some situations have preferred orientations. For instance, a hexagonal plate in the atmosphere with thickness much smaller than the diameter has a tendency to be in horizontal orientation (e.g., Ono, 1969). In these situations, Euler angle $\gamma$ is in random orientation, but the angle $\beta$ is approximately fixed due to the earlier tendency. The average $\alpha$ is equivalent to the azimuthal angle average over the scattering phase matrix in a fixed orientation and usually is frequently encountered in light scattering. The partially random orientations with respect to Euler angle $\alpha$ or to Euler angles $\alpha$ and $\gamma$ are discussed in this section.

The extinction and scattering cross sections in partially random orientations with respect to Euler angle $\alpha$ can be given by

$$\langle C_{\text{ext}}\rangle(\beta,\gamma) = \frac{1}{8\pi^2}\int_0^{2\pi} d\alpha \left\{ \frac{-1}{k^2|\mathbf{E}_0|^2} \sum_{n=1}^{\infty}\sum_{m=-n}^{n} Re\left[a_{mn}p_{mn}^* + b_{mn}q_{mn}^*\right] \right\}, \tag{3.4.90a}$$

$$\langle C_{\text{sca}}\rangle(\beta,\gamma) = \frac{1}{2\pi}\int_0^{2\pi} d\alpha \left\{ \frac{1}{k^2|\mathbf{E}_0|^2} \sum_{n=1}^{\infty}\sum_{m=-n}^{n} \left[|p_{mn}|^2 + |q_{mn}|^2\right] \right\}, \tag{3.4.90b}$$

and the extinction and scattering cross sections in partially random orientations with respect to Euler angles $\alpha$ and $\gamma$ are given by

$$\langle C_{\text{ext}}\rangle(\beta) = \frac{1}{8\pi^2}\int_0^{2\pi} d\alpha \int_0^{2\pi} d\gamma \left\{ \frac{-1}{k^2|\mathbf{E}_0|^2} \sum_{n=1}^{\infty}\sum_{m=-n}^{n} Re\left[a_{mn}p_{mn}^* + b_{mn}q_{mn}^*\right] \right\}, \tag{3.4.91a}$$

$$\langle C_{\text{sca}}\rangle(\beta) = \frac{1}{4\pi^2}\int_0^{2\pi} d\alpha \int_0^{2\pi} d\gamma \left\{ \frac{1}{k^2|\mathbf{E}_0|^2} \sum_{n=1}^{\infty}\sum_{m=-n}^{n} \left[|p_{mn}|^2 + |q_{mn}|^2\right] \right\}. \tag{3.4.91b}$$

Using the same steps as described in Section 3.4.3, the extinction cross section in partially random orientations can be given by

$$\begin{aligned}
\langle C_{\text{ext}}\rangle(\beta,\gamma) = -\frac{2\pi}{k^2} Re \Bigg\{ & \sum_{n=1}^{\infty}\sum_{n'=1}^{\infty} \sqrt{(2n+1)(2n'+1)}\, i^{n-n'} \sum_{m_1=-n}^{n}\sum_{m_2=-n'}^{n'} \exp\left[i(m_1-m_2)\gamma\right] \\
& \cdot \Bigg[ \frac{d_{1m_1}^n(\beta)d_{1m_2}^{n'}(\beta) + d_{1m_1}^n(\beta)d_{1m_2}^{n'}(\beta)}{2} \left(\mathrm{T}_{m_1nm_2n'}^{11} + \mathrm{T}_{m_1nm_2n'}^{22}\right)^* \\
& + \frac{d_{1m_1}^n(\beta)d_{1m_2}^{n'}(\beta) - d_{1m_1}^n(\beta)d_{1m_2}^{n'}(\beta)}{2} \left(\mathrm{T}_{m_1nm_2n'}^{12} + \mathrm{T}_{m_1nm_2n'}^{21}\right)^* \Bigg] \Bigg\},
\end{aligned} \tag{3.4.92}$$

$$\begin{aligned}
\langle C_{\text{ext}}\rangle(\beta) = -\frac{2\pi}{k^2} Re \Bigg\{ & \sum_{n=1}^{\infty}\sum_{n'=1}^{\infty} \sqrt{(2n+1)(2n'+1)}\, i^{n-n'} \\
& \cdot \sum_{m=-\min(n,n')}^{\min(n,n')} \Bigg[ \frac{d_{1m}^n(\beta)d_{1m}^{n'}(\beta) + d_{-1m}^n(\beta)d_{-1m}^{n'}(\beta)}{2} \left(\mathrm{T}_{mnmn'}^{11} + \mathrm{T}_{mnmn'}^{22}\right)^* \\
& + \frac{d_{1m}^n(\beta)d_{1m}^{n'}(\beta) - d_{-1m}^n(\beta)d_{-1m}^{n'}(\beta)}{2} \left(\mathrm{T}_{mnmn'}^{12} + \mathrm{T}_{mnmn'}^{21}\right)^* \Bigg] \Bigg\}.
\end{aligned} \tag{3.4.93}$$

The scattering cross section in partially random orientations can also be straightforwardly obtained using the same steps. However, it is not given here because the expressions are tedious.

Partially random orientation with respect to Euler angle $\alpha$ is implemented by

$$\mathbf{P}^{\text{CP}}(\Theta;\beta,\gamma) = \frac{4\pi}{k^2\langle C_{\text{sca}}\rangle}\frac{1}{2\pi}\int_0^{2\pi} d\alpha \mathbb{F}^{\text{CP}}(\Theta;\alpha,\beta,\gamma), \tag{3.4.94}$$

and partially random orientation with respect to Euler angles $\alpha$ and $\gamma$ is implemented by

$$\mathbf{P}^{\text{CP}}(\Theta;\beta) = \frac{4\pi}{k^2\langle C_{\text{sca}}\rangle}\frac{1}{4\pi^2}\int_0^{2\pi} d\alpha \int_0^{2\pi} d\gamma \mathbb{F}^{\text{CP}}(\Theta;\alpha,\beta,\gamma). \tag{3.4.95}$$

The scattering phase matrix in the circular polarization representation is given in Eq. (3.4.50). The scattering phase matrix in partially random orientation usually has 16 independent elements as follows:

$$\mathbf{P} = \begin{pmatrix} P_{11} & P_{12} & P_{13} & P_{14} \\ P_{21} & P_{22} & P_{23} & P_{24} \\ P_{31} & P_{32} & P_{33} & P_{34} \\ P_{41} & P_{42} & P_{43} & P_{44} \end{pmatrix}. \tag{3.4.96}$$

The corresponding scattering phase matrix in the circular polarization representation is

$$\mathbf{P}^{CP} = \frac{1}{2}\begin{pmatrix} (P_{22}+P_{33})-(P_{23}-P_{32})i & (P_{21}+P_{24})+(P_{34}+P_{31})i & (P_{21}-P_{24})-(P_{34}-P_{31})i & (P_{22}-P_{33})+(P_{23}+P_{32})i \\ (P_{12}+P_{42})-(P_{13}+P_{43})i & P_{11}+P_{44}+P_{14}+P_{41} & P_{11}-P_{44}-P_{14}+P_{41} & (P_{12}+P_{42})+(P_{13}+P_{43})i \\ (P_{12}-P_{42})-(P_{13}-P_{43})i & P_{11}-P_{44}+P_{14}-P_{41} & P_{11}+P_{44}-P_{14}-P_{41} & (P_{12}-P_{42})+(P_{13}-P_{43})i \\ (P_{22}-P_{33})-(P_{23}+P_{32})i & (P_{21}+P_{24})-(P_{34}+P_{31})i & (P_{21}-P_{24})+(P_{34}-P_{31})i & (P_{22}+P_{33})+(P_{23}-P_{32})i \end{pmatrix}. \tag{3.4.97}$$

The following symmetry relations can be obtained from Eq. (3.4.97):

$$P_{22}^{CP} = \left(P_{-2-2}^{CP}\right)^*, \; P_{2-2}^{CP} = \left(P_{-22}^{CP}\right)^*, \; P_{20}^{CP} = \left(P_{-20}^{CP}\right)^*, \tag{3.4.98a}$$

$$P_{2-0}^{CP} = \left(P_{-2-0}^{CP}\right)^*, \; P_{02}^{CP} = \left(P_{0-2}^{CP}\right)^*, \; P_{-02}^{CP} = \left(P_{-0-2}^{CP}\right)^*, \tag{3.4.98b}$$

where elements $P_{00}^{CP}$, $P_{0-0}^{CP}$, $P_{-00}^{CP}$, and $P_{-0-0}^{CP}$ are real and the others are usually complex. The scattering phase matrix in the circular polarization representation can be given in the same way as Eq. (3.4.54):

$$P_{pq}^{CP}(\Theta) = \sum_{s=0}^{\infty} g_{pq}^s d_{pq}^s(\Theta), \;\; p,q = 2,0,-0,-2, \tag{3.4.99}$$

where the independent expansion coefficients are $g_{00}^s$, $g_{0-0}^s$, $g_{-00}^s$, $g_{-0-0}^s$, $g_{22}^s$, $g_{02}^s$, $g_{-02}^s$, $g_{-22}^s$, $g_{-20}^s$, and $g_{-2-0}^s$. Also, $g_{00}^s$, $g_{0-0}^s$, $g_{-00}^s$, and $g_{-0-0}^s$ are real and the other quantities are usually complex. The expansion coefficients have the following expressions:

$$g_{00}^s = \sum_{n=1}^{\infty} \sum_{j=\max(1,|n-s|)}^{n+s} h_{snj}^+ C_{n1s0}^{j1} \sum_{m=-\min(n,j)}^{\min(n,j)} C_{nms0}^{jm} D_{mnj}^{00}, \tag{3.4.100a}$$

$$g_{0-0}^s = \sum_{n=1}^{\infty} \sum_{j=\max(1,|n-s|)}^{n+s} h_{snj}^- C_{n1s0}^{j1} \sum_{m=-\min(n,j)}^{\min(n,j)} C_{nms0}^{jm} D_{mnj}^{0-0}, \tag{3.4.100b}$$

$$g^s_{-00}=\sum_{n=1}^{\infty}\sum_{j=\max(1,|n-s|)}^{n+s}h^-_{snj}C^{j1}_{n1s0}\sum_{m=-\min(n,j)}^{\min(n,j)}C^{jm}_{nms0}D^{-00}_{mnj}, \tag{3.4.100c}$$

$$g^s_{-0-0}=\sum_{n=1}^{\infty}\sum_{j=\max(1,|n-s|)}^{n+s}h^+_{snj}C^{j1}_{n1s0}\sum_{m=-\min(n,j)}^{\min(n,j)}C^{jm}_{nms0}D^{-0-0}_{mnj}, \tag{3.4.100d}$$

$$g^s_{-20}=\sum_{n=1}^{\infty}\sum_{j=\max(1,|n-s|)}^{n+s}h^-_{snj}C^{j1}_{n-1s2}\sum_{m=-\min(n,j)}^{\min(n,j)}C^{jm}_{nms0}D^{-20}_{mnj}, \tag{3.4.100e}$$

$$g^s_{-2-0}=\sum_{n=1}^{\infty}\sum_{j=\max(1,|n-s|)}^{n+s}h^+_{snj}C^{j1}_{n-1s2}\sum_{m=-\min(n,j)}^{\min(n,j)}C^{jm}_{nms0}D^{-2-0}_{mnj}, \tag{3.4.100f}$$

$$g^s_{22}=\sum_{n=1}^{\infty}\sum_{j=\max(1,|n-s|)}^{n+s}h^+_{snj}C^{j1}_{n-1s2}\sum_{m=-\min(n,j-2)}^{\min(n,j+2)}C^{j2-m}_{n-ms2}D^{22}_{mnj}, \tag{3.4.100g}$$

$$g^s_{-22}=\sum_{n=1}^{\infty}\sum_{j=\max(1,|n-s|)}^{n+s}h^-_{snj}C^{j1}_{n-1s2}\sum_{m=-\min(n,j-2)}^{\min(n,j+2)}C^{j2-m}_{n-ms2}D^{-22}_{mnj}, \tag{3.4.100h}$$

$$g^s_{02}=\sum_{n=1}^{\infty}\sum_{j=\max(1,|n-s|)}^{n+s}h^+_{snj}C^{j1}_{n1s0}\sum_{m=-\min(n,j-2)}^{\min(n,j+2)}C^{j2-m}_{n-ms2}D^{02}_{mnj}, \tag{3.4.100i}$$

$$g^s_{-02}=\sum_{n=1}^{\infty}\sum_{j=\max(1,|n-s|)}^{n+s}h^-_{snj}C^{j1}_{n1s0}\sum_{m=-\min(n,j-2)}^{\min(n,j+2)}C^{j2-m}_{n-ms2}D^{-02}_{mnj}, \tag{3.4.100j}$$

where the coefficients $D$ for partially random orientation with respect to Euler angle $\alpha$, as given in Eq. (3.4.94), are

$$D^{00}_{mnj}=\sum_{n_1=|m-1|}^{\infty}\sum_{j_1=|m-1|}^{\infty}(2n_1+1)(2j_1+1)\sum_{\kappa=-n_1}^{n_1}\sum_{\kappa_1=-j_1}^{j_1}\Big[d^{n_1}_{-m+1,\kappa}(\beta)d^{j_1}_{-m+1,\kappa_1}(\beta)\cdot\exp[-i(\kappa-\kappa_1)\gamma](-1)^{\kappa+\kappa_1}B^3_{\kappa mnn_1}\left(B^3_{\kappa_1 mjj_1}\right)^*\Big], \tag{3.4.101a}$$

$$D^{0-0}_{mnj}=\sum_{n_1=|m-1|}^{\infty}\sum_{j_1=|m-1|}^{\infty}(2n_1+1)(2j_1+1)\sum_{\kappa=-n_1}^{n_1}\sum_{\kappa_1=-j_1}^{j_1}\Big[d^{n_1}_{m-1,-\kappa}(\beta)d^{j_1}_{m-1,-\kappa_1}(\beta)\cdot\exp[i(\kappa-\kappa_1)\gamma](-1)^{\kappa+\kappa_1}B^2_{\kappa mnn_1}\left(B^2_{\kappa_1 mjj_1}\right)^*\Big], \tag{3.4.101b}$$

$$D_{mnj}^{-00} = \sum_{n_1=|m-1|}^{\infty}\sum_{j_1=|m-1|}^{\infty}(2n_1+1)(2j_1+1)\sum_{\kappa=-n_1}^{n_1}\sum_{\kappa_1=-j_1}^{j_1}\Big[d_{-m+1,\kappa}^{n_1}(\beta)d_{-m+1,\kappa_1}^{j_1}(\beta)$$
$$\cdot\exp[-i(\kappa-\kappa_1)\gamma](-1)^{\kappa+\kappa_1}B_{\kappa m n n_1}^{4}\left(B_{\kappa_1 m j j_1}^{4}\right)^{*}\Big],$$
(3.4.101c)

$$D_{mnj}^{-0-0} = \sum_{n_1=|m-1|}^{\infty}\sum_{j_1=|m-1|}^{\infty}(2n_1+1)(2j_1+1)\sum_{\kappa=-n_1}^{n_1}\sum_{\kappa_1=-j_1}^{j_1}\Big[d_{m-1,-\kappa}^{n_1}(\beta)d_{m-1,-\kappa_1}^{j_1}(\beta)$$
$$\cdot\exp[i(\kappa-\kappa_1)\gamma](-1)^{\kappa+\kappa_1}B_{\kappa m n n_1}^{1}\left(B_{\kappa_1 m j j_1}^{1}\right)^{*}\Big],$$
(3.4.101d)

$$D_{mnj}^{-20} = \sum_{n_1=|m-1|}^{\infty}\sum_{j_1=|m-1|}^{\infty}(2n_1+1)(2j_1+1)\sum_{\kappa=-n_1}^{n_1}\sum_{\kappa_1=-j_1}^{j_1}\Big[d_{-m+1,\kappa}^{n_1}(\beta)d_{-m+1,\kappa_1}^{j_1}(\beta)$$
$$\cdot\exp[-i(\kappa-\kappa_1)\gamma](-1)^{\kappa+\kappa_1}B_{\kappa m n n_1}^{3}\left(B_{\kappa_1 m j j_1}^{4}\right)^{*}\Big],$$
(3.4.101e)

$$D_{mnj}^{-2-0} = \sum_{n_1=|m-1|}^{\infty}\sum_{j_1=|m-1|}^{\infty}(2n_1+1)(2j_1+1)\sum_{\kappa=-n_1}^{n_1}\sum_{\kappa_1=-j_1}^{j_1}\Big[d_{m-1,-\kappa}^{n_1}(\beta)d_{m-1,-\kappa_1}^{j_1}(\beta)$$
$$\cdot\exp[i(\kappa-\kappa_1)\gamma](-1)^{\kappa+\kappa_1}B_{\kappa m n n_1}^{2}\left(B_{\kappa_1 m j j_1}^{1}\right)^{*}\Big],$$
(3.4.101f)

$$D_{mnj}^{22} = \sum_{n_1=|m-1|}^{\infty}\sum_{j_1=|m-1|}^{\infty}(2n_1+1)(2j_1+1)\sum_{\kappa=-n_1}^{n_1}\sum_{\kappa_1=-j_1}^{j_1}\Big[d_{m-1,-\kappa}^{n_1}(\beta)d_{m-1,\kappa_1}^{j_1}(\beta)$$
$$\cdot\exp[i(\kappa+\kappa_1)\gamma](-1)^{\kappa+\kappa_1}B_{\kappa m n n_1}^{1}\left(B_{\kappa_1 2-m j j_1}^{3}\right)^{*}\Big],$$
(3.4.101g)

$$D_{mnj}^{-22} = \sum_{n_1=|m-1|}^{\infty}\sum_{j_1=|m-1|}^{\infty}(2n_1+1)(2j_1+1)\sum_{\kappa=-n_1}^{n_1}\sum_{\kappa_1=-j_1}^{j_1}\Big[d_{m-1,-\kappa}^{n_1}(\beta)d_{m-1,\kappa_1}^{j_1}(\beta)$$
$$\cdot\exp[i(\kappa+\kappa_1)\gamma](-1)^{\kappa+\kappa_1}B_{\kappa m n n_1}^{2}\left(B_{\kappa_1 2-m j j_1}^{4}\right)^{*}\Big],$$
(3.4.101h)

$$D_{mnj}^{02} = \sum_{n_1=|m-1|}^{\infty}\sum_{j_1=|m-1|}^{\infty}(2n_1+1)(2j_1+1)\sum_{\kappa=-n_1}^{n_1}\sum_{\kappa_1=-j_1}^{j_1}\Big[d_{m-1,-\kappa}^{n_1}(\beta)d_{m-1,\kappa_1}^{j_1}(\beta)$$
$$\cdot\exp[i(\kappa+\kappa_1)\gamma](-1)^{\kappa+\kappa_1}B_{\kappa m n n_1}^{2}\left(B_{\kappa_1 2-m j j_1}^{3}\right)^{*}\Big],$$
(3.4.101i)

$$D_{mnj}^{-02} = \sum_{n_1=|m-1|}^{\infty} \sum_{j_1=|m-1|}^{\infty} (2n_1+1)(2j_1+1) \sum_{\kappa=-n_1}^{n_1} \sum_{\kappa_1=-j_1}^{j_1} \left[ d_{m-1,-\kappa}^{n_1}(\beta) d_{m-1,\kappa_1}^{j_1}(\beta) \cdot \exp[i(\kappa+\kappa_1)\gamma](-1)^{\kappa+\kappa_1} B_{\kappa mnn_1}^{1} \left( B_{\kappa_1 2-mjj_1}^{4} \right)^{*} \right], \tag{3.4.101j}$$

and the coefficients $D$ for partially random orientation with respect to Euler angles $\alpha$ and $\gamma$, given in Eq. (3.4.95), are

$$D_{mnj}^{00} = \sum_{n_1=|m-1|}^{\infty} \sum_{j_1=|m-1|}^{\infty} (2n_1+1)(2j_1+1) \sum_{\kappa=-n_1}^{n_1} \left[ d_{-m+1,\kappa}^{n_1}(\beta) d_{-m+1,\kappa}^{j_1}(\beta) B_{\kappa mnn_1}^{3} \left( B_{\kappa mjj_1}^{3} \right)^{*} \right], \tag{3.4.102a}$$

$$D_{mnj}^{0-0} = \sum_{n_1=|m-1|}^{\infty} \sum_{j_1=|m-1|}^{\infty} (2n_1+1)(2j_1+1) \sum_{\kappa=-n_1}^{n_1} \left[ d_{m-1,-\kappa}^{n_1}(\beta) d_{m-1,-\kappa}^{j_1}(\beta)\, B_{\kappa mnn_1}^{2} \left( B_{\kappa mjj_1}^{2} \right)^{*} \right], \tag{3.4.102b}$$

$$D_{mnj}^{-00} = \sum_{n_1=|m-1|}^{\infty} \sum_{j_1=|m-1|}^{\infty} (2n_1+1)(2j_1+1) \sum_{\kappa=-n_1}^{n_1} \left[ d_{-m+1,\kappa}^{n_1}(\beta) d_{-m+1,\kappa}^{j_1}(\beta) B_{\kappa mnn_1}^{4} \left( B_{\kappa mjj_1}^{4} \right)^{*} \right], \tag{3.4.102c}$$

$$D_{mnj}^{-0-0} = \sum_{n_1=|m-1|}^{\infty} \sum_{j_1=|m-1|}^{\infty} (2n_1+1)(2j_1+1) \sum_{\kappa=-n_1}^{n_1} \left[ d_{m-1,-\kappa}^{n_1}(\beta) d_{m-1,-\kappa}^{j_1}(\beta) B_{\kappa mnn_1}^{1} \left( B_{\kappa mjj_1}^{1} \right)^{*} \right], \tag{3.4.102d}$$

$$D_{mnj}^{-20} = \sum_{n_1=|m-1|}^{\infty} \sum_{j_1=|m-1|}^{\infty} (2n_1+1)(2j_1+1) \sum_{\kappa=-n_1}^{n_1} \left[ d_{-m+1,\kappa}^{n_1}(\beta) d_{-m+1,\kappa}^{j_1}(\beta) B_{\kappa mnn_1}^{3} \left( B_{\kappa mjj_1}^{4} \right)^{*} \right], \tag{3.4.102e}$$

$$D_{mnj}^{-2-0} = \sum_{n_1=|m-1|}^{\infty} \sum_{j_1=|m-1|}^{\infty} (2n_1+1)(2j_1+1) \sum_{\kappa=-n_1}^{n_1} \left[ d_{m-1,-\kappa}^{n_1}(\beta) d_{m-1,-\kappa}^{j_1}(\beta) B_{\kappa mnn_1}^{2} \left( B_{\kappa mjj_1}^{1} \right)^{*} \right], \tag{3.4.102f}$$

$$D_{mnj}^{22} = \sum_{n_1=|m-1|}^{\infty} \sum_{j_1=|m-1|}^{\infty} (2n_1+1)(2j_1+1) \sum_{\kappa=-n_1}^{n_1} \left[ d_{m-1,-\kappa}^{n_1}(\beta) d_{m-1,-\kappa}^{j_1}(\beta) B_{\kappa mnn_1}^{1} \left( B_{-\kappa 2-mjj_1}^{3} \right)^{*} \right], \tag{3.4.102g}$$

$$D_{mnj}^{-22} = \sum_{n_1=|m-1|}^{\infty} \sum_{j_1=|m-1|}^{\infty} (2n_1+1)(2j_1+1) \sum_{\kappa=-n_1}^{n_1} \left[ d_{m-1,-\kappa}^{n_1}(\beta) d_{m-1,-\kappa}^{j_1}(\beta) B_{\kappa mnn_1}^{2} \left( B_{-\kappa 2-mjj_1}^{4} \right)^{*} \right], \tag{3.4.102h}$$

$$D_{mnj}^{02} = \sum_{n_1=|m-1|}^{\infty} \sum_{j_1=|m-1|}^{\infty} (2n_1+1)(2j_1+1) \sum_{\kappa=-n_1}^{n_1} \left[ d_{m-1,-\kappa}^{n_1}(\beta) d_{m-1,-\kappa}^{j_1}(\beta) B_{\kappa mnn_1}^{2} \left( B_{-\kappa 2-mjj_1}^{3} \right)^* \right], \tag{3.4.102i}$$

$$D_{mnj}^{-02} = \sum_{n_1=|m-1|}^{\infty} \sum_{j_1=|m-1|}^{\infty} (2n_1+1)(2j_1+1) \sum_{\kappa=-n_1}^{n_1} \left[ d_{m-1,-\kappa}^{n_1}(\beta) d_{m-1,-\kappa}^{j_1}(\beta) B_{\kappa mnn_1}^{1} \left( B_{-\kappa 2-mjj_1}^{4} \right)^* \right]. \tag{3.4.102j}$$

Coefficients $B$ are given in Eq. (3.4.48b).

The scattering phase matrix elements in the linear representation can be expanded using the Wigner-d function as follows:

$$\mathrm{P}_{11}(\Theta) = \sum_{s=0}^{\infty} \alpha_1^s d_{00}^s(\Theta), \tag{3.4.103a}$$

$$\mathrm{P}_{22}(\Theta) + \mathrm{P}_{33}(\Theta) = \sum_{s=2}^{\infty} \left(\alpha_2^s + \alpha_3^s\right) d_{22}^s(\Theta), \tag{3.4.103b}$$

$$\mathrm{P}_{22}(\Theta) - \mathrm{P}_{33}(\Theta) = \sum_{s=2}^{\infty} \left(\alpha_2^s - \alpha_3^s\right) d_{2-2}^s(\Theta), \tag{3.4.103c}$$

$$\mathrm{P}_{44}(\Theta) = \sum_{s=0}^{\infty} \alpha_4^s d_{00}^s(\Theta), \tag{3.4.103d}$$

$$\mathrm{P}_{12}(\Theta) = \sum_{s=2}^{\infty} \beta_1^s d_{02}^s(\Theta), \tag{3.4.103e}$$

$$\mathrm{P}_{21}(\Theta) = \sum_{s=2}^{\infty} \gamma_1^s d_{02}^s(\Theta), \tag{3.4.103f}$$

$$\mathrm{P}_{34}(\Theta) = \sum_{s=2}^{\infty} \beta_2^s d_{02}^s(\Theta), \tag{3.4.103g}$$

$$\mathrm{P}_{43}(\Theta) = \sum_{s=2}^{\infty} \gamma_2^s d_{02}^s(\Theta), \tag{3.4.103h}$$

$$\mathrm{P}_{13}(\Theta) = \sum_{s=2}^{\infty} \beta_3^s d_{02}^s(\Theta), \tag{3.4.103i}$$

$$\mathrm{P}_{31}(\Theta) = \sum_{s=2}^{\infty} \gamma_3^s d_{02}^s(\Theta), \tag{3.4.103j}$$

$$P_{23}(\Theta) - P_{32}(\Theta) = \sum_{s=2}^{\infty} (\beta_4^s - \gamma_4^s) d_{22}^s(\Theta), \tag{3.4.103k}$$

$$P_{23}(\Theta) + P_{32}(\Theta) = \sum_{s=2}^{\infty} (\beta_4^s + \gamma_4^s) d_{2-2}^s(\Theta), \tag{3.4.103l}$$

$$P_{14}(\Theta) = \sum_{s=0}^{\infty} \beta_5^s d_{00}^s(\Theta), \tag{3.4.103m}$$

$$P_{41}(\Theta) = \sum_{s=0}^{\infty} \gamma_5^s d_{00}^s(\Theta), \tag{3.4.103n}$$

$$P_{24}(\Theta) = \sum_{s=2}^{\infty} \beta_6^s d_{02}^s(\Theta), \tag{3.4.103o}$$

$$P_{42}(\Theta) = \sum_{s=2}^{\infty} \gamma_6^s d_{02}^s(\Theta), \tag{3.4.103p}$$

where $\alpha$, $\beta$, and $\gamma$ correspond to the expansion coefficients of the diagonal, the upper off-diagonal, and the lower off-diagonal elements of the scattering phase matrix and should be distinguished from the three rotational Euler angles, and

$$\alpha_1^s = \frac{1}{2}\left(g_{00}^s + g_{0-0}^s + g_{-00}^s + g_{-0-0}^s\right), \tag{3.4.104a}$$

$$\alpha_2^s = Re\left[g_{22}^s\right] + Re\left[g_{-22}^s\right], \tag{3.4.104b}$$

$$\alpha_3^s = Re\left[g_{22}^s\right] - Re\left[g_{-22}^s\right], \tag{3.4.104c}$$

$$\alpha_4^s = \frac{1}{2}\left(g_{00}^s - g_{0-0}^s - g_{-00}^s + g_{-0-0}^s\right), \tag{3.4.104d}$$

$$\beta_1^s = Re\left[g_{02}^s\right] + Re\left[g_{-02}^s\right], \tag{3.4.104e}$$

$$\gamma_1^s = Re\left[g_{-20}^s\right] + Re\left[g_{-2-0}^s\right], \tag{3.4.104f}$$

$$\beta_2^s = Im\left[g_{-2-0}^s\right] - Im\left[g_{-20}^s\right], \tag{3.4.104g}$$

$$\gamma_2^s = Im\left[g_{-02}^s\right] - Im\left[g_{02}^s\right], \tag{3.4.104h}$$

$$\beta_3^s = -Im\left[g_{02}^s\right] - Im\left[g_{-02}^s\right], \tag{3.4.104i}$$

$$\gamma_3^s = -Im\left[g_{-20}^s\right] - Im\left[g_{-2-0}^s\right], \tag{3.4.104j}$$

$$\beta_4^s = -Im[g_{22}^s] - Im[g_{-22}^s], \tag{3.4.104k}$$

$$\gamma_4^s = Im[g_{22}^s] - Im[g_{-22}^s], \tag{3.4.104l}$$

$$\beta_5^s = \frac{1}{2}(g_{00}^s - g_{0-0}^s + g_{-00}^s - g_{-0-0}^s), \tag{3.4.104m}$$

$$\gamma_5^s = \frac{1}{2}(g_{00}^s + g_{0-0}^s - g_{-00}^s - g_{-0-0}^s), \tag{3.4.104n}$$

$$\beta_6^s = Re[g_{-20}^s] - Re[g_{-2-0}^s], \tag{3.4.104o}$$

$$\gamma_6^s = Re[g_{02}^s] - Re[g_{-02}^s]. \tag{3.4.104p}$$

Even though the scattering phase matrix of a particle in partially random orientations can be analytically obtained, the computation is much more time-consuming and laborious than the computation with completely random orientations because extra summations are needed for the partially random orientation condition.

# Invariant-imbedding T-matrix method

# 4

Light scattering by spherical particles can be analytically solved using the Lorenz-Mie theory (van de Hulst, 1957; Bohren and Huffman, 1983). Moreover, the scattering by a spherical particle or a spherical shell has been extensively studied (e.g., Aden and Kerker, 1951; Kattawar and Hood, 1976; Wiscombe, 1980; Toon and Ackerman, 1981; Hovenac and Lock, 1992; Nussenzveig, 1992; Gouesbet and Grehan, 2011). The Lorenz-Mie theory can solve the scattering of light by a spherical particle for the entire practical size range. Importantly, the cpu time usage for the calculation in the case of a spherical particle is so minimal that it does not pose a concern given the present computational resources. For nonspherical particles, however, only scattering by small particles can be obtained using numerically accurate methods. Here, a numerically accurate method refers to a method that gives converged results with a sufficiently fine resolution. In Section 2.3, several frequently used accurate methods, such as FDTD and DDA, are introduced for the computation of the scattering of light by nonspherical particles. With an increasing particle size, not only do the cpu time usage and the computer memory demand significantly increase, but the convergence rates for these methods also substantially degrade. Moreover, the single-scattering properties of a particle calculated by any numerically accurate method depend on the incident direction with respect to the orientation of the scattering particle. Accordingly, the scattering properties of a particle under the random orientation condition can only be obtained by numerically averaging the results associated with different orientations. The number of orientations necessary for convergence drastically increases with increased particle size. Consequently, these numerically accurate methods are only efficient for light scattering by small particles.

The T-matrix method, however, is a semianalytical method expressed in terms of vector spherical wave functions. The T-matrix of a particle is also independent of the incident direction. Accordingly, the random orientation condition can be analytically accounted for once the T-matrix is given. The T-matrix concept is given in Chapter 3, including introduction of the vector spherical wave function and analytical computation for random orientations using the T-matrix. A commonly used method, the extended boundary condition method (EBCM), which uses a surface integral equation and matrix inversion, is introduced in Section 3.3.2. The EBCM has been extensively used to calculate light scattering by particles with axial symmetry, such as spheroids, cylinders, and Chebyshev particles (Mishchenko et al., 2002). Over a period of time, the EBCM was almost a synonym of the "T-matrix method." In principle, the EBCM is a general theory to calculate the T-matrix and can be applied to any symmetric or

**Invariant Imbedding T-matrix Method for Light Scattering by Nonspherical and Inhomogeneous Particles**
**https://doi.org/10.1016/B978-0-12-818090-7.00004-8**

asymmetric particle, but this method is usually applied to the computation for particles with axial symmetry (e.g., Mugnai and Wiscombe, 1986; Mishchenko and Travis, 1998; Havemann and Baran, 2001). One reason for this is that effort of computing the T-matrix of a particle with axial symmetry significantly reduces due to the T-matrix symmetry (Schulz et al., 1999; Kahnert et al., 2001). Another reason is related to numerical stability. The computation of the T-matrix includes a matrix inversion process, that is, $\mathbf{T} = -(\mathrm{Rg}\mathbf{Q})\mathbf{Q}^{-1}$. Matrices Rg**Q** and **Q** are both calculated from the surface integrals introduced in Section 3.3.2. For nonspherical particles with large size parameters or largely deformed geometries, such as those having extreme aspect ratios, the eigenvalues of matrix **Q** differ by many orders of magnitude, and the matrix **Q** becomes ill-conditioned due to the fact that the surface integrals introduce the spherical Hankel function of the first kind in different arguments into the matrix, and instability of the matrix inversion process becomes worse if the dimension of the matrix **Q** becomes larger.

Several efforts have been made to improve the numerical stability of the T-matrix computation. A perturbation approach was developed for the application of EBCM, which replaces the matrix inversion problem with an implicit T-matrix equation, and the implicit equation can be solved iteratively (Kahnert and Rother, 2011). However, the perturbation method only works for the particle geometry slightly deviating from a reference geometry. A straightforward improvement for the numerical stability results from using the extended precision variables to replace double precision variables in the computation and inversion of matrix **Q** (Mishchenko and Travis, 1994). Computations using extended precision may be roughly five to eight times slower than using double precision, but the applicable particle sizes can be two or three times larger if extended precision is used. Another method uses the unitary property of the T-matrix introduced in Section 3.4.1 and iteratively obtains stable results for strongly deformed geometries (e.g., Lakhtakia et al., 1984), where a large difference between the smallest and largest dimensions is needed to characterize the particle. However, this unitary equation is only applicable to particles without absorption. A method using discrete sources can also be implemented in the EBCM to improve numerical stability for prolate and oblate particles. One method is called the iterative EBCM (IEBCM) (Iskander et al., 1983, 1989), and another is the null field with discrete sources (e.g., Doicu et al., 2006). However, although the numerical stability is attained by such methods, the computational complexity and cpu time increase compared with the original EBCM. An alternative method is required not only to retain numerical stability but also to preserve numerical efficiency, so the method can be applied to particles with regular or extreme geometries and large size parameters.

The invariant-imbedding T-matrix method (IITM) is a volume integral method that uses the invariant-imbedding technique to iteratively obtain the T-matrix of an arbitrary particle. The volume integral equation of the electric field has the form of the Fredholm equation of the second kind, which is usually well conditioned (Press et al., 2007). The IITM shows numerical stability in matrix inversion and can be used to obtain the T-matrix of particles with large size parameters or extreme geometries. The invariant-imbedding technique was introduced by Ambarzumian (1943). One of the most famous applications of the invariant-imbedding technique is reported in *Radiative Transfer* by Chandrasekhar (1960). The general concept of invariant imbedding is discussed in detail in *An Introduction to Invariant Imbedding* by Bellman and Wing (1975). The concept of "invariant imbedding" was first applied in light scattering to obtain the T-matrix of an arbitrary particle by Johnson (1988). Unfortunately, this paper received little attention from the light scattering community and was cited only several times in the literature before 2013. Johnson's paper was fortunately revisited, and the IITM was implemented as a powerful light scattering computational program by Bi et al. (2013). The IITM shows stability not only for particles with axial symmetry but also for particles without axial symmetry (Bi and Yang, 2014; Yang et al., 2019). Moreover, it is straightforward to extend IITM from homogeneous to inhomogeneous particles and from particles with symmetry to particles without symmetry. The current computational range of the IITM has been applied to particle geometries with axial symmetry, finite-fold rotational symmetry (such as a hexagonal column with sixfold rotational symmetry), or even with no symmetry (Bi and Yang, 2014; Liu et al., 2015; Sun et al., 2017; Yang et al., 2019).

This section presents a comprehensive introduction to the IITM. Section 4.1 briefly derives the volume integral equation from Maxwell's equations and reorganizes the expansion of the electromagnetic fields using the vector spherical wave functions. In addition, the free space dyadic Green function is revisited using eigenfunction expansions and can easily be generalized to other sets of eigenfunction expansions, such as expansion using both the outgoing and incoming vector spherical wave functions used in the Debye series study. The invariant-imbedding technique is introduced in Section 4.2. Section 4.3 gives the concept of the IITM, and discussions about several aspects of the IITM are given in Section 4.4.

## 4.1 Electromagnetic volume integral equation

### 4.1.1 Volume integral equation

Here, only dielectric particles are considered, and the particles are assumed to be isotropic and have a linear response to an applied field. Maxwell's equations in such a medium for time-harmonic fields can be given as follows:

$$\nabla \times \mathbf{E}\left(\vec{r}\right) = i\omega\mu\mathbf{H}\left(\vec{r}\right), \tag{4.1.1a}$$

$$\nabla \times \mathbf{H}\left(\vec{r}\right) = -i\omega\varepsilon\mathbf{E}\left(\vec{r}\right), \tag{4.1.1b}$$

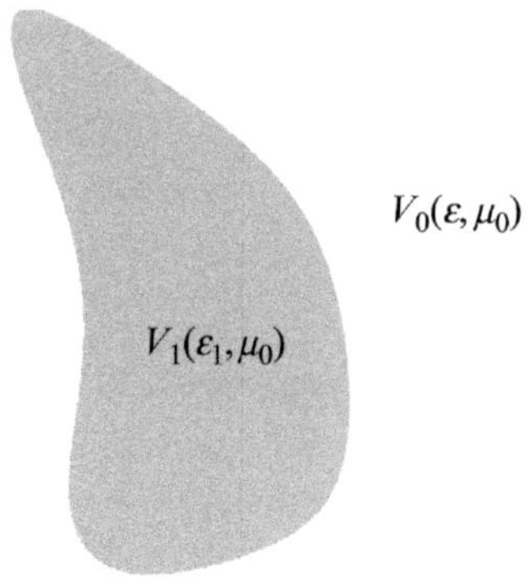


**Fig. 4.1** A dielectric scattering particle. The region occupied by the particle is denoted as $V_1$, while the exterior region is denoted as $V_0$.

where $\varepsilon$ and $\mu$ are the permittivity and the permeability in the medium, respectively. These are the general equations for the electromagnetic field in a medium. For light scattering of a particle in a surrounding medium as in Fig. 4.1, the permittivity of the particle is $\varepsilon_1\left(\vec{r}\right)$, which may vary inside the particle, and the permittivity of the surrounding medium is $\varepsilon$. Also, we assume that both the particle and the surrounding medium are nonferromagnetic and thus $\mu_1 = \mu = \mu_0$.

According to Maxwell's equations in Eq. (4.1.1), we have the following equations:

$$\nabla \times \nabla \times \mathbf{E}\left(\vec{r}\right) - k^2\mathbf{E}\left(\vec{r}\right) = 0, \quad \vec{r} \in \mathrm{V}_0, \tag{4.1.2a}$$

$$\nabla \times \nabla \times \mathbf{E}\left(\vec{r}\right) - k_1^2\mathbf{E}\left(\vec{r}\right) = 0, \quad \vec{r} \in \mathrm{V}_1, \tag{4.1.2b}$$

$$\mathbf{E}\left(\vec{r}\right) = \mathbf{E}^{\mathrm{inc}}\left(\vec{r}\right) + \mathbf{E}^{\mathrm{sca}}\left(\vec{r}\right), \tag{4.1.3}$$

where $k = \omega\sqrt{\varepsilon\mu_0}$ and $k_1 = \omega\sqrt{\varepsilon_1\left(\vec{r}\right)\mu_0}$; the space is divided into the region occupied by the particle $\mathrm{V}_1$ and the region exterior to the particle $\mathrm{V}_0$. It is evident that the incident field always satisfies the equation

$$\nabla \times \nabla \times \mathbf{E}^{\mathrm{inc}}\left(\vec{r}\right) - k^2\mathbf{E}^{\mathrm{inc}}\left(\vec{r}\right) = 0. \tag{4.1.4}$$

Eq. (4.1.2) can be rewritten in a compact form (Mishchenko et al., 2002):

$$\nabla \times \nabla \times \mathbf{E}\left(\vec{r}\right) - k^2\mathbf{E}\left(\vec{r}\right) = \mathbf{j}\left(\vec{r}\right), \tag{4.1.5}$$

where

$$\mathbf{j}(\vec{r}) = \begin{cases} 0, & \vec{r} \in \mathrm{V}_0, \\ k^2\left[\varepsilon_1(\vec{r})/\varepsilon - 1\right]\mathbf{E}(\vec{r}), & \vec{r} \in \mathrm{V}_1. \end{cases} \tag{4.1.6}$$

The free space dyadic Green function satisfies the following equation:

$$\nabla \times \nabla \times \overleftrightarrow{G}(\vec{r}, \vec{r}') - k^2\overleftrightarrow{G}(\vec{r}, \vec{r}') = \overleftrightarrow{\mathrm{I}}\delta(\vec{r} - \vec{r}'), \tag{4.1.7}$$

where the operator $\nabla$ acts only on the vector $\vec{r}$. We use the vector Green theorem to derive the volume integral equation. It is rewritten here as (Morse and Feshbach, 1953; Tsang et al., 2000; Mishchenko et al., 2002)

$$\int_{\mathrm{V}} d\mathrm{V}[\mathbf{A} \cdot (\nabla \times \nabla \times \mathbf{B}) - \mathbf{B} \cdot (\nabla \times \nabla \times \mathbf{A})] = \oint_S dS\hat{n}_s \cdot (\mathbf{B} \times \nabla \times \mathbf{A} - \mathbf{A} \times \nabla \times \mathbf{B}), \tag{4.1.8}$$

where volume $V$ is enclosed by surface $S$ and $\mathbf{A}$ and $\mathbf{B}$ are arbitrary vectors. Using Eq. (3.3.19), the identity of the incident field at infinity is

$$\mathbf{E}^{\mathrm{inc}}(\vec{r}') = -\oint_{S_\infty} dS\hat{n}_s \cdot \left\{\left[\nabla \times \mathbf{E}^{\mathrm{inc}}(\vec{r})\right] \times \overleftrightarrow{G}(\vec{r}, \vec{r}') + \mathbf{E}^{\mathrm{inc}}(\vec{r}) \times \left[\nabla \times \overleftrightarrow{G}(\vec{r}, \vec{r}')\right]\right\}. \tag{4.1.9}$$

Next, we can insert the total electric field as the vector $\mathbf{A}$ as

$$\mathbf{A} = \mathbf{E}(\vec{r}), \tag{4.1.10a}$$

and $\mathbf{B}$ in this case is

$$\mathbf{B} = \overleftrightarrow{G}(\vec{r}, \vec{r}') \cdot \vec{a}, \tag{4.1.10b}$$

Substituting Eq. (4.1.10) into Eq. (4.1.8) and integrating it over all space, the vector Green theorem applied to Eqs. (4.1.5)–(4.1.7) and Eq. (4.1.9) can give the following expression:

$$\mathbf{E}(\vec{r}') = \mathbf{E}^{\mathrm{inc}}(\vec{r}') + \int_{\mathrm{V}_1} \mathbf{j}(\vec{r}) \cdot \overleftrightarrow{G}(\vec{r}, \vec{r}')d^3\vec{r}$$

$$-\oint_{S_\infty} dS\hat{n}_s \cdot \left\{ \left[ \nabla \times \mathbf{E}^{\text{sca}}(\vec{r}) \right] \times \overleftrightarrow{G}(\vec{r}, \vec{r}') + \mathbf{E}^{\text{sca}}(\vec{r}) \times \left[ \nabla \times \overleftrightarrow{G}(\vec{r}, \vec{r}') \right] \right\}.$$

(4.1.11)

Using the Sommerfeld radiation condition with respect to the dyadic Green function given in Eq. (3.3.22), the surface integration at infinity in Eq. (4.1.11) goes to zero. Using the symmetric property of the dyadic Green function given in Eq. (3.3.5), Eq. (4.1.11) gives the volume integral equation for the electric field:

$$\mathbf{E}(\vec{r}) = \mathbf{E}^{\text{inc}}(\vec{r}) + \int_{V_1} u(\vec{r}') \overleftrightarrow{G}(\vec{r}, \vec{r}') \cdot \mathbf{E}(\vec{r}') d^3\vec{r}', \tag{4.1.12a}$$

$$\mathbf{E}^{\text{sca}}(\vec{r}) = \int_{V_1} u(\vec{r}') \overleftrightarrow{G}(\vec{r}, \vec{r}') \cdot \mathbf{E}(\vec{r}') d^3\vec{r}', \tag{4.1.12b}$$

where $u(\vec{r}') = k^2 \left[ \varepsilon_1(\vec{r}') / \varepsilon - 1 \right]$.

### 4.1.2 Equivalence between volume integral equation and surface integral equation

In Section 3.3.2, the surface integral equation of the electric field is given in Eq. (3.3.23) using the vector Green theorem and is reorganized as follows:

$$\mathbf{E}^{\text{sca}}(\vec{r}) = \oint_S dS'\hat{n}_s \times \left\{ i\omega\mu_0 \left[ \mathbf{H}(\vec{r}') \times \overleftrightarrow{G}(\vec{r}', \vec{r}) \right] + \mathbf{E}(\vec{r}') \times \left[ \nabla' \times \overleftrightarrow{G}(\vec{r}', \vec{r}) \right] \right\},$$

(4.1.13)

where the field point $\vec{r}$ is the region outside the scattering particle, that is, $\vec{r} \in V_0$; $S$ encloses the scattering particle and $\hat{n}_s$ is the outward unit normal to the surface; and $\nabla'$ acts on the variable $\vec{r}'$. Both the volume integral equation shown in Eq. (4.1.12) and surface integral equation shown in Eq. (4.1.13) are derived from Maxwell's equations and the vector Green theorem, which should be equivalent.

Using Gauss's theorem, the surface integral term in Eq. (4.1.13) becomes the volume integral as follows:

$$\mathbf{E}^{\text{sca}}(\vec{r}) = \int_{V_1} d^3\vec{r}' \nabla' \cdot \left\{ i\omega\mu_0 \left[ \mathbf{H}(\vec{r}') \times \overleftrightarrow{G}(\vec{r}', \vec{r}) \right] + \mathbf{E}(\vec{r}') \times \left[ \nabla' \times \overleftrightarrow{G}(\vec{r}', \vec{r}) \right] \right\}.$$

(4.1.14)

This divergence operator identity will be useful in what follows:

$$\nabla\cdot(\mathbf{A}\times\mathbf{B}) = (\nabla\times\mathbf{A})\cdot\mathbf{B} - \mathbf{A}\cdot(\nabla\times\mathbf{B}). \tag{4.1.15}$$

Using Maxwell's equations in Eqs. (4.1.1), (4.1.2) and Eq. (4.1.7), Eq. (4.1.14) can be further transformed:

$$\mathbf{E}^{\text{sca}}\left(\vec{r}\right) = \int_{V_1} d^3\vec{r}'\left\{k_1^2\mathbf{E}\left(\vec{r}'\right)\cdot\overleftrightarrow{G}\left(\vec{r}',\vec{r}\right) - k^2\mathbf{E}\left(\vec{r}'\right)\cdot\overleftrightarrow{G}\left(\vec{r}',\vec{r}\right) - \mathbf{E}\left(\vec{r}'\right)\delta\left(\vec{r}-\vec{r}'\right)\right\}. \tag{4.1.16}$$

Since the field point is outside region $V_1$, the integral over $\mathbf{E}\left(\vec{r}'\right)\delta\left(\vec{r}-\vec{r}'\right)$ becomes zero. Also, the free space dyadic Green function has the symmetry property:

$$\overleftrightarrow{G}\left(\vec{r}',\vec{r}\right) = \overleftrightarrow{G}^T\left(\vec{r},\vec{r}'\right). \tag{4.1.17}$$

Eq. (4.1.16) becomes the volume integral equation:

$$\mathbf{E}^{\text{sca}}\left(\vec{r}\right) = \int_V k^2(\varepsilon_1/\varepsilon - 1)\overleftrightarrow{G}\left(\vec{r},\vec{r}'\right)\cdot\mathbf{E}\left(\vec{r}'\right)d^3\vec{r}'. \tag{4.1.18}$$

The equivalence between the volume integral equation and the surface integral equation is proved using the Green theorem.

In the far field or $r\to\infty$, the free space dyadic Green function is given in Eq. (3.3.8). Also, for an arbitrary vector $\mathbf{A}$, one has

$$\left(\overleftrightarrow{I} - \hat{r}\hat{r}\right)\cdot\mathbf{A} = -\hat{r}\times\hat{r}\times\mathbf{A}. \tag{4.1.19}$$

The volume integral and the surface integral equations can be given in the far field as

$$\mathbf{E}^{\text{sca}}\left(\vec{r}\right)\Big|_{r\to\infty} = \frac{\exp(ikr)}{-ikr}\frac{ik^3}{4\pi}\int_{V_1}\left(1-m^2\right)\left(\overleftrightarrow{I} - \hat{r}\hat{r}\right)\cdot\mathbf{E}\left(\vec{r}'\right)\exp\left(-ik\hat{r}\cdot\vec{r}'\right)d^3\vec{r}', \tag{4.1.20}$$

$$\mathbf{E}^{\text{sca}}\left(\vec{r}\right)\Big|_{r\to\infty} = \frac{\exp(ikr)}{-ikr}\frac{k^2}{4\pi}\hat{r}\times\oint_S d^2\vec{r}'\left\{\left[\hat{n}_s\times\mathbf{E}\left(\vec{r}'\right)\right] - \frac{\omega\mu_0}{k}\hat{r}\times\left[\hat{n}_s\times\mathbf{H}\left(\vec{r}'\right)\right]\right\}, \tag{4.1.21}$$

where $m^2 = \varepsilon_1/\varepsilon$ is the square of the relative refractive index of the particle.

### 4.1.3 Free space dyadic green function revisited

The free space dyadic Green function is given using the so-called Ohm-Rayleigh method in Section 3.3.1. The method first expands a corresponding $\delta$ function using vector spherical wave functions and then finds the expansion of the free space dyadic Green function. This method is used to get the expansion of the free space dyadic Green function in various coordinate systems, such as Cartesian, cylindrical, and spherical coordinates (Tai, 1994). Contour integration is used to get the corresponding expansions. When the expansion eigenfunctions change, the contour integration may be complicated. Here, we want to use a general technique to expand the Green function so the eigenfunctions can be any combination as long as their Wronskian is not equal to zero, which is introduced by Tai (1994). Since the expansion of the scalar Green function is straightforward using this technique, we shall start with the expansion of the scalar Green function and then find the expansion of the dyadic Green function using the relation between the scalar and the dyadic Green functions. Note that only the expansion in spherical coordinates is discussed in this book and the expansion in other coordinate systems can be found in Morse and Feshbach (1953) and Tai (1994). The direct expansion of the dyadic Green function using this general technique can also be found in Morse and Feshbach (1953).

For convenience, the scalar Green function and its expression given in Eq. (3.3.1) are restated here. The scalar Green function for the scalar Helmholtz equation satisfies the equation

$$\nabla^2 g\left(\vec{r}, \vec{r}'\right) + k^2 g\left(\vec{r}, \vec{r}'\right) = -\delta\left(\vec{r} - \vec{r}'\right), \tag{4.1.22}$$

The explicit expression for the scalar Green function is ( Jackson, 1998)

$$g\left(\vec{r}, \vec{r}'\right) = \frac{\exp\left[ik\left|\vec{r} - \vec{r}'\right|\right]}{4\pi\left|\vec{r} - \vec{r}'\right|}. \tag{4.1.23}$$

We will first expand the scalar Green function using scalar spherical wave functions and then discuss how to expand the free space dyadic Green function in terms of the scalar expansions. A general technique for computing the scalar Green function is given in Chapter 7 of Morse and Feshbach (1953). Using this technique, not only can the Green function in different coordinate systems be expanded, but the Green function can also have different basis functions. Here, the scalar Green function is given in spherical coordinates using spherical Bessel functions and spherical Hankel functions of the first kind as the basis set. Since two angular coordinates $(\theta, \varphi)$ have a finite domain, the spherical harmonics function associated with angular coordinates forms a complete set of two-dimensional eigenfunctions, and the eigenvalues are discrete. The Green function can be expanded in terms of the spherical harmonics, and the inhomogeneous equation related to the radial coordinate can be solved using the two solutions of the corresponding homogeneous equations. The spherical harmonics satisfy the orthogonality relation:

$$\left(\gamma'_{mn}\right)^2 \int\limits_{4\pi} \mathbb{Y}^*_{n'm'}(\theta,\varphi)\mathbb{Y}_{nm}(\theta,\varphi)\sin\theta d\theta d\varphi = \delta_{nn'}\delta_{mm'}, \tag{4.1.24a}$$

$$\left(\gamma'_{mn}\right)^2 = \frac{2n+1}{4\pi}\frac{(n-m)!}{(n+m)!}. \tag{4.1.24b}$$

The spherical harmonics are defined in Eqs. (3.2.11), (3.2.12). The scalar Green function can be expanded using the spherical harmonics and the undetermined functions as

$$g\left(\vec{r}\,,\,\vec{r}'\right) = \sum_{n=0}^{\infty}\sum_{m=-n}^{n} \mathrm{R}_n(r,r')\mathbb{X}_{mn}(\theta',\varphi')\mathbb{Y}_{nm}(\theta,\varphi), \tag{4.1.25}$$

where the functions $\mathrm{R}_n(r,r')$ and $\mathbb{X}_{nm}(\theta',\varphi')$ are undetermined. The radial function $\mathrm{R}_n(r,r')$ can be shown later to be only related to index $n$. Substituting Eq. (4.1.25) into Eq. (4.1.22) and using Eqs. (3.2.5), (3.2.8a), the equation for the scalar Green function is

$$\sum_{n=0}^{\infty}\sum_{m=-n}^{n} \mathbb{X}_{nm}(\theta',\varphi')\mathbb{Y}_{nm}(\theta,\varphi)\left\{\left[\frac{1}{r^2}\frac{\partial}{\partial r}r^2\frac{\partial}{\partial r}\mathrm{R}_n(r,r')\right] + \left[k^2 - \frac{n(n+1)}{r^2}\right]\mathrm{R}_n(r,r')\right\}$$
$$= -\frac{\delta(r-r')\delta(\theta-\theta')\delta(\varphi-\varphi')}{r^2\sin\theta}. \tag{4.1.26}$$

Using the orthogonality of spherical harmonics in Eq. (4.1.24), we can decouple Eq. (4.1.26) into two equations:

$$\mathbb{X}_{nm}(\theta',\varphi') = \left(\gamma'_{mn}\right)^2\mathbb{Y}^*_{nm}(\theta',\varphi'), \tag{4.1.27}$$

$$\left[\frac{\partial^2}{\partial(kr)^2} + \frac{2}{(kr)}\frac{\partial}{\partial(kr)}\right]\mathrm{R}_n(r,r') + \left[1 - \frac{n(n+1)}{(kr)^2}\right]\mathrm{R}_n(r,r') = -\frac{\delta(r-r')}{(kr)^2}. \tag{4.1.28}$$

Eq. (4.1.28) is an inhomogeneous equation for the radial function, and the homogeneous equation is actually the spherical Bessel equation based on Eq. (3.2.8b). Before giving the solution of Eq. (4.1.28), let us first discuss the general form of a second-order ordinary differential equation (ODE). Homogeneous and corresponding inhomogeneous second-order ODEs are stated as

$$\frac{d^2y(x)}{dx^2} + p(x)\frac{dy(x)}{dx} + q(x) = 0, \tag{4.1.29a}$$

$$\frac{d^2y(x)}{dx^2}+p(x)\frac{dy(x)}{dx}+q(x)=r(x), \tag{4.1.29b}$$

and $y_1(x)$ and $y_2(x)$ are two linearly independent solutions, that is, the Wronskian is not zero, or

$$\Delta(y_1,y_2)=y_1y_2'-y_2y_1'\neq 0. \tag{4.1.30}$$

Accordingly, the solution of the inhomogeneous second-order ODE can be explicitly expressed using the two independent solutions from the homogeneous second-order ODE (Morse and Feshbach, 1953):

$$y(x)=y_2(x)\int^x dt\frac{r(t)y_1(t)}{\Delta(y_1(t),y_2(t))}+y_1(x)\int_x dt\frac{r(t)y_2(t)}{\Delta(y_1(t),y_2(t))}, \tag{4.1.31}$$

where the lower and upper limit choices depend on the two independent solutions $y_1(x)$ and $y_2(x)$ and the boundary condition of $y(x)$. Note that the solution in Eq. (4.1.31) exists only when the coefficient of the second-order derivative in Eq. (4.1.29) is unity. Comparing Eq. (4.1.28) and Eq. (4.1.29), the inhomogeneous term here is the right-hand side of Eq. (4.1.28). Accordingly, the solution of Eq. (4.1.28) can be given as

$$\mathrm{R}_n(r,r')=z_{2n}(r)\int_a^r du\frac{-\delta(u-r')z_{1n}(u)}{(kr')^2\Delta(z_{1n}(u),z_{2n}(u))}+z_{1n}(r)\int_r^b du\frac{-\delta(u-r')z_{2n}(u)}{(kr')^2\Delta(z_{1n}(u),z_{2n}(u))}, \tag{4.1.32}$$

where the choices that the first limit is less than $r$ and the second limit is larger than $r$ are allowed because boundary conditions are not determined. For $r>r'$, the second integral in Eq. (4.1.32) becomes zero, while for $r<r'$, the first integral becomes zero. Accordingly, the solution in radial coordinates can be written as

$$\mathrm{R}_n(r,r')=-\frac{k}{(kr')^2\Delta(z_{1n}(r'),z_{2n}(r'))}\begin{cases} z_{1n}(r)z_{2n}(r'); & r\leq r', \\ z_{2n}(r)z_{1n}(r'); & r\geq r', \end{cases} \tag{4.1.33}$$

where the property $k\delta(kx)=\delta(x)$ since $k>0$ is used. For our light scattering formulation, the boundary condition requires the radial function to be finite at the origin ($r=0$) and also to the radiation condition at infinity ($r\to\infty$) as given in Eq. (3.2.35). As a result, the following two independent homogeneous solutions exist:

$$z_{1n}(r)=j_n(kr),\quad z_{2n}(r)=h_n^{(1)}(kr), \tag{4.1.34}$$

and the corresponding Wronskian is given in Eq. (3.2.32b) as

$$\Delta\left(j_n(kr), h_n^{(1)}(kr)\right) = \frac{i}{(kr)^2}. \tag{4.1.35}$$

Consequently, the radial solution can be explicitly written as

$$\mathrm{R}_n(r, r') = -ik\begin{cases} j_n(kr)h_n^{(1)}(kr'); & r \le r', \\ h_n^{(1)}(kr)j_n(kr'); & r \ge r'. \end{cases} \tag{4.1.36}$$

Then,

$$\begin{aligned} g\left(\vec{r}, \vec{r}'\right) &= -ik\sum_{n=0}^{\infty}\sum_{m=-n}^{n}\left(\gamma'_{mn}\right)^2\mathbb{Y}^*_{nm}(\theta', \varphi')\mathbb{Y}_{nm}(\theta, \varphi)\begin{cases} j_n(kr)h_n^{(1)}(kr'); & r \le r' \\ h_n^{(1)}(kr)j_n(kr'); & r \ge r' \end{cases} \\ &= -ik\sum_{n=0}^{\infty}\frac{2n+1}{4\pi}\sum_{m=-n}^{n}(-1)^m\begin{cases} \mathrm{Rg}\psi_{mn}(kr, \theta, \varphi)\psi_{-mn}(kr', \theta', \varphi'); & r \le r', \\ \psi_{mn}(kr, \theta, \varphi)\mathrm{Rg}\psi_{-mn}(kr', \theta', \varphi'); & r \ge r'. \end{cases} \end{aligned} \tag{4.1.37}$$

The definition of the dyadic Green function in Eq. (4.1.7) is from the wave equation of the electric field in Eq. (4.1.5). Only in this section do we denote it as the free space electric dyadic Green function $\overleftrightarrow{G}_e\left(\vec{r}, \vec{r}'\right)$ for distinction. Correspondingly, it is straightforward to define the dyadic Green function associated with the wave equation of the magnetic field as (Tai, 1994)

$$\nabla\times\nabla\times\overleftrightarrow{G}_m\left(\vec{r}, \vec{r}'\right) - k^2\overleftrightarrow{G}_m\left(\vec{r}, \vec{r}'\right) = \nabla\times\left[\overleftrightarrow{\mathrm{I}}\delta\left(\vec{r} - \vec{r}'\right)\right], \tag{4.1.38}$$

where $\overleftrightarrow{G}_m\left(\vec{r}, \vec{r}'\right)$ is the free space magnetic dyadic Green function. The explicit expressions for the two dyadic Green functions using the scalar Green function are

$$\overleftrightarrow{G}_e\left(\vec{r}, \vec{r}'\right) = \left(\overleftrightarrow{\mathrm{I}} + \frac{1}{k^2}\nabla\nabla\right)g\left(\vec{r}, \vec{r}'\right), \tag{4.1.39a}$$

$$\overleftrightarrow{G}_m\left(\vec{r}, \vec{r}'\right) = \nabla\times\left[\overleftrightarrow{\mathrm{I}}g\left(\vec{r}, \vec{r}'\right)\right]. \tag{4.1.39b}$$

After some algebraic steps, the free space magnetic dyadic Green function has this symmetry property:

$$\overleftrightarrow{G}_m^T\left(\vec{r}', \vec{r}\right) = \overleftrightarrow{G}_m\left(\vec{r}, \vec{r}'\right). \tag{4.1.40}$$

It is evident that the two dyadic Green functions are related by

$$\nabla \times \overleftrightarrow{G}_e\left(\vec{r}, \vec{r}'\right) = \overleftrightarrow{G}_m\left(\vec{r}, \vec{r}'\right), \tag{4.1.41a}$$

$$\nabla \times \overleftrightarrow{G}_m\left(\vec{r}, \vec{r}'\right) - \overleftrightarrow{\mathrm{I}}\delta\left(\vec{r} - \vec{r}'\right) = k^2\overleftrightarrow{G}_e\left(\vec{r}, \vec{r}'\right). \tag{4.1.41b}$$

We have two facts to point out in relation to introducing the free space magnetic dyadic Green function in this section. The scalar Green function is expanded using the scalar spherical wave functions, and the explicit expression for $\overleftrightarrow{G}_e\left(\vec{r}, \vec{r}'\right)$ contains the gradient operator in Eq. (4.1.39a). The longitudinal component of the vector spherical wave function (Rg)$\mathbf{L}_{mn}$ would inevitably appear in the process of expanding $\overleftrightarrow{G}_e\left(\vec{r}, \vec{r}'\right)$, even though the final expansion may not explicitly contain the longitudinal component. Moreover, the longitudinal component is associated with the field point in the source region. The curl operator in Eq. (4.1.39b) for $\overleftrightarrow{G}_m\left(\vec{r}, \vec{r}'\right)$ automatically excludes the longitudinal component. Consequently, the derivation of $\overleftrightarrow{G}_m\left(\vec{r}, \vec{r}'\right)$ is much easier than the one of $\overleftrightarrow{G}_e\left(\vec{r}, \vec{r}'\right)$. A comparison of the degree of complexity using the two methods is shown in Chapter 5 of Tai (1994). The expansion of the free space dyadic Green function using the Ohm-Rayleigh method in Section 3.3.1 contains an isolated singular term in Eq. (3.3.14), which can be given in terms of $\overleftrightarrow{G}_m\left(\vec{r}, \vec{r}'\right)$ and Eq. (4.1.41b). For convenience, the cases $r<r'$ and $r>r'$ are denoted by superscript minus and plus signs, respectively. Therefore, we can rewrite the scalar Green function as follows:

$$g^-\left(\vec{r}, \vec{r}'\right) = ik\sum_{n=0}^{\infty}\sum_{m=-n}^{n}\frac{2n+1}{4\pi}(-1)^{m+1}\mathrm{Rg}\psi_{mn}(kr,\theta,\varphi)\psi_{-mn}(kr',\theta',\varphi'), \tag{4.1.42a}$$

$$g^+\left(\vec{r}, \vec{r}'\right) = ik\sum_{n=0}^{\infty}\sum_{m=-n}^{n}\frac{2n+1}{4\pi}(-1)^{m+1}\psi_{mn}(kr,\theta,\varphi)\mathrm{Rg}\psi_{-mn}(kr',\theta',\varphi'), \tag{4.1.42b}$$

$$g\left(\vec{r}, \vec{r}'\right) = g^+\left(\vec{r}, \vec{r}'\right)U(r-r') + g^-\left(\vec{r}, \vec{r}'\right)U(r'-r), \tag{4.1.43}$$

where the step function $U$ is defined as

$$U(r-r') = \begin{cases} 1, & r>r' \\ 0, & r<r' \end{cases}, \tag{4.1.44a}$$

$$U(r'-r)=\begin{cases}0, & r>r'\\ 1, & r<r'\end{cases},\tag{4.1.44b}$$

$$\nabla U(r-r')=-\nabla U(r'-r)=\hat{r}\delta(r-r').\tag{4.1.45}$$

Using the explicit expression for the free space magnetic dyadic Green function in Eq. (4.1.39b) and Eq. (4.1.43), we have the expression

$$\overleftrightarrow{G}_m\left(\vec{r},\vec{r}'\right)=\overleftrightarrow{G}_m^+\left(\vec{r},\vec{r}'\right)U(r-r')+\overleftrightarrow{G}_m^-\left(\vec{r},\vec{r}'\right)U(r'-r)$$
$$+\hat{r}\times\overleftrightarrow{\mathrm{I}}\delta(r-r')\left(g^+\left(\vec{r},\vec{r}'\right)-g^-\left(\vec{r},\vec{r}'\right)\right).\tag{4.1.46}$$

$$\overleftrightarrow{G}_m^{\pm}\left(\vec{r},\vec{r}'\right)=\nabla\times\left[\overleftrightarrow{\mathrm{I}}g^{\pm}\left(\vec{r},\vec{r}'\right)\right].\tag{4.1.47}$$

The scalar Green function at $r=r'$ is continuous as a result of the reciprocity shown in Eq. (3.3.2). Consequently, the free space magnetic dyadic Green function in Eq. (4.1.46) is reduced to

$$\overleftrightarrow{G}_m\left(\vec{r},\vec{r}'\right)=\overleftrightarrow{G}_m^+\left(\vec{r},\vec{r}'\right)U(r-r')+\overleftrightarrow{G}_m^-\left(\vec{r},\vec{r}'\right)U(r'-r).\tag{4.1.48}$$

Then, one can have

$$\nabla\times\overleftrightarrow{G}_m\left(\vec{r},\vec{r}'\right)=\left[\nabla\times\overleftrightarrow{G}_m^+\left(\vec{r},\vec{r}'\right)\right]U(r-r')+\left[\nabla\times\overleftrightarrow{G}_m^-\left(\vec{r},\vec{r}'\right)\right]U(r'-r)$$
$$+\delta(r-r')\left\{\hat{r}\times\left[\overleftrightarrow{G}_m^+\left(\vec{r},\vec{r}'\right)-\overleftrightarrow{G}_m^-\left(\vec{r},\vec{r}'\right)\right]\right\}.\tag{4.1.49}$$

The tangential discontinuity of the Green function at $r=r'$ can be stated as (Tai, 1994, Section 4.1 and p. 128)

$$\hat{r}\times\left[\overleftrightarrow{G}_m^+\left(\vec{r},\vec{r}'\right)-\overleftrightarrow{G}_m^-\left(\vec{r},\vec{r}'\right)\right]=\left(\overleftrightarrow{\mathrm{I}}-\hat{r}\hat{r}\right)\delta(\Omega-\Omega'),\tag{4.1.50}$$

where $\delta(\Omega-\Omega')$ is a two-dimensional delta function at the spherical surface $r=r'$. Using Eq. (4.1.41b) and Eqs. (4.1.49), (4.1.50), the free space electric dyadic Green function can be given as

$$\overleftrightarrow{G}_e\left(\vec{r},\vec{r}'\right)=\frac{1}{k^2}\left[\nabla\times\overleftrightarrow{G}_m^+\left(\vec{r},\vec{r}'\right)\right]U(r-r')+\frac{1}{k^2}\left[\nabla\times\overleftrightarrow{G}_m^-\left(\vec{r},\vec{r}'\right)\right]U(r'-r)$$
$$-\frac{\hat{r}\hat{r}\delta\left(\vec{r}-\vec{r}'\right)}{k^2},\tag{4.1.51a}$$

$$\delta\left(\vec{r}-\vec{r}'\right)=\delta(r-r')\delta(\Omega-\Omega'), \tag{4.1.51b}$$

where $\delta\left(\vec{r}-\vec{r}'\right)$ is a three-dimensional delta function.

Accordingly, the electric dyadic Green function can be obtained from the magnetic dyadic Green function and the isolated singular term. The explicit expression for $\overleftrightarrow{G}_m^{\pm}\left(\vec{r},\vec{r}'\right)$ in Eq. (4.1.47) can be written as

$$\overleftrightarrow{G}_m^{\pm}\left(\vec{r},\vec{r}'\right)=\vec{G}_m^{\pm(x)}\left(\vec{r},\vec{r}'\right)\hat{x}+\vec{G}_m^{\pm(y)}\left(\vec{r},\vec{r}'\right)\hat{y}+\vec{G}_m^{\pm(z)}\left(\vec{r},\vec{r}'\right)\hat{z}, \tag{4.1.52a}$$

$$\vec{G}_m^{\pm(c)}=\nabla\times\left[g^{\pm}\left(\vec{r},\vec{r}'\right)\hat{c}\right],\ \hat{c}=\hat{x},\ \hat{y},\ \hat{z}, \tag{4.1.52b}$$

where $\hat{x}$, $\hat{y}$, and $\hat{z}$ are the unit vectors in Cartesian coordinates and

$$\hat{x}=\ \sin\theta\cos\varphi\hat{r}+\cos\theta\cos\varphi\hat{\theta}-\sin\varphi\hat{\varphi}, \tag{4.1.53a}$$

$$\hat{y}=\ \sin\theta\sin\varphi\hat{r}+\cos\theta\sin\varphi\hat{\theta}+\cos\varphi\hat{\varphi}, \tag{4.1.53b}$$

$$\hat{z}=\ \cos\theta\hat{r}-\sin\theta\hat{\theta}. \tag{4.1.53c}$$

In addition to the standard vector spherical wave functions, we can define the following ones for convenience in expanding the free space dyadic Green function:

$$\mathrm{M}_{mn}^{(c)}=\gamma_{mn}\frac{\nabla\times(\Psi_{mn}\hat{c})}{k}=\frac{1}{k}\nabla\times\mathrm{N}_{mn}^{(c)}, \tag{4.1.54a}$$

$$\mathrm{N}_{mn}^{(c)}=\gamma_{mn}\frac{\nabla\times\nabla\times(\Psi_{mn}\hat{c})}{k^2}=\frac{1}{k}\nabla\times\mathrm{M}_{mn}^{(c)}, \tag{4.1.54b}$$

$$\mathrm{L}_{mn}^{(c)}=\gamma'_{mn}\frac{1}{k}(\nabla\Psi_{mn})\cdot\hat{c}=\mathrm{L}_{mn}\cdot\hat{c}, \tag{4.1.54c}$$

where the definitions of $\gamma_{mn}$ and $\gamma'_{mn}$ are given in Eq. (3.2.39) and the type of the spherical Bessel function used in $\mathrm{M}_{mn}^{(c)}$, $\mathrm{N}_{mn}^{(c)}$, or $\mathrm{L}_{mn}^{(c)}$ depends on $\Psi_{mn}$ and can be spherical Bessel functions $j_n$, spherical Neumann functions $y_n$, or spherical Hankel functions of the first kind $h_n^{(1)}$ or the second kind $h_n^{(2)}$. For instance, $\Psi_{mn}$ uses $h_n^{(1)}$ in Eq. (4.1.54), that is, $\Psi_{mn}=\psi_{mn}$, then $\mathrm{M}_{mn}^{(c)}=\mathbf{M}_{mn}^{(c)}$, $\mathrm{N}_{mn}^{(c)}=\mathbf{N}_{mn}^{(c)}$, and $\mathrm{L}_{mn}^{(c)}=\mathbf{L}_{mn}^{(c)}$. Using the recurrence relations of the spherical Bessel functions listed in Eq. (3.2.29) and the associated Legendre polynomials listed in Eq. (3.2.24), after some straightforward though tedious algebra, the defined VSWF in Eq. (4.1.54) can be presented using the standard VSWF (Tai, 1994):

$$\frac{\mathrm{A}_{mn}^{(x)}}{\gamma_{mn}} = \frac{i}{2n(n+1)}\left[\frac{\mathrm{B}_{m+1n}}{\gamma_{m+1n}} + (n+m)(n-m+1)\frac{\mathrm{B}_{m-1n}}{\gamma_{m-1n}}\right]$$
$$-\frac{1}{2(n+1)(2n+1)}\left[\frac{\mathrm{A}_{m+1n+1}}{\gamma_{m+1n+1}} - (n-m+1)(n-m+2)\frac{\mathrm{A}_{m-1n+1}}{\gamma_{m-1n+1}}\right]$$
$$+\frac{1}{2n(2n+1)}\left[\frac{\mathrm{A}_{m+1n-1}}{\gamma_{m+1n-1}} - (n+m-1)(n+m)\frac{\mathrm{A}_{m-1n-1}}{\gamma_{m-1n-1}}\right], \tag{4.1.55a}$$

$$\frac{\mathrm{A}_{mn}^{(y)}}{\gamma_{mn}} = \frac{1}{2n(n+1)}\left[\frac{\mathrm{B}_{m+1n}}{\gamma_{m+1n}} - (n+m)(n-m+1)\frac{\mathrm{B}_{m-1n}}{\gamma_{m-1n}}\right]$$
$$+\frac{i}{2(n+1)(2n+1)}\left[\frac{\mathrm{A}_{m+1n+1}}{\gamma_{m+1n+1}} + (n-m+1)(n-m+2)\frac{\mathrm{A}_{m-1n+1}}{\gamma_{m-1n+1}}\right]$$
$$-\frac{i}{2n(2n+1)}\left[\frac{\mathrm{A}_{m+1n-1}}{\gamma_{m+1n-1}} + (n+m-1)(n+m)\frac{\mathrm{A}_{m-1n-1}}{\gamma_{m-1n-1}}\right], \tag{4.1.55b}$$

$$\frac{\mathrm{A}_{mn}^{(z)}}{\gamma_{mn}} = \frac{im}{n(n+1)}\frac{\mathrm{B}_{mn}}{\gamma_{mn}} + \frac{1}{2n+1}\left[\frac{n-m+1}{n+1}\frac{\mathrm{A}_{mn+1}}{\gamma_{mn+1}} + \frac{n+m}{n}\frac{\mathrm{A}_{mn-1}}{\gamma_{mn-1}}\right], \tag{4.1.55c}$$

where

$$\begin{cases} \text{if A} = \text{M}, \ \ \text{B} = \text{N}, \\ \text{if A} = \text{N}, \ \ \text{B} = \text{M}, \end{cases} \tag{4.1.56}$$

and the $\mathrm{M}_{mn}$ and $\mathrm{N}_{mn}$ are defined in Eq. (3.2.100). Also (Tai, 1994),

$$\frac{\mathrm{L}_{mn}^{(x)}}{\gamma'_{mn}} = \frac{1}{2(2n+1)}[\Psi_{m+1n+1} - (n-m+1)(n-m+2)\Psi_{m-1n+1}]$$
$$+\frac{1}{2(2n+1)}[\Psi_{m+1n-1} - (n+m-1)(n+m)\Psi_{m-1n-1}], \tag{4.1.57a}$$

$$\frac{\mathrm{L}_{mn}^{(y)}}{\gamma'_{mn}} = -\frac{i}{2(2n+1)}[\Psi_{m+1n+1} + (n-m+1)(n-m+2)\Psi_{m-1n+1}]$$
$$-\frac{i}{2(2n+1)}[\Psi_{m+1n-1} + (n+m-1)(n+m)\Psi_{m-1n-1}], \tag{4.1.57b}$$

$$\frac{\mathrm{L}_{mn}^{(z)}}{\gamma'_{mn}} = -\frac{n-m+1}{2n+1}\Psi_{mn+1} + \frac{n+m}{2n+1}\Psi_{mn-1}. \tag{4.1.57c}$$

In Eqs. (4.1.55), (4.1.57), $m$ can be any integer: positive, negative, or zero. Similar formulas for $\mathbf{M}^{(c)}$ and $\mathbf{N}^{(c)}$ are given by Tai (1994), but different definitions are used.

Using $r<r'$ as an example, the corresponding free space magnetic dyadic Green function can be expressed as (Tai, 1994)

$$\begin{aligned}\vec{G}_m^{-(c)}\left(\vec{r},\vec{r}'\right) &= ik^2\sum_{n=0}^{\infty}\sum_{m=-n}^{n}\frac{\mathrm{Rg}\mathbf{M}_{mn}^{(c)}(kr,\theta,\varphi)}{\gamma_{mn}}\frac{2n+1}{4\pi}(-1)^{m+1}\psi_{-mn}(kr',\theta',\varphi')\\ &= ik^2\sum_{n=0}^{\infty}\sum_{m=-n}^{n}\frac{(-1)^m}{4\pi}\left[\frac{\mathrm{Rg}\mathbf{M}_{mn}(kr,\theta,\varphi)}{\gamma_{mn}}\alpha_{mn}^{(c)}+\frac{\mathrm{Rg}\mathbf{N}_{mn}(kr,\theta,\varphi)}{\gamma_{mn}}\beta_{mn}^{(c)}\right],\\ c&=x,y,z, \end{aligned}\tag{4.1.58}$$

where

$$\begin{aligned}\alpha_{mn}^{(x)} = \frac{1}{2n}&\left[\psi'_{-m+1n-1}-(n-m-1)(n-m)\psi'_{-m-1n-1}\right]\\ &-\frac{1}{2(n+1)}\left[\psi'_{-m+1n+1}-(n+m+1)(n+m+2)\psi'_{-m-1n+1}\right],\end{aligned}\tag{4.1.59a}$$

$$\begin{aligned}\alpha_{mn}^{(y)} = -\frac{i}{2n}&\left[\psi'_{-m+1n-1}+(n-m-1)(n-m)\psi'_{-m-1n-1}\right]\\ &+\frac{i}{2(n+1)}\left[\psi'_{-m+1n+1}+(n+m+1)(n+m+2)\psi'_{-m-1n+1}\right],\end{aligned}\tag{4.1.59b}$$

$$\alpha_{mn}^{(z)} = \frac{n-m}{n}\psi'_{-mn-1}+\frac{n+m+1}{n+1}\psi'_{-mn+1},\tag{4.1.59c}$$

$$\beta_{mn}^{(x)} = i\frac{2n+1}{2n(n+1)}\left[\psi'_{-m+1n}+(n+m+1)(n-m)\psi'_{-m-1n}\right],\tag{4.1.60a}$$

$$\beta_{mn}^{(y)} = \frac{2n+1}{2n(n+1)}\left[\psi'_{-m+1n}-(n+m+1)(n-m)\psi'_{-m-1n}\right],\tag{4.1.60b}$$

$$\beta_{mn}^{(z)} = -i\frac{2n+1}{n(n+1)}m\psi'_{-mn}.\tag{4.1.60c}$$

The primed arguments of $\psi$ in Eqs. (4.1.59), (4.1.60) are suppressed and are indicated by $\psi'$ instead. Using Eq. (4.1.52), the free space magnetic dyadic Green function can be written as

$$\begin{aligned}\overleftrightarrow{G}_m^{-}\left(\vec{r},\vec{r}'\right) = ik^2\sum_{n=1}^{\infty}\sum_{m=-n}^{n}\frac{(-1)^m}{4\pi}&\left\{\frac{\mathrm{Rg}\mathbf{M}_{mn}(kr,\theta,\varphi)}{\gamma_{mn}}\left[\alpha_{mn}^{(x)}\hat{x}+\alpha_{mn}^{(y)}\hat{y}+\alpha_{mn}^{(z)}\hat{z}\right]\right.\\ &\left.+\frac{\mathrm{Rg}\mathbf{N}_{mn}(kr,\theta,\varphi)}{\gamma_{mn}}\left[\beta_{mn}^{(x)}\hat{x}+\beta_{mn}^{(y)}\hat{y}+\beta_{mn}^{(z)}\hat{z}\right]\right\}.\end{aligned}\tag{4.1.61}$$

Based on the definitions of Eq. (4.1.54) and the standard vector spherical wave functions, we have the following equations:

$$\frac{\mathbf{M}_{mn}}{\gamma_{mn}}\cdot\hat{c}=\frac{\mathbf{M}_{mn}^{(c)}}{\gamma_{mn}}\cdot\left(k\vec{r}\right), \tag{4.1.62a}$$

$$\frac{\mathbf{N}_{mn}}{\gamma_{mn}}\cdot\hat{c}=\frac{\mathbf{N}_{mn}^{(c)}}{\gamma_{mn}}\cdot\left(k\vec{r}\right)+2\frac{\mathbf{L}_{mn}^{(c)}}{\gamma'_{mn}}, \tag{4.1.62b}$$

$$\frac{\mathbf{M}_{mn}}{\gamma_{mn}}\cdot\left(k\vec{r}\right)=0, \tag{4.1.63a}$$

$$\frac{\mathbf{N}_{mn}}{\gamma_{mn}}\cdot\left(k\vec{r}\right)=n(n+1)\psi_{mn}. \tag{4.1.63b}$$

Using the aforementioned equations, we can have

$$\beta_{mn}^{(x)}\hat{x}+\beta_{mn}^{(y)}\hat{y}+\beta_{mn}^{(z)}\hat{z}=\frac{2n+1}{n(n+1)}\frac{\mathbf{M}_{-mn}(kr',\theta',\varphi')}{\gamma_{-mn}}, \tag{4.1.64a}$$

$$\alpha_{mn}^{(x)}\hat{x}+\alpha_{mn}^{(y)}\hat{y}+\alpha_{mn}^{(z)}\hat{z}=\frac{2n+1}{n(n+1)}\frac{\mathbf{N}_{-mn}(kr',\theta',\varphi')}{\gamma_{-mn}}. \tag{4.1.64b}$$

The free space magnetic dyadic Green function at $r<r'$ can be rewritten as

$$\begin{aligned}\overleftrightarrow{G}_m^{-}\left(\vec{r},\vec{r}'\right)=ik^2\sum_{n=1}^{\infty}\sum_{m=-n}^{n}(-1)^m[&\mathrm{Rg}\mathbf{M}_{mn}(kr,\theta,\varphi)\mathbf{N}_{-mn}(kr',\theta',\varphi')\\&+\mathrm{Rg}\mathbf{N}_{mn}(kr,\theta,\varphi)\mathbf{M}_{-mn}(kr',\theta',\varphi')].\end{aligned} \tag{4.1.65}$$

The expansion at $r>r'$ following the same steps can be given as

$$\begin{aligned}\overleftrightarrow{G}_m^{+}\left(\vec{r},\vec{r}'\right)=ik^2\sum_{n=1}^{\infty}\sum_{m=-n}^{n}(-1)^m[&\mathbf{M}_{mn}(kr,\theta,\varphi)\mathrm{Rg}\mathbf{N}_{-mn}(kr',\theta',\varphi')\\&+\mathbf{N}_{mn}(kr,\theta,\varphi)\mathrm{Rg}\mathbf{M}_{-mn}(kr',\theta',\varphi')].\end{aligned} \tag{4.1.66}$$

Using the relation between the free space electric and magnetic dyadic Green function given in Eq. (4.1.51), the free space electric dyadic Green function can be written as

$$\begin{aligned}&\overleftrightarrow{G}_e\left(\vec{r},\vec{r}'\right)=-\hat{r}\hat{r}\frac{\delta\left(\vec{r}-\vec{r}'\right)}{k^2}+ik\sum_{n=1}^{\infty}\sum_{m=-n}^{n}(-1)^m\\&\begin{cases}\mathbf{M}_{mn}(kr,\theta,\varphi)\mathrm{Rg}\mathbf{M}_{-mn}(kr',\theta',\varphi')+\mathbf{N}_{mn}(kr,\theta,\varphi)\mathrm{Rg}\mathbf{N}_{-mn}(kr',\theta',\varphi'), & r>r',\\ \mathrm{Rg}\mathbf{M}_{mn}(kr,\theta,\varphi)\mathbf{M}_{-mn}(kr',\theta',\varphi')+\mathrm{Rg}\mathbf{N}_{mn}(kr,\theta,\varphi)\mathbf{N}_{-mn}(kr',\theta',\varphi'), & r<r'.\end{cases}\end{aligned} \tag{4.1.67}$$

This section presents a general technique to obtain the expansion of the free space dyadic Green function. The crucial step is to choose two independent solutions of the spherical Bessel equation as shown in Eq. (3.1.34). It can be any combination, such as $(j_n(kr), y_n(kr))$, $(h_n^{(1)}(kr), h_n^{(2)}(kr))$, or $(j_n(kr), h_n^{(1)}(kr))$. Using the corresponding Wronskian of the chosen two functions as shown in Eq. (4.1.35), the scalar Green function can be obtained as shown in Eq. (4.1.37). Then, the free space magnetic dyadic Green function can be obtained using the scalar Green function as shown in Eqs. (4.1.38)–(4.1.66). Finally, the free space electric Green function can be obtained after some algebra series as shown in Eq. (4.1.67). In this section, we use $(j_n(kr), h_n^{(1)}(kr))$ as the basis to give the expansion. Using other independent combinations can be easily realized using the same steps because all the definition and recurrence relations are the same for different spherical Bessel functions.

### 4.1.4 Vector spherical wave function expansion

Several quantities related to the angular functions and the radial functions will be introduced, so the discussion can be performed in a compact way. By analyzing the angular functions in the form of vector spherical wave functions, the following angular quantities are defined as

$$\begin{aligned}\mathbf{X}_{mn}(\theta,\varphi) &= (\mathrm{A}_{mn}(\theta,\varphi)\ \mathrm{B}_{mn}(\theta,\varphi)\ \mathrm{C}_{mn}(\theta,\varphi)) \\ &= (-1)^m \beta_n \exp(im\varphi)\begin{pmatrix} 0 & 0 & \sqrt{n(n+1)}d_{0m}^n(\theta) \\ i\pi_{mn}(\theta) & \tau_{mn}(\theta) & 0 \\ -\tau_{mn}(\theta) & i\pi_{mn}(\theta) & 0 \end{pmatrix},\end{aligned} \tag{4.1.68}$$

where

$$\beta_n = \sqrt{\frac{2n+1}{4\pi n(n+1)}}, \tag{4.1.69}$$

$$\mathrm{A}_{mn}(\theta,\varphi) = (-1)^m \beta_n \exp(im\varphi)\begin{pmatrix} 0 \\ i\pi_{mn}(\theta) \\ -\tau_{mn}(\theta) \end{pmatrix}, \tag{4.1.70a}$$

$$\mathrm{B}_{mn}(\theta,\varphi) = (-1)^m \beta_n \exp(im\varphi)\begin{pmatrix} 0 \\ \tau_{mn}(\theta) \\ i\pi_{mn}(\theta) \end{pmatrix}, \tag{4.1.70b}$$

$$\mathrm{C}_{mn}(\theta,\varphi) = (-1)^m \beta_n \exp(im\varphi)\begin{pmatrix} \sqrt{n(n+1)}d_{0m}^n(\theta) \\ 0 \\ 0 \end{pmatrix}, \tag{4.1.70c}$$

where the real functions $\pi_{mn}(\theta)$ and $\tau_{mn}(\theta)$ and their symmetric relations are defined in Eq. (3.2.59), the real function $d_{0m}^{n}(\theta)$ is defined in Eq. (3.2.79a), and the corresponding symmetric relations are defined in Eq. (3.2.78). Using the symmetric relations, one can obtain

$$\mathrm{A}_{mn}^{*} = (-1)^{m}\mathrm{A}_{-mn}, \mathrm{B}_{mn}^{*} = (-1)^{m}\mathrm{B}_{-mn}, \mathrm{C}_{mn}^{*} = (-1)^{m}\mathrm{C}_{-mn}, \tag{4.1.71a}$$

$$\mathbf{X}_{mn}^{*} = (-1)^{m}\mathbf{X}_{-mn}. \tag{4.1.71b}$$

Using Eq. (3.2.59) and Eq. (3.3.38) for the functions $\pi_{mn}(\theta)$ and $\tau_{mn}(\theta)$ and using Eq. (3.2.76a) for the function $d_{0m}^{n}(\theta)$, their orthogonality relations are

$$\frac{1}{2}\int_{0}^{\pi} \sin\theta d\theta[\pi_{mn}(\theta)\pi_{mn'}(\theta) + \tau_{mn}(\theta)\tau_{mn'}(\theta)] = \frac{n(n+1)}{2n+1}\delta_{nn'}, \tag{4.1.72a}$$

$$\frac{1}{2}\int_{0}^{\pi} \sin\theta d\theta[\pi_{mn}(\theta)\tau_{mn'}(\theta) + \tau_{mn}(\theta)\pi_{mn'}(\theta)] = 0, \tag{4.1.72b}$$

$$\frac{1}{2}\int_{0}^{\pi} \sin\theta d\theta d_{0m}^{n}(\theta) d_{0m}^{n'}(\theta) = \frac{1}{2n+1}\delta_{nn'}. \tag{4.1.72c}$$

Moreover, the radial quantities can be defined as

$$\mathbf{H}_{n}(kr) = \begin{pmatrix} h_a(n;kr) & 0 \\ 0 & h_b(n;kr) \\ 0 & h_c(n;kr) \end{pmatrix}, \tag{4.1.73a}$$

$$\mathbf{J}_{n}(kr) = \begin{pmatrix} j_a(n;kr) & 0 \\ 0 & j_b(n;kr) \\ 0 & j_c(n;kr) \end{pmatrix}, \tag{4.1.73b}$$

where

$$\begin{pmatrix} h_a(n;kr) \\ h_b(n;kr) \\ h_c(n;kr) \end{pmatrix} = \begin{pmatrix} h_n^{(1)}(kr) \\ \dfrac{1}{kr}\dfrac{d\left[krh_n^{(1)}(kr)\right]}{d(kr)} \\ \sqrt{n(n+1)}\dfrac{h_n^{(1)}(kr)}{kr} \end{pmatrix}, \tag{4.1.74a}$$

$$\begin{pmatrix} j_a(n;kr) \\ j_b(n;kr) \\ j_c(n;kr) \end{pmatrix} = \begin{pmatrix} j_n(kr) \\ \dfrac{1}{kr}\dfrac{d[krj_n(kr)]}{d(kr)} \\ \sqrt{n(n+1)}\dfrac{j_n(kr)}{kr} \end{pmatrix}. \tag{4.1.74b}$$

The regular and outgoing vector spherical wave functions can be written in a compact way as follows:

$$\begin{aligned}
&(\mathrm{Rg}\mathbf{M}_{mn}(kr,\theta,\varphi)\ \ \mathrm{Rg}\mathbf{N}_{mn}(kr,\theta,\varphi)) \\
&\quad = (\mathrm{A}_{mn}(\theta,\varphi)j_a(n;kr)\ \ \mathrm{B}_{mn}(\theta,\varphi)j_b(n;kr)+\mathrm{C}_{mn}(\theta,\varphi)j_c(n;kr)) \\
&\quad = \mathbf{X}_{mn}(\theta,\varphi)\mathbf{J}_n(kr),
\end{aligned} \tag{4.1.75a}$$

$$\begin{aligned}
&(\mathbf{M}_{mn}(kr,\theta,\varphi)\ \ \mathbf{N}_{mn}(kr,\theta,\varphi)) \\
&\quad = (\mathrm{A}_{mn}(\theta,\varphi)h_a(n;kr)\ \ \mathrm{B}_{mn}(\theta,\varphi)h_b(n;kr)+\mathrm{C}_{mn}(\theta,\varphi)h_c(n;kr)) \\
&\quad = \mathbf{X}_{mn}(\theta,\varphi)\mathbf{H}_n(kr).
\end{aligned} \tag{4.1.75b}$$

The free space Dyadic green function in Eq. (4.1.67) can also be written in a compact way:

$$\overleftrightarrow{G}\left(\vec{r},\vec{r}'\right) = -\hat{r}\hat{r}\frac{\delta\left(\vec{r}-\vec{r}'\right)}{k^2} + \overleftrightarrow{G}_0\left(\vec{r},\vec{r}'\right), \tag{4.1.76a}$$

$$\overleftrightarrow{G}_0\left(\vec{r},\vec{r}'\right) = ik\sum_{n=1}^{\infty}\sum_{m=-n}^{n}(-1)^m\begin{cases}\left(\mathbf{M}_{mn}(kr,\theta,\varphi)\ \ \mathbf{N}_{mn}(kr,\theta,\varphi)\right)\begin{pmatrix}\mathrm{Rg}\mathbf{M}_{-mn}(kr',\theta',\varphi')\\ \mathrm{Rg}\mathbf{N}_{-mn}(kr',\theta',\varphi')\end{pmatrix}, & r>r',\\ \left(\mathrm{Rg}\mathbf{M}_{mn}(kr,\theta,\varphi)\ \ \mathrm{Rg}\mathbf{N}_{mn}(kr,\theta,\varphi)\right)\begin{pmatrix}\mathbf{M}_{-mn}(kr',\theta',\varphi')\\ \mathbf{N}_{-mn}(kr',\theta',\varphi')\end{pmatrix}, & r<r'.\end{cases} \tag{4.1.76b}$$

The expansion for $\overleftrightarrow{G}_0\left(\vec{r},\vec{r}'\right)$ has the symmetric property, $\overleftrightarrow{G}_0^T\left(\vec{r}',\vec{r}\right) = \overleftrightarrow{G}_0\left(\vec{r},\vec{r}'\right)$, as stated in Eq. (3.3.5) and a result of reciprocity (Tai, 1994; Tsang et al., 2000). Using Eqs. (4.1.68), (4.1.75), the free space Dyadic green function can be written in a simple form

$$\overleftrightarrow{G}_0\left(\vec{r},\vec{r}'\right) = \sum_{n=1}^{\infty}\sum_{m=-n}^{n}\mathbf{X}_{mn}(\theta,\varphi)\cdot\mathrm{G}_n(r,r')\cdot\mathbf{X}_{mn}^{\dagger}(\theta',\varphi'), \tag{4.1.77}$$

where the superscript "†" represents the matrix transpose and complex conjugate, and

$$\mathrm{G}_n(r,r') = ik\begin{cases}\mathbf{H}_n(kr)\mathbf{J}_n^T(kr'), & r>r',\\ \mathbf{J}_n(kr)\mathbf{H}_n^T(kr'), & r<r.\end{cases} \tag{4.1.78}$$

$\mathrm{G}_n(r,r')$ is discontinuous at $r=r'$ and $\mathrm{G}_n^T(r',r)=\mathrm{G}_n(r,r')$.

## 4.2 Concept of the invariant-imbedding T-matrix method

Spherical coordinates are used in this book to obtain the T-matrix method. Accordingly, the imbedding process is performed in the radial coordinate in the form of infinitesimal spherical shells. Consequently, we need to integrate the angular coordinates and find the invariant-imbedding formula associated with the radial coordinate.

### 4.2.1 Radial integral

Using the volume integral equation associated with the electric field in (4.1.12a) and the free space dyadic Green function (4.1.76a), the volume integral equation can be transformed as

$$\left[\overleftrightarrow{I}+\frac{u\left(\vec{r}\right)\hat{r}\hat{r}}{k^2}\right]\cdot\mathbf{E}\left(\vec{r}\right)=\mathbf{E}^{\mathrm{inc}}\left(\vec{r}\right)+\int_{\mathrm{V}_1}u\left(\vec{r}'\right)\overleftrightarrow{G}_0\left(\vec{r},\vec{r}'\right)\cdot\mathbf{E}\left(\vec{r}'\right)d^3\vec{r}'. \tag{4.2.1}$$

The singular term in the Green function is considered in Eq. (4.2.1). Accordingly, we can define the electric field excluding the contribution from the source region as $\mathbf{E}_0\left(\vec{r}\right)$ and

$$\mathbf{E}_0\left(\vec{r}\right)=\left[\overleftrightarrow{I}+\frac{u\left(\vec{r}\right)\hat{r}\hat{r}}{k^2}\right]\cdot\mathbf{E}\left(\vec{r}\right)=\left[\mathbf{Z}\left(\vec{r}\right)\right]^{-1}\cdot\mathbf{E}\left(\vec{r}\right). \tag{4.2.2}$$

The coefficient $\mathbf{Z}$ is written in matrix form in Eq. (4.2.2) using spherical coordinates and can be given as

$$\mathbf{Z}\left(\vec{r}\right)=\begin{pmatrix}\varepsilon/\varepsilon_1\left(\vec{r}\right) & 0 & 0\\ 0 & 1 & 0\\ 0 & 0 & 1\end{pmatrix}. \tag{4.2.3}$$

It is evident that $\mathbf{E}_0\left(\vec{r}\right)\equiv\mathbf{E}\left(\vec{r}\right)$, if $r>r_c$, where $r_c$ is the radius of the smallest circumscribed sphere of the scattering particle as shown in Fig. 4.2. Replacing the electric field $\mathbf{E}\left(\vec{r}\right)$ with the electric field $\mathbf{E}_0\left(\vec{r}\right)$ in Eq. (4.2.1), the volume integral equation can be rearranged as

$$\mathbf{E}_0\left(\vec{r}\right)=\mathbf{E}^{\mathrm{inc}}\left(\vec{r}\right)+\int_{\mathrm{V}_1}u\left(\vec{r}'\right)\overleftrightarrow{G}_0\left(\vec{r},\vec{r}'\right)\cdot\mathbf{Z}\left(\vec{r}'\right)\cdot\mathbf{E}_0\left(\vec{r}'\right)d^3\vec{r}'. \tag{4.2.4}$$

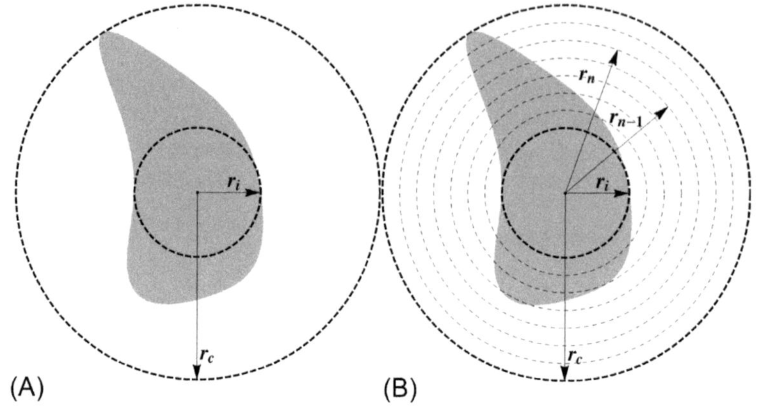


**Fig. 4.2** Schematic figure for the invariant-imbedding method. In panel (A), the shaded region is the dielectric scattering particle, and $r_i$ and $r_c$ are the radii of the largest inscribed and the smallest circumscribed spheres of the particle. In panel (B), the region where the radius is between $r_i$ and $r_c$ can be divided into spherical shells, and light scattering can be computed using the invariant-imbedding technique. $r_{n-1}$ and $r_n$ are the radii of layers $(n-1)$ and $n$, respectively.

In Eq. (3.2.51a), the incident field can be expanded using the regular vector spherical wave functions. Substituting Eq. (4.1.77) into Eq. (4.2.4) and using a single incident component ($\text{Rg}\mathbf{M}_{mn}$ $\text{Rg}\mathbf{N}_{mn}$), we can rewrite the aforementioned equation as

$$\left(\mathbf{E}_0\left(\vec{r}\right)\right)_{m'n'} = \mathbf{X}_{m'n'}(\Omega)\mathbf{J}_{n'}(kr) + \sum_{n=1}^{\infty}\sum_{m=-n}^{n}\mathbf{X}_{mn}(\Omega)\cdot\int_0^{r_c}\left\{\{dr'r'^2\text{G}_n(r,r')\right.$$
$$\left.\cdot\left[\int_{4\pi} d\Omega' u\left(\vec{r}'\right)\mathbf{X}_{mn}^{\dagger}(\Omega')\cdot\mathbf{Z}\left(\vec{r}'\right)\cdot\left(\mathbf{E}_0\left(\vec{r}'\right)\right)_{m'n'}\right]\right\}, \tag{4.2.5}$$

where the angular arguments $(\theta, \varphi)$ are replaced by the argument $\Omega$ for convenience.

We can organize all the angular-related functions into a new function:

$$\mathbf{F}_{mnm'n'}(r) = r^2\int_{4\pi} d\Omega\left[u\left(\vec{r}\right)\mathbf{X}_{mn}^{\dagger}(\Omega)\cdot\mathbf{Z}\left(\vec{r}\right)\cdot\left(\mathbf{E}_0\left(\vec{r}\right)\right)_{m'n'}\right]. \tag{4.2.6}$$

Accordingly, Eq. (4.2.5) can be rewritten as

$$\left(\mathbf{E}_0\left(\vec{r}\right)\right)_{m'n'} = \mathbf{X}_{m'n'}(\Omega)\mathbf{J}_{n'}(kr) + \sum_{n=1}^{\infty}\sum_{m=-n}^{n}\mathbf{X}_{mn}(\Omega)\cdot\int_0^{r_c} dr'\text{G}_n(r,r')\cdot\mathbf{F}_{mnm'n'}(r'). \tag{4.2.7}$$

When $r > r_c$, this satisfies the relation $r > r'$. Substituting Eq. (4.1.78) into Eq. (4.2.7), we can obtain

$$\left(\mathbf{E}_0\left(\vec{r}\right)\right)_{m'n'} = \mathbf{X}_{m'n'}(\Omega)\mathbf{J}_{n'}(kr) + \sum_{n=1}^{\infty}\sum_{m=-n}^{n} \mathbf{X}_{mn}(\Omega)\cdot\mathbf{H}_n(kr)ik\int_0^{r_c} dr'\mathbf{J}_n^T(kr')\cdot\mathbf{F}_{mnm'n'}(r'). \tag{4.2.8}$$

We can only consider a single incident component $(\text{Rg}\mathbf{M}_{mn}\ \ \text{Rg}\mathbf{N}_{mn})$ in Eq. (4.2.5). Using Eq. (4.1.75b), $\mathbf{X}_{mn}(\Omega)\mathbf{H}_n(kr)$ in the second term of Eq. (4.2.8) means a single component associated with the outgoing vector spherical wave functions. The physical meaning of Eq. (4.2.8) is evident. The electric field in Eq. (4.2.8) is composed of two terms, specifically, the incident field and the scattered field. According to the definition of the T-matrix in Eq. (3.2.53), the T-matrix can be represented as the following integral:

$$\mathbf{T}_{mnm'n'} = ik\int_0^{r_c} dr'\mathbf{J}_n^T(kr')\cdot\mathbf{F}_{mnm'n'}(r'). \tag{4.2.9}$$

Eq. (4.2.8) is physically simplified to the following form:

$$\left(\mathbf{E}_0\left(\vec{r}\right)\right)_{m'n'} = \mathbf{X}_{m'n'}(\Omega)\mathbf{J}_{n'}(kr) + \sum_{n=1}^{\infty}\sum_{m=-n}^{n} \mathbf{X}_{mn}(\Omega)\cdot\mathbf{H}_n(kr)\mathbf{T}_{mnm'n'}. \tag{4.2.10}$$

The electric field inside the smallest circumscribed sphere of the scattering particle is still necessary according to Eqs. (4.2.6), (4.2.9). Using Eqs. (4.2.5), (4.2.6), the angular integral function $\mathbf{F}_{mnm'n'}$ has the following relation:

$$\mathbf{F}_{mnm'n'}(r) = \mathbf{U}_{mnm'n'}(r)\mathbf{J}_{n'}(kr) + \sum_{n''=1}^{\infty}\sum_{m''=-n''}^{n''} \mathbf{U}_{mnm''n''}(r)\cdot\int_0^{r_c} dr'\mathrm{G}_{n''}(r,r')\cdot\mathbf{F}_{m''n''m'n'}(r'), \tag{4.2.11}$$

where

$$\mathbf{U}_{mnm'n'}(r) = r^2\int d\Omega u\left(\vec{r}\right)\mathbf{X}_{mn}^{\dagger}(\Omega)\cdot\mathbf{Z}(r,\Omega)\cdot\mathbf{X}_{m'n'}(\Omega). \tag{4.2.12}$$

Eq. (4.2.9) and Eqs. (4.2.11), (4.2.12) are the key equations to obtain the T-matrix of a dielectric scattering particle using the invariant-imbedding T-matrix method.

Eqs. (4.2.5)–(4.2.12) earlier are given in component form for clarity. For succinctness, we next use matrix notation to replace all the equations and quantities. As stated in Section 3.2.3, the indices $(m,n)$ are replaced by one index l. Once the truncation term $N$ associated with the vector spherical wave functions is determined, the matrix or vector dimension $L$ associated with the index $l$ is given in Eq. (3.2.54). Accordingly, matrices $\mathbf{J}$ and $\mathbf{H}$ have dimension $3L\times 2L$ and are block diagonal; matrix $\mathbf{F}$ has dimension $3L \times 2L$; matrices $\mathbf{U}$ and G have dimension $3L \times 3L$, and G is block diagonal; and the dimension of the T-matrix $\mathbf{T}$ is $2L \times 2L$. Eqs. (4.2.9), (4.2.11) can be written in compact matrix forms as

$$\mathbf{T} = ik\int_0^{r_c} dr'\mathbf{J}^T(kr')\mathbf{F}(r'), \tag{4.2.13}$$

$$\mathbf{F}(r) = \mathbf{U}(r)\mathbf{J}(kr) + \mathbf{U}(r)\int_0^{r_c} dr'\mathrm{G}(r, r')\mathbf{F}(r'). \tag{4.2.14}$$

### 4.2.2 Invariant-imbedding method: Differential form and difference form

The invariant-imbedding method is comprehensively introduced in general formulas and applications by Bellman and Wing (1975). Here, we only present the invariant-imbedding technique according to the T-matrix computations given in Eqs. (4.2.13), (4.2.14) in our convenience. The upper limit $r_c$ in Eqs. (4.2.13), (4.2.14) is a constant. If we consider light scattering by a particle whose smallest circumscribed sphere radius is $r_n$, the upper limit now is a variable, $0 \le r_n \le r_c$. The integral forms in Eqs. (4.2.13), (4.2.14) remain unchanged with the increase of the radial length, which is called the principle of invariance. We can imbed the radial length from $r_n$ to $r_n + \Delta r$ and find the relation between $\mathbf{T}(r_n + \Delta r)$ and $\mathbf{T}(r_n)$ or the expression for $\frac{d\mathbf{T}(r_n)}{dr_n}$. The imbedding method is similar to a perturbation method so that we can obtain the perturbation equation $\frac{d\mathbf{T}(r_n)}{dr_n}$. The final T-matrix can be obtained by iteratively imbedding the radial length $r_n$ from 0 to $r_c$, as shown in Fig. 4.3. The invariant-imbedding method transforms a boundary condition problem to an initial condition problem. In the

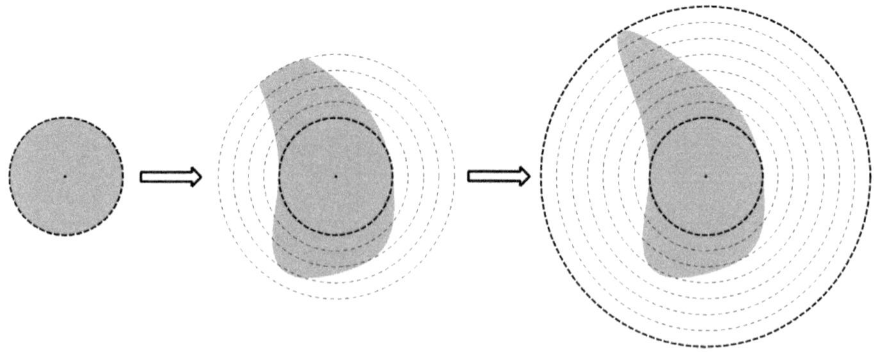

**Fig. 4.3** Invariant-imbedding steps. The invariant-imbedding algorithm starts from the inscribed sphere of a scattering particle and keeps imbedding spherical shells until the whole particle is included in the imbedded region.

following part, we will give two invariant-imbedding formulas for the T-matrix: the derivative form $\frac{d\mathbf{T}(r_n)}{dr_n}$, which is derived from the integral equation, and the difference form for $\mathbf{T}(r_n+\Delta r)$ and $\mathbf{T}(r_n)$, which is derived from the quadrature expressions associated with the integral equations. Moreover, we will prove that the difference form is reduced to the derivative form when $\Delta r \to 0$.

For clarity, we rewrite Eqs. (4.2.13), (4.2.14) and change the upper limit from the constant $r_c$ to an intermediate variable $r_n$ as

$$\mathbf{T}(r_n) = ik\int_0^{r_n} dr'\mathbf{J}^T(kr')\mathbf{F}(r_n|\,r'), \tag{4.2.15}$$

$$\mathbf{F}(r_n|\,r) = \mathbf{U}(r)\mathbf{J}(kr) + \mathbf{U}(r)\int_0^{r_n} dr'\mathrm{G}(r,r')\mathbf{F}(r_n|\,r'), \tag{4.2.16}$$

where the $r_n$ dependence in the matrix $\mathbf{F}$ is explicitly shown. It is apparent that Eq. (4.2.16) is a Fredholm integral equation of the second kind, where the forcing term $\mathbf{U}(r)\mathbf{J}(r)$ and the integral kernel $\mathrm{G}(r,r')$ are known. A general invariant-imbedding algorithm for the Fredholm integral equation of the second kind is given in Section 12.9 by Bellman and Wing (1975). Since our integral kernel in Eq. (4.1.78) and the forcing term are explicitly known, the invariant-imbedding form for the T-matrix is presented in the following part.

If we differentiate Eq. (4.2.16) with respect to the variable $r_n$, we can obtain

$$\frac{\partial\mathbf{F}(r_n|\,r)}{\partial r_n} = \mathbf{U}(r)\mathbf{J}(kr)ik\mathbf{H}^T(kr_n)\mathbf{F}(r_n|\,r_n) + \mathbf{U}(r)\int_0^{r_n} dr'\mathrm{G}(r,r')\frac{\partial\mathbf{F}(r_n|\,r')}{\partial r_n}, \tag{4.2.17}$$

where the condition $r<r_n$ for $\mathrm{G}(r,r_n)$ is used in the first term of the right-hand side. Comparing the forcing term and the integral kernel in Eqs. (4.2.16), (4.2.17), it is evident that

$$\frac{\partial\mathbf{F}(r_n|\,r)}{\partial r_n} = \mathbf{F}(r_n|\,r)ik\mathbf{H}^T(kr_n)\mathbf{F}(r_n|\,r_n). \tag{4.2.18}$$

Using Eq. (4.2.16), we can obtain

$$\mathbf{F}(r_n|\,r_n) = \mathbf{U}(r_n)[\mathbf{J}(kr_n) + \mathbf{H}(kr_n)\mathbf{T}(r_n)], \tag{4.2.19}$$

where the condition $r_n>r'$ for $\mathrm{G}(r_n,r')$ is used in the second term of the right-hand side and Eq. (4.2.15) is also used. Differentiating Eq. (4.2.15) with respect to the variable $r_n$, we can obtain

$$\frac{d\mathbf{T}(r_n)}{dr_n} = ik\mathbf{J}^T(kr_n)\mathbf{F}(r_n|\,r_n) + ik\int_0^{r_n} dr'\mathbf{J}^T(kr')\frac{\partial\mathbf{F}(r_n|\,r')}{\partial r_n}. \tag{4.2.20}$$

Using Eq. (4.2.15), Eqs. (4.2.18), (4.2.19), Eq. (4.2.20) can be presented as

$$\frac{d\mathbf{T}(r_n)}{dr_n} = ik\left[\mathbf{J}^T(kr_n) + \mathbf{T}(r_n)\mathbf{H}^T(kr_n)\right]\mathbf{U}(r_n)[\mathbf{J}(kr_n) + \mathbf{H}(kr_n)\mathbf{T}(r_n)]. \tag{4.2.21}$$

Eq. (4.2.21) is the invariant-imbedding equation for the T-matrix in a derivative form ( Johnson, 1988). A similar treatment with respect to the derivative form is also given by Doicu and Wriedt (2018).

Before obtaining the invariant-imbedding equation for the T-matrix in a difference form, we first need to discretize the integrations into quadrature expressions for Eqs. (4.2.13), (4.2.14) as shown in (e.g., Press et al., 2007, p. 989)

$$\mathbf{T} = ik\sum_{j=1}^{n_0}\omega_j\mathbf{J}^T(kr_j)\mathbf{F}(r_j), \tag{4.2.22}$$

$$\mathbf{F}(r_i) = \mathbf{U}(r_i)\mathbf{J}(kr_i) + \mathbf{U}(r_i)\sum_{j=1}^{n_0}\omega_j\mathrm{G}(r_i, r_j)\mathbf{F}(r_j), \quad i = 1,\ldots,n_0, \tag{4.2.23}$$

where $n_0$ is the total order of discretization, or total layers, $\{r_j\}$ are the abscissas, and $\{\omega_j\}$ are the corresponding weights of a quadrature rule. Eq. (4.2.23) is actually a set of linear algebraic equations associated with function $\mathbf{F}(r_i)$. The solution can be algebraically obtained by solving the linear equations because the algebraic matrix for the Fredholm equation of the second kind is usually well conditioned (Press et al., 2007, p. 990). However, the dimension of the matrix or vector is too large to directly compute since Eq. (4.2.23) itself is a matrix equation. Alternatively, the imbedding form is employed to iteratively obtain the solution. Because the imbedding form is derived from the Fredholm equation of the second kind, the matrix expression in the imbedding form still remains well conditioned. When the layer number is $n$, where $1 \le n \le n_0$, the corresponding T-matrix and radial functions are the same as the ones in Eqs. (4.2.22), (4.2.23) except that the upper limits of the summations have to be $n$ instead of $n_0$, due to the principle of invariance. We can rewrite Eqs. (4.2.22), (4.2.23) for a changing layer number $n$ and explicitly present the layer number dependence as follows:

$$\mathbf{T}(n) = ik\sum_{j=1}^{n}\omega_j\mathbf{J}^T(kr_j)\mathbf{F}(n|\, r_j), \tag{4.2.24}$$

$$\mathbf{F}(n|\, r_i) = \mathbf{U}(r_i)\mathbf{J}(kr_i) + \mathbf{U}(r_i)\sum_{j=1}^{n}\omega_j\mathrm{G}(r_i, r_j)\mathbf{F}(n|\, r_j), \quad i = 1,\ldots,n, \tag{4.2.25}$$

where $\mathbf{T}(n) = \mathbf{T}(r_n)$. The difference form of the invariant-imbedding method for the T-matrix was proposed by Johnson (1988). Using similar steps to the derivation of the derivative form, the difference form is given in the following part.

Using Eqs. (4.1.78), (4.2.25), we can get the expression for $\mathbf{F}(n|r_n)$ when $i=n$ as follows:

$$\mathbf{F}(n|\,r_n) = \omega_n^{-1}\mathbf{Q}(r_n)\left[\mathbf{J}(kr_n) + \mathbf{H}(kr_n)ik\sum_{j=1}^{n-1}\omega_j\mathbf{J}^T(kr_n)\mathbf{F}(n|\,r_j)\right], \tag{4.2.26}$$

where

$$\mathbf{Q}(r_n) = \omega_n[I - \omega_n\mathbf{U}(kr_n)\mathrm{G}(r_n, r_n)]^{-1}\mathbf{U}(r_n). \tag{4.2.27}$$

Then, we can get the expression for $\mathbf{F}(n|r_i)$ when $1 \le i < n$ as follows:

$$\mathbf{F}(n|\,r_i) = \mathbf{U}(r_i)\mathbf{J}(kr_i) + \mathbf{U}(r_i)\sum_{j=1}^{n-1}\omega_j\mathrm{G}(r_i, r_j)\mathbf{F}(n|\,r_j) + \mathbf{U}(r_i)\mathbf{J}(kr_i)\omega_n ik\mathbf{H}^T(kr_n)\mathbf{F}(n|\,r_n), \tag{4.2.28a}$$

or,

$$\mathbf{F}(n|\,r_i) = \mathbf{U}(r_i)\mathbf{J}(kr_i)\left[\mathbf{I} + \omega_n ik\mathbf{H}^T(kr_n)\mathbf{F}(n|\,r_n)\right] + \mathbf{U}(r_i)\sum_{j=1}^{n-1}\omega_j\mathrm{G}(r_i, r_j)\mathbf{F}(n|\,r_j). \tag{4.2.28b}$$

Comparing the forcing term in Eq. (4.2.28b) and the forcing term in Eq. (4.2.25) and also using the expression for $\mathbf{F}(n-1|r_i)$ in Eq. (4.2.25), it is evident that we can define the following relation:

$$\mathbf{F}(n|\,r_i) = \mathbf{F}(n-1|\,r_i)(\mathbf{I} + \mathrm{p}_n), \tag{4.2.29}$$

where

$$\mathrm{p}_n = \omega_n ik\mathbf{H}^T(kr_n)\mathbf{F}(n|\,r_n). \tag{4.2.30}$$

Using Eqs. (4.2.24), (4.2.29), Eq. (4.2.26) can be rewritten as

$$\mathbf{F}(n|\,r_n) = \omega_n^{-1}\mathbf{Q}(r_n)[\mathbf{J}(kr_n) + \mathbf{H}(kr_n)\mathbf{T}(n-1)(\mathbf{I} + \mathrm{p}_n)], \tag{4.2.31}$$

Using Eqs. (4.2.30), (4.2.31), we can explicitly write down

$$\mathrm{p}_n = \mathbf{Q}_{hj}(r_n) + \mathbf{Q}_{hh}(r_n)\mathbf{T}(n-1)(\mathbf{I} + \mathrm{p}_n), \tag{4.2.32a}$$

or,

$$\mathbf{I} + \mathrm{p}_n = [\mathbf{I} - \mathbf{Q}_{hh}(r_n)\mathbf{T}(n-1)]^{-1}\left[\mathbf{I} + \mathbf{Q}_{hj}(r_n)\right], \tag{4.2.32b}$$

where

$$\mathbf{Q}_{hj} = ik\mathbf{H}^T\mathbf{QJ}, \tag{4.2.33}$$

$$\mathbf{Q}_{hh} = ik\mathbf{H}^T\mathbf{QH}. \tag{4.2.34}$$

Accordingly, the T-matrix for $n$ layers in Eq. (4.2.24) can be written as

$$\mathbf{T}(n) = ik\omega_n\mathbf{J}^T(kr_n)\mathbf{F}(n|\, r_n) + \mathbf{T}(n-1)[\mathbf{I} + \mathrm{p}_n]. \tag{4.2.35}$$

Using Eqs. (4.2.31), (4.2.32), Eq. (4.2.35) can be written in an iterative form as

$$\mathbf{T}(n) = \mathbf{Q}_{jj}(r_n) + \left[\mathbf{I} + \mathbf{Q}_{jh}(r_n)\right]\mathbf{T}(n-1)[\mathbf{I} - \mathbf{Q}_{hh}(r_n)\mathbf{T}(n-1)]^{-1}\left[\mathbf{I} + \mathbf{Q}_{hj}(r_n)\right], \tag{4.2.36}$$

where

$$\mathbf{Q}_{jj} = ik\mathbf{J}^T\mathbf{QJ}, \tag{4.2.37}$$

$$\mathbf{Q}_{jh} = ik\mathbf{J}^T\mathbf{QH}. \tag{4.2.38}$$

Eq. (4.2.36) is the invariant-imbedding T-matrix method in difference form. The difference form directly gives a recurrence relation, which is computationally efficient, so the difference form is used in the following T-matrix computation.

Since all $\mathbf{Q}$-related functions are only related to the quadrature point $r_n$, they can be treated as the contributions of the reflection and transmission associated with the corresponding single imbedding layer. $\mathbf{Q}_{jj}(r_n)$ represents the exterior to exterior reflection of the single layer and $\mathbf{Q}_{hh}(r_n)$ the interior to interior reflection of the single layer; $\mathbf{I}+\mathbf{Q}_{hj}(r_n)$ represents the transmission from the exterior to interior direction and $\mathbf{I}+\mathbf{Q}_{jh}(r_n)$ the transmission of the single layer from the interior to exterior direction. Moreover, $\mathbf{T}(n-1)$ represents the reflection from all the ($n$-1) layers. The matrix inversion in Eq. (4.2.36) can be decomposed into summations using a Taylor expansion as

$$[\mathbf{I} - \mathbf{Q}_{hh}(r_n)\mathbf{T}(n-1)]^{-1} = \mathbf{I} + \sum_{p=1}^{\infty}[\mathbf{Q}_{hh}(r_n)\mathbf{T}(n-1)]^p. \tag{4.2.39}$$

The physical interpretation of the difference form can be given in terms of the scattering orders between the single layer and the interior layers as follows:

First-order scattering: $\mathbf{Q}_{jj}(r_n)$
Second-order scattering: $[\mathbf{I}+\mathbf{Q}_{jh}(r_n)]\mathbf{T}(n-1)[\mathbf{I}+\mathbf{Q}_{hj}(r_n)]$
Third-order scattering: $[\mathbf{I}+\mathbf{Q}_{jh}(r_n)]\mathbf{T}(n-1)\mathbf{Q}_{hh}(r_n)\mathbf{T}(n-1)[\mathbf{I}+\mathbf{Q}_{hj}(r_n)]$
...
$p$th-order scattering: $[\mathbf{I}+\mathbf{Q}_{jh}(r_n)]\mathbf{T}(n-1)[\mathbf{Q}_{hh}(r_n)\mathbf{T}(n-1)]^p[\mathbf{I}+\mathbf{Q}_{hj}(r_n)]$
...

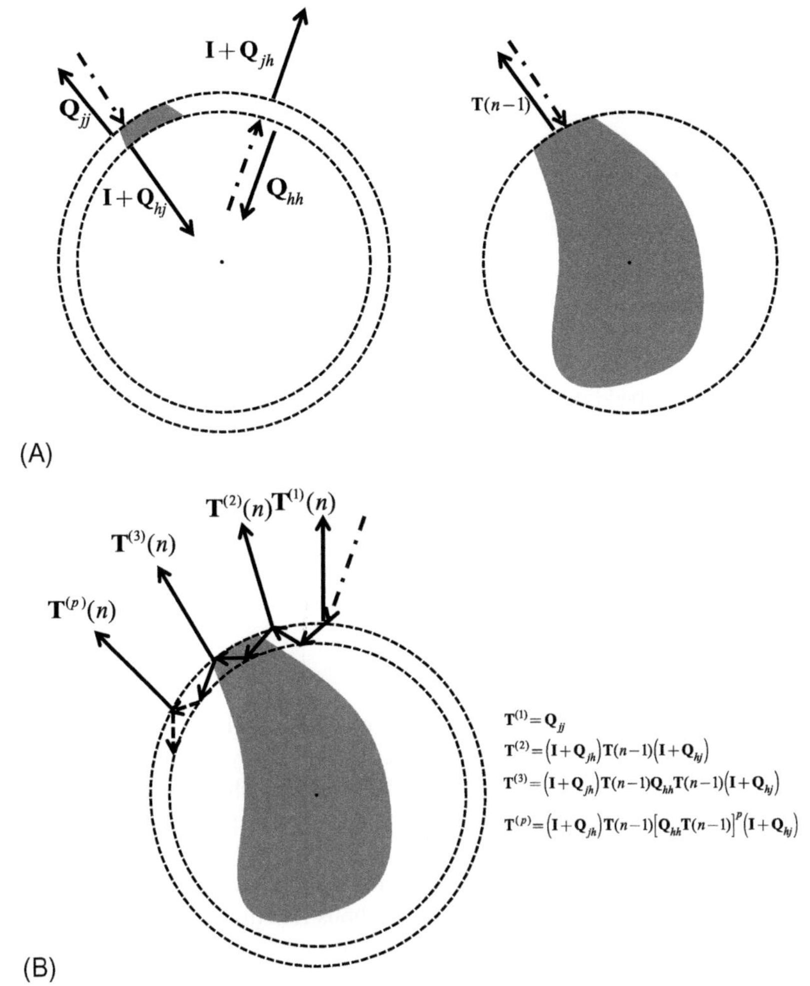


**Fig. 4.4** Physical interpretation of the invariant-imbedding expression of the T-matrix. The left panel of (A) shows reflection and transmission for a single layer for exterior and interior incidence, and the right panel shows the reflection for the layers inside the single layer. (B) A schematic of the physical interpretation. The dash-dot arrows represent the incident light.

which are also shown in Fig. 4.4.

Now, we prove that the difference form is the same as the derivative form in the limit $\Delta r \rightarrow 0$. We use the layer depth between layer $n$-1 and layer $n$ as the weight, that is,

$$\omega_n = \Delta r = r_n - r_{n-1}. \tag{4.2.40}$$

Only retaining terms in which are first order in $\Delta r$, we can obtain the following expressions:

$$\mathbf{Q}(r_n) \approx \Delta r \mathbf{U}(r_n), \tag{4.2.41}$$

$$\mathbf{Q}_{hj} \approx ik\Delta r \mathbf{H}^T \mathbf{U}\mathbf{J}, \quad \mathbf{Q}_{hh} \approx ik\Delta r \mathbf{H}^T \mathbf{U}\mathbf{H}, \tag{4.2.42a}$$

$$\mathbf{Q}_{jj} \approx ik\Delta r \mathbf{J}^T \mathbf{U}\mathbf{J}, \quad \mathbf{Q}_{jh} \approx ik\Delta r \mathbf{J}^T \mathbf{U}\mathbf{H}, \tag{4.2.42b}$$

$$[\mathbf{I} - \mathbf{Q}_{hh}(r_n)\mathbf{T}(n-1)]^{-1} \approx \mathbf{I} + ik\Delta r \mathbf{H}^T(r_n)\mathbf{U}(r_n)\mathbf{H}(r_n)\mathbf{T}(n-1). \tag{4.2.43}$$

Substituting Eqs. (4.2.41)–(4.2.43) into Eq. (4.2.36) and retaining first-order terms in $\Delta r$, the invariant-imbedding equation in the derivative form can be obtained in the limit $\Delta r \to 0$ as follows ( Johnson, 1988):

$$\begin{aligned} \frac{d\mathbf{T}(n)}{dr} &= \lim_{\Delta r \to 0} \frac{\mathbf{T}(n) - \mathbf{T}(n-1)}{\Delta r} \\ &= ik\left[\mathbf{J}^T(r_n) + \mathbf{T}(r_n)\mathbf{H}^T(r_n)\right]\mathbf{U}(r_n)[\mathbf{J}(r_n) + \mathbf{H}(r_n)\mathbf{T}(r_n)]. \end{aligned} \tag{4.2.44}$$

### 4.2.3 T-matrix computation

As stated in Eq. (4.1.78), function $\mathrm{G}(r,r')$ is discontinuous when $r$ approaches $r'$ from above or below. The integration in Eq. (4.2.14) is replaced by a quadrature expression in Eq. (4.2.23). The quadrature method should be carefully chosen because high-order quadrature methods may give worse results than low-order quadrature methods when applied to an integrand with a discontinuity (Farrington, 1961), here caused by $\mathrm{G}(r,r')$. Therefore, the trapezoidal rule is used in computation to give the quadrature expression in Eq. (4.2.23). Accordingly, the discontinuous function $\mathrm{G}(r,r')$ at $r = r'$ can be stated as the average of two limiting values approaching $r'$ from above and below:

$$\mathrm{G}_n(r,r) = \frac{ik}{2}\left[\mathbf{H}_n(kr)\mathbf{J}_n^T(kr) + \mathbf{J}_n(kr)\mathbf{H}_n^T(kr)\right]. \tag{4.2.45}$$

In the imbedding process, a recurrence relation with respect to the T-matrix is given in Eq. (4.2.36). As stated in Section 3.3.3, the T-matrix of a homogeneous spherical particle can be analytically obtained using the Lorenz-Mie theory, and the coefficients are given in Eq. (3.3.41). To optimally save on computational time, the initial value for Eq. (4.2.36) is taken to be the T-matrix at $r = r_i$ as shown in Fig. 4.3, and the imbedding or iterative region is from $r_i$ to $r_c$.

Eq. (4.2.36) is the equation for iteratively computing the T-matrix of an arbitrary particle. As shown for the physical interpretation in Fig. 4.4, the T-matrix $\mathbf{T}(n)$ associated with $n$ layers is determined by the T-matrix $\mathbf{T}(n-1)$ associated with $(n-1)$ layers and $\mathbf{Q}_{jj}$, $\mathbf{Q}_{jh}$, $\mathbf{Q}_{hj}$, and $\mathbf{Q}_{hh}$, which only depend on the characteristics of the $n$th layer. Radial functions $\mathbf{H}(kr_n)$, $\mathbf{J}(kr_n)$, and $\mathrm{G}(r_n,r_n)$ only depend on the radius of the $n$th layer and not on any particle properties, such as shape and refractive index. Accordingly, the only particle-related quantity is the matrix $\mathbf{U}(r_n)$, and using its definition in Eq. (4.2.12), it can be explicitly expressed as

$$\mathbf{U}_{mnm'n'}(r) = k^2 r^2 (-1)^{m+m'} \sqrt{\frac{(2n+1)(2n'+1)}{n(n+1)n'(n'+1)}}$$
$$\cdot \frac{1}{4\pi}\int_0^{\pi} \sin\theta d\theta \int_0^{2\pi} d\varphi \left\{ \exp[i(m'-m)\varphi] \left( \frac{\varepsilon_1(r,\Omega)}{\varepsilon} - 1 \right) \mathrm{u}_{mnm'n'}(r,\Omega) \right\}, \tag{4.2.46}$$

where

$$\mathrm{u}_{mnm'n'}(r,\Omega) = \begin{pmatrix} \begin{matrix}\pi_{mn}(\theta)\pi_{m'n'}(\theta) \\ +\tau_{mn}(\theta)\tau_{m'n'}(\theta)\end{matrix} & \begin{matrix}-i[\pi_{mn}(\theta)\tau_{m'n'}(\theta) \\ +\tau_{mn}(\theta)\pi_{m'n'}(\theta)]\end{matrix} & 0 \\ \begin{matrix}i[\pi_{mn}(\theta)\tau_{m'n'}(\theta) \\ +\tau_{mn}(\theta)\pi_{m'n'}(\theta)]\end{matrix} & \begin{matrix}\pi_{mn}(\theta)\pi_{m'n'}(\theta) \\ +\tau_{mn}(\theta)\tau_{m'n'}(\theta)\end{matrix} & 0 \\ 0 & 0 & \dfrac{\sqrt{n(n+1)n'(n'+1)}d^n_{0m}(\theta)d^{n'}_{0m'}(\theta)}{[\varepsilon_1(r,\Omega)/\varepsilon]} \end{pmatrix}, \tag{4.2.47}$$

where $\varepsilon_1(r,\Omega)$ and $\varepsilon$ are the permittivity of the scattering particle and the surrounding medium, respectively, and the refractive index $m$ of the scattering particle relative to the medium has the relation $m^2(r,\Omega) = \varepsilon_1(r,\Omega)/\varepsilon$. All symmetries associated with the T-matrix caused by the geometric symmetries of the scattering particle introduced in Section 3.4.2 depend on the symmetries of matrix $\mathbf{U}$ because the only shape-related quantities in the recurrence relation of Eq. (4.2.26) are all involved in matrix $\mathbf{U}$. Moreover, the permittivity $\varepsilon_1(r,\Omega)$ in matrix $\mathbf{U}$ is the only shape-related quantity. The application of matrix symmetries can reduce the computational burdens. Especially, axially or discrete rotational symmetries can decouple the T-matrix, so the computation can be significantly reduced. We will discuss the expressions of matrix $\mathbf{U}$ under some shape symmetries. The argument $\theta$ for functions $\pi_{mn}$, $\tau_{mn}$, and $d^n_{0m}$ and the arguments $(r,\Omega)$ or $(r,\theta,\varphi)$ for the permittivity $\varepsilon_1$ in the following expressions are suppressed for clarity.

If the scattering particle has horizontal mirror symmetry, the particle permittivity is described as

$$\varepsilon_1(r,\pi-\theta,\varphi) = \varepsilon_1(r,\theta,\varphi). \tag{4.2.48}$$

Using the symmetric relations in Eq. (4.1.72), matrix $\mathbf{U}$ can be written as

$$\mathbf{U}_{mnm'n'}(r) = k^2 r^2 (-1)^{m+m'} \sqrt{\frac{(2n+1)(2n'+1)}{n(n+1)n'(n'+1)}}$$
$$\cdot \frac{1}{4\pi}\int_0^{\frac{\pi}{2}} \sin\theta d\theta \int_0^{2\pi} d\varphi \left\{ \exp[i(m'-m)\varphi] \left( \frac{\varepsilon_1}{\varepsilon} - 1 \right) \tilde{\mathrm{u}}_{mnm'n'}(r,\Omega) \right\}, \tag{4.2.49}$$

$$\widetilde{\mathrm{u}}_{mnm'n'} = \begin{pmatrix} c_{mnm'n'} \left[\pi_{mn}\pi_{m'n'} + \tau_{mn}\tau_{m'n'}\right] & -i\widetilde{c}_{mnm'n'} \left[\pi_{mn}\tau_{m'n'} + \tau_{mn}\pi_{m'n'}\right] & 0 \\ i\widetilde{c}_{mnm'n'} \left[\pi_{mn}\tau_{m'n'} + \tau_{mn}\pi_{m'n'}\right] & c_{mnm'n'} \left[\pi_{mn}\pi_{m'n'} + \tau_{mn}\tau_{m'n'}\right] & 0 \\ 0 & 0 & c_{mnm'n'} \dfrac{\sqrt{n(n+1)n'(n'+1)}d^{n}_{0m}d^{n'}_{0m'}}{[\varepsilon_1/\varepsilon]} \end{pmatrix}, \tag{4.2.50}$$

where

$$c_{mnm'n'} = 1+(-1)^{n+m+n'+m'}, \tag{4.2.51a}$$

$$\widetilde{c}_{mnm'n'} = 1-(-1)^{n+m+n'+m'}. \tag{4.2.51b}$$

If the scattering particle has vertical mirror symmetry, we generally assume the line of symmetry to be along the direction specified by azimuthal angle $\varphi_0$ (or $\varphi_0+\pi$), and the particle permittivity can be described as

$$\varepsilon_1(r,\theta,2\pi+2\varphi_0-\varphi) = \varepsilon_1(r,\theta,\varphi). \tag{4.2.52}$$

Matrix $\mathbf{U}$ can be expressed as

$$\begin{aligned} \mathbf{U}_{mnm'n'}(r) = {} & k^2r^2(-1)^{m+m'}\sqrt{\frac{(2n+1)(2n'+1)}{n(n+1)n'(n'+1)}}\exp\left[i(m'-m)\varphi_0\right] \\ & \cdot\frac{1}{2\pi}\int_0^{\pi}\sin\theta d\theta\int_{\varphi_0}^{\varphi_0+\pi}d\varphi\left\{\cos\left[i(m'-m)(\varphi-\varphi_0)\right]\left(\frac{\varepsilon_1(r,\Omega)}{\varepsilon}-1\right)\mathrm{u}_{mnm'n'}(r,\Omega)\right\}, \end{aligned} \tag{4.2.53}$$

where $\mathrm{u}_{mnm'n'}$ is given in Eq. (4.2.47).

If the scattering particle has $N$-fold ($N$ is a finite integer) rotational symmetry, the particle permittivity satisfies the relation

$$\varepsilon_1(r,\theta,\varphi+j2\pi/N) = \varepsilon_1(r,\theta,\varphi),\ \ j = 1,2,\ldots,N-1. \tag{4.2.54}$$

Matrix $\mathbf{U}$ can be partially decoupled as

$$\begin{aligned} \mathbf{U}_{mnm'n'}(r) = {} & \delta_{(m+lN)m'}k^2r^2(-1)^{lN}\sqrt{\frac{(2n+1)(2n'+1)}{n(n+1)n'(n'+1)}} \\ & \cdot\frac{N}{4\pi}\int_0^{\pi}\sin\theta d\theta\int_0^{\frac{2\pi}{N}}d\varphi\left\{\exp\left[ilN\varphi\right]\left(\frac{\varepsilon_1}{\varepsilon}-1\right)\mathrm{u}_{mnm'n'}\right\}, \end{aligned} \tag{4.2.55}$$

where $l$ is an integer, that is, if mod($m' - m$,$N$) = 0 (mod is the modular arithmetic function), $\delta_{(m+lN)m'} = 1$; $lN = l*N$. Accordingly, matrix **U** can be decomposed into $N$ components, and the components can be formally described as

$$(\mathbf{U}_{mnm'n'})_j, j = 0,1,2,\ldots,N-1 \cdot \begin{cases} \mathrm{mod}\,(m'-m,N) = 0, \\ \mathrm{mod}\,(m,N) = j. \end{cases} \tag{4.2.56}$$

If the scattering particle has axial symmetry, the particle permittivity is azimuthally independent. The matrix **U** can be decoupled with component $m$ as

$$\mathbf{U}_{mnm'n'}(r) = \delta_{mm'}k^2r^2\sqrt{\frac{(2n+1)(2n'+1)}{n(n+1)n'(n'+1)}}\frac{1}{2}\int_0^\pi \sin\theta d\theta\left[\left(\frac{\varepsilon_1(r,\theta)}{\varepsilon}-1\right)\mathrm{u}_{mnmn'}(r,\theta)\right], \tag{4.2.57}$$

where

$$\mathrm{u}_{mnmn'}(r,\theta) = \begin{pmatrix} \pi_{mn}(\theta)\pi_{mn'}(\theta) + \tau_{mn}(\theta)\tau_{mn'}(\theta) & -i[\pi_{mn}(\theta)\tau_{mn'}(\theta) + \tau_{mn}(\theta)\pi_{mn'}(\theta)] & 0 \\ i[\pi_{mn}(\theta)\tau_{mn'}(\theta) + \tau_{mn}(\theta)\pi_{mn'}(\theta)] & \pi_{mn}(\theta)\pi_{mn'}(\theta) + \tau_{mn}(\theta)\tau_{mn'}(\theta) & 0 \\ 0 & 0 & \dfrac{\sqrt{n(n+1)n'(n'+1)}d^n_{0m}(\theta)d^{n'}_{0m}(\theta)}{[\varepsilon_1(r,\theta)/\varepsilon]} \end{pmatrix}. \tag{4.2.58}$$

Accordingly, the matrix **U** can be decomposed into an infinite number of components as

$$(\mathbf{U}_{mnmn'})_m, \quad m = 0, \pm 1, \pm 2, \ldots \tag{4.2.59}$$

For a particle with spherical symmetry, the particle permittivity is independent of the angular variables. Using the symmetry given in Eq. (4.1.72), matrix **U** is completely decoupled as

$$\mathbf{U}_{mnm'n'}(r) = \delta_{mm'}\delta_{nn'}k^2r^2\left(\frac{\varepsilon_1(r)}{\varepsilon}-1\right)\mathrm{diag}\left(1 \;\; 1 \;\; \frac{\varepsilon}{\varepsilon_1(r)}\right). \tag{4.2.60}$$

There are similar discussions for matrix **U** in Bi et al. (2013).

The layer-related matrices $\mathbf{H}(kr_n)$, $\mathbf{J}(kr_n)$, and $\mathrm{G}(r_n, r_n)$ are radial functions and block diagonal. Accordingly, if matrix **U** is decoupled such as when it is associated with a particle with $N$-fold rotational symmetry or axial symmetry, the T-matrix also has the same decoupled property. The symmetric property of the T-matrix associated with the particle shape is comprehensively discussed in Section 3.4.2.

We have discussed the decoupled T-matrix here according to the symmetry of matrix **U**. If the particle has $N$-fold rotational symmetry, the recurrence relation for the T-matrix can be decomposed into $N$ components as

$$(\mathbf{T}(n))_j = (\mathbf{Q}_{jj}(r_n))_j + \left[\mathbf{I} + (\mathbf{Q}_{jh}(r_n))_j\right](\mathbf{T}(n-1))_j \\ \left[\mathbf{I} - (\mathbf{Q}_{hh}(r_n))_j(\mathbf{T}(n-1))_j\right]^{-1}\left[\mathbf{I} + (\mathbf{Q}_{hj}(r_n))_j\right], \tag{4.2.61}$$

where the $j$th component is defined as $\text{mod}(m-m',N)=0$ and $j=\text{mod}(m,N)$ in Eq. (4.2.56). As stated in Section 3.4.2, only components for $j=0, 1, 2, \ldots, \text{int}[N/2]$ are independent, and the remaining components can be obtained using the T-matrix symmetry. If the particle has axial symmetry, the recurrence relation for the T-matrix can be decomposed into infinite components according to index $m$ as

$$(\mathbf{T}(n))_m = (\mathbf{Q}_{jj}(r_n))_m + \left[\mathbf{I} + (\mathbf{Q}_{jh}(r_n))_m\right](\mathbf{T}(n-1))_m \\ \left[\mathbf{I} - (\mathbf{Q}_{hh}(r_n))_m(\mathbf{T}(n-1))_m\right]^{-1}\left[\mathbf{I} + (\mathbf{Q}_{hj}(r_n))_m\right], \tag{4.2.62}$$

where component $m$ is defined as the one in Eq. (4.2.57). Only components for $m=0$, 1, 2, ... are independent, and the remaining components can be obtained using the T-matrix symmetry. If the particle has spherical symmetry, the recurrence relation for the T-matrix is completely decoupled as

$$(\mathbf{T}(n))_n = (\mathbf{Q}_{jj}(r_n))_n + \left[\mathbf{I} + (\mathbf{Q}_{jh}(r_n))_n\right](\mathbf{T}(n-1))_n \\ \left[\mathbf{I} - (\mathbf{Q}_{hh}(r_n))_n(\mathbf{T}(n-1))_n\right]^{-1}\left[\mathbf{I} + (\mathbf{Q}_{hj}(r_n))_n\right], \tag{4.2.63}$$

where all quantities are $m$ independent and $n=1, 2, \ldots$; all the **Q** matrices and the T-matrix are $2\times 2$ diagonal matrices, for instance, $(\mathbf{T}(n))_n = \begin{pmatrix} \mathrm{T}_n^{11}(n) & 0 \\ 0 & \mathrm{T}_n^{22}(n) \end{pmatrix}$; and the analytical expressions are given in Eq. (3.3.41).

### 4.2.4 Spherical validation of differential form using Lorenz-Mie coefficients

Similarly to Eq. (4.2.63), the derivative form of the IITM with respect to a spherical particle given in Eq. (4.2.21) can be explicitly given by

$$\frac{d\mathrm{T}_n^{11}(r)}{dr} = ik(kr)^2(m^2-1)\left(j_a + \mathrm{T}_n^{11}(r)h_a\right)^2, \tag{4.2.64a}$$

$$\frac{d\mathrm{T}_n^{22}(r)}{dr} = ik(kr)^2(m^2-1)\left[\left(j_b + \mathrm{T}_n^{22}(r)h_b\right)^2 + \left(j_c + \mathrm{T}_n^{22}(r)h_c\right)^2/m^2\right], \tag{4.2.64b}$$

where $k$ is the wave number in the surrounding medium; $m$ here is the refractive index relative to the surrounding medium, that is, $m^2 = \varepsilon_1(r)/\varepsilon$; $n$ is the index to denote the expansion of VSWF; and the functions $j$ and $h$ are defined in terms of the spherical Bessel function and the spherical Hankel function of the first kind as in Eq. (4.1.74), respectively.

For a homogeneous spherical particle with relative refractive index $m$ and radius $r$, the Lorenz-Mie coefficients are given in Eq. (3.3.40), and the corresponding Riccati-Bessel functions are given in Eq. (3.3.41). We use the conventional notation $x = kr$ for spherical particles. After some algebra, the derivatives of the Lorenz-Mie coefficients can be given by

$$\frac{db_n(x)}{dx} = i(1-m^2)(\zeta_n(x) - b_n(x)\xi_n(x))^2. \tag{4.2.65a}$$

$$\frac{da_n(x)}{dx} = i(1-m^2)\left[(\zeta_n'(x) - a_n(x)\xi_n'(x))^2 + n(n+1)\left(j_n(x) - a_n(x)h_n^{(1)}(x)\right)^2/m^2\right], \tag{4.2.65b}$$

It is evident that

$$\begin{cases} \zeta_n(x) = xj_a(x), & \xi_n(x) = xh_a(x); \\ \zeta_n'(x) = xj_b(x), & \xi_n'(x) = xh_b(x); \\ \sqrt{n(n+1)}j_n(x) = j_c(x), & \sqrt{n(n+1)}h_n^{(1)}(x) = h_c(x). \end{cases} \tag{4.2.66}$$

For a spherical particle, $\mathrm{T}_n^{11} = -b_n$, and $\mathrm{T}_n^{22} = -a_n$. Accordingly, Eq. (4.2.65) is the same as Eq. (4.2.64). This verifies the differential form of the IITM for a spherical particle using the Lorenz-Mie coefficients.

## 4.3 Application of the IITM to arbitrary particle morphologies

### 4.3.1 Stability and memory restriction

The key equation for computing the T-matrix is Eq. (4.2.14) with respect to function $\mathbf{F}(r)$, which is a Fredholm equation of the second kind. The integral can be approximated by the choice of a quadrature rule in Eq. (4.2.23). The algebraic solution associated with $n_0$ quadrature points is usually well conditioned because a diagonally dominant matrix inversion is encountered (Press et al., 2007, Chapter 19). The equation in Eq. (4.2.23) is itself a matrix equation. The matrix of a matrix equation, however, invokes super large matrices, which can become extremely time-consuming for large particles. Alternatively, the imbedding algorithm is employed to iteratively get the solution, as shown in Eq. (4.2.36). Moreover, the invariant-imbedding presentation of the T-matrix remains well conditioned. Accordingly, IITM can be applied to arbitrary particle morphologies.

Even without the concern of an ill-conditioned matrix, the efficiency of the IITM depends on the memory and computational time requirements. The matrix computations involve matrix inversion, multiplication, and addition. A large matrix requires both large memory requirements and long computational times for the matrix operations. Apparently, we can have the following relations using the properties of the spherical Bessel function introduced in Section 3.2.1, as follows:

$$(\omega_n \mathbf{U}\mathrm{G})_{mnm'n'} \xrightarrow{n,n'\to\infty} 0, \tag{4.3.1}$$

$$(\mathbf{Q}_{jj})_{mnm'n'},\ (\mathbf{Q}_{jh})_{mnm'n'},\ (\mathbf{Q}_{hh})_{mnm'n'},\ (\mathbf{Q}_{hj})_{mnm'n'},\mathbf{T}_{mnm'n'} \xrightarrow{n,n'\to\infty} 0. \tag{4.3.2}$$

Accordingly, the matrices in the recurrence relation can be truncated at the term related to the current imbedding radius. Naturally, the truncation term increases with the imbedding process. We assume that the truncation term for the vector spherical wave functions is $N$, that is,

$$\sum_{n=1}^{\infty}\sum_{m=-n}^{n} \to \sum_{n=1}^{N}\sum_{m=-n}^{n}. \tag{4.3.3}$$

As described in Section 3.2.2, index $(m, n)$ can be replaced by index $l$ on the basis of a unique correspondence. For clarity, we only consider the maximum truncation term $N_0$ associated with the radius of the smallest circumscribed sphere of the scattering particle. The dimension of the vector spherical wave function is $L=N_0(N_0+2)$. For an arbitrary particle morphology, the dimensions of matrices $\mathbf{U}$ and G are $3L\times 3L$, and the dimensions of matrices $\mathbf{H}$ and $\mathbf{J}$ are $3L\times 2L$. The dimensions of matrices $\mathbf{Q}_{jj}$, $\mathbf{Q}_{jh}$, $\mathbf{Q}_{hh}$, $\mathbf{Q}_{hj}$, and $\mathbf{T}$ are $2L\times 2L$. For a particle with $N$-fold rotational symmetry, however, the dimensions of the matrices are roughly decoupled as $L/N$. Note that the decoupled dimension depends on the remainder in Eq. (4.2.56) and is not exactly $L/N$. Accordingly, the dimensions for (int$[N/2]+1$)-independent T-matrix components are roughly $2\frac{L}{N}\times 2\frac{L}{N}$. For a particle with axial symmetry, the matrix is decoupled according to index $m$. Since only index values $m>0$ are independent, the dimension for $m=0$ is $N_0$, and the dimension for $m=1,2,\ldots,N_0$ is $(N_0-m+1)$. For instance, the dimensions for T-matrix component $m=0$ is $2N_0\times 2N_0$, and the dimensions for T-matrix components with $m>0$ are $2(N_0-m+1)\times 2(N_0-m+1)$, where the details are given in Table 4.1. Fig. 4.5 shows the memory requirements of the T-matrix associated with the size parameter. The truncation term $N_0$ uses $x$+$4.05x^{1/3}+8$ (where $x=kr_c$) as an example to show the requirements. We will discuss the truncation term at the end of this section. The T-matrix is a complex matrix, and double precision is assumed in the computation. With an increased particle size, the memory requirements for the T-matrix significantly increase for particles with arbitrary morphologies. The memory requirements in the computation are much larger than shown in Fig. 4.5 because the transitional matrices, such as $\mathbf{Q}_{jj}$, $\mathbf{Q}_{jh}$, $\mathbf{Q}_{hh}$, $\mathbf{Q}_{hj}$, and $\mathbf{U}$, have memory requirements similar to the T-matrix.

**Table 4.1** Matrix dimension and total element number for an arbitrary morphology, finite-fold symmetric morphology, and axially symmetric morphology.

| $L=N_0(N_0+2)$ | | **U** | $\mathbf{Q}_{jj}$, $\mathbf{Q}_{jh}$, $\mathbf{Q}_{hh}$, $\mathbf{Q}_{hj}$, **T** |
|---|---|---|---|
| Arbitrary morphology | Dimension | $3L\times 3L$ | $2L\times 2L$ |
| | Independent elements | $9L^2$ | $4L^2$ |
| *N*-Fold symmetric morphology | Dimension | $3\frac{L}{N}\times 3\frac{L}{N}$ | $2\frac{L}{N}\times 2\frac{L}{N}$ |
| | Independent elements | $9\frac{L^2}{N^2}(\mathrm{int}[N/2]+1)$ | $4\frac{L^2}{N^2}(\mathrm{int}[N/2]+1)$ |
| Axially symmetric morphology | Dimension for $m=1, 2, \ldots, N_0$ | $3(N_0-m+1)\times 3(N_0-m+1)$ | $2(N_0-m+1)\times 2(N_0-m+1)$ |
| | Dimension for $m=0$ | $3N_0\times 3N_0$ | $2N_0\times 2N_0$ |
| | Independent elements | $9\left[N_0^2+\frac{N_0(N_0+1)(2N_0+1)}{6}\right]$ | $4\left[N_0^2+\frac{N_0(N_0+1)(2N_0+1)}{6}\right]$ |

Note that $N$ indicates the morphology symmetry and $N_0$ is the truncation term in the expansion using the vector spherical wave functions.

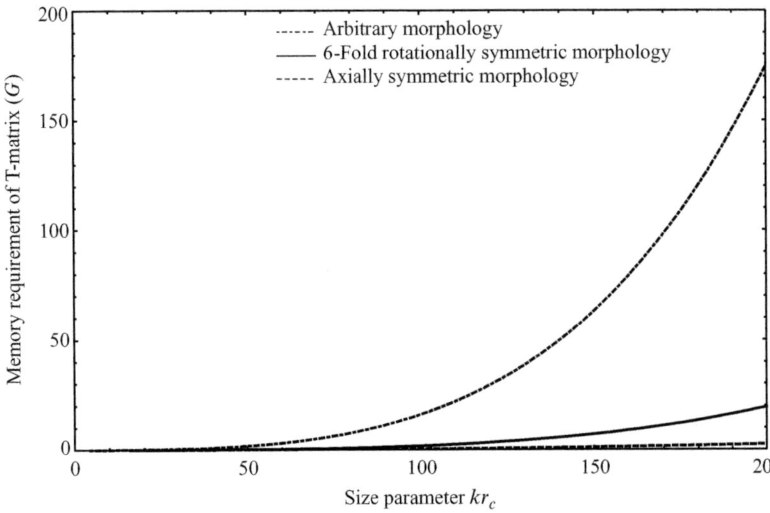


**Fig. 4.5** The memory requirement for the T-matrix associated with the size parameter in the unit of gigabytes. The truncation term $N_0$ uses $x+4.05x^{1/3}+8$ (where $x=kr_c$) as an example to show the memory requirement. The memory requirements at $x=200$ for arbitrary morphology, sixfold symmetric morphology, and axially symmetric morphology are roughly 175G, 19G, and 2.5G, respectively.

According to the analysis of the memory requirement for the T-matrix, two parallelization methods, Open Multi-Processing (OpenMP) and Message Passing Interface (MPI), are used to reduce the computational time and the memory restriction. OpenMP supports shared-memory programming. Even though it cannot reduce the memory restriction, the computational time can be significantly reduced. Also,

OpenMP is easy to implement. On the other hand, MPI supports distributed-memory programming. Not only can it distribute the memory requirements to multiple nodes to significantly reduce memory restrictions, but also it can reduce the computational time. However, during the computation of the MPI, message passing in different processes is usually necessary, which can take time and increase the complexity of implementation. Based on the shape symmetry and the particle size, we use different parallelizations to improve the computational efficiency as shown in Table 4.2. For arbitrary morphologies, OpenMP is used for small particles (the estimated size parameter is $x = kr_c < 50$). For large particles, MPI with message passing has to be used to reduce the memory requirements for each node. For finite-fold symmetric morphologies, the T-matrix is decomposed into several components shown in Eq. (4.2.56). We assign the independent components to different nodes and use OpenMP for the computation of the corresponding component. Accordingly, OpenMP and MPI without message passing are used for sizes $x = kr_c < 100$. For large particles, MPI with message passing during the recurrence relation computation must be used. For axially symmetric morphologies, the T-matrix is decomposed into components according to index $m$. Each $m$-component can be assigned to a process, and no message passing is necessary during the recurrence computation. Since the required memory for axially symmetric morphologies is much smaller than for other morphologies, MPI without message passing can be usually used to calculate the T-matrix for intermediate particle sizes.

### 4.3.2 Extension to arbitrary morphology

The stability of the invariant-imbedding method enables extensive applications not only for axially symmetric morphologies, finite-fold symmetric morphologies, and arbitrary morphologies but also for both homogeneous and inhomogeneous particles. As mentioned in Section 4.2.3, the imbedding process starts at radius $r_i$ of the largest inscribed sphere of the scattering particle and ends at radius $r_c$ of the smallest circumscribed sphere of the scattering particle as shown in Fig. 4.2A. Then, we can determine the quadrature points from radius $r_i$ to radius $r_c$, so the intersected area between the scattering particle and the spherical shell at certain quadrature points can be determined. Consequently, the angular integrations in matrix $\mathbf{U}$ can be computed so that matrices $\mathbf{Q}_{jj}$, $\mathbf{Q}_{jh}$, $\mathbf{Q}_{hj}$, and $\mathbf{Q}_{hh}$ can be obtained in terms of the spherical Bessel functions. The T-matrix can be updated step by step using the recurrence relation until the

**Table 4.2** Parallelization methods used for different morphologies at different sizes.

| | Arbitrary morphology | $N$-Fold symmetric morphology | Axially symmetric morphology |
|---|---|---|---|
| $x < 50$ | OpenMP | MPI + OpenMP | MPI |
| $50 < x < 100$ | MPI | MPI + OpenMP | MPI |
| $100 < x$ | MPI | MPI | MPI |

The numbers given in the table are rough or estimated.

quadrature point reaches radius $r_c$. Technically, we need to find the starting radius $r_i$, the ending radius $r_c$, and the intersected area between the scattering particle and the spherical shell at a quadrature point to compute matrix **U**. The angular integrations in matrix **U** as given in Section 4.2.3 are computed using Gaussian quadrature for both the zenith (cos$\theta$) and azimuthal ($\varphi$) integration. For simple shapes, such as a spheroid, a cylinder, or a hexagonal column, the starting and ending radii and the intersected area at certain quadrature points can be analytically given. For a complex shape, these parameters are not analytically obtained. We will discuss how to proceed with the invariant-imbedding algorithm for a complex shape with a given origin. Without loss of generality, we can assume that the complex shape is constructed by faceted meshes. Each facet can be specified by vertices and a normal direction. Moreover, the faceted meshes are assumed to not intersect. Similar discretization using faceted meshes has been used extensively to emulate a complex shape in many fields such as in technology and computer sciences.

#### 4.3.2.1 Inscribed sphere and circumscribed sphere of a complex shape

For a given origin at a complex shape, the radius $r_c$ of the smallest circumscribed sphere of the morphology can be easily obtained by comparing the distances between the origin and the vertices. In contrast, obtaining radius $r_i$ is not as apparent as obtaining radius $r_c$. We can first find the minimum distance between the origin and one facet (a polygon) and then exhaust all the facets to find the least distance. The following gives the steps to obtain the minimum distance from the origin to a facet.

The equation for a plane including a facet or a polygon can be formally expressed as

$$ax + by + cz = d. \tag{4.3.4}$$

We assume that point $p_0$ is in the plane given in Eq. (4.3.4) and has the minimum distance from the origin. In Cartesian coordinates, the coordinates of point $p_0$ can be expressed as

$$p_0 = \frac{d}{a^2 + b^2 + c^2}(a, b, c). \tag{4.3.5}$$

Whether the point is on the boundary or not can be easily determined, and if the point is not on the boundary, whether the point is inside or outside the polygon can be determined using the ray-casting algorithm (Roth, 1982). If the point is on the boundary or inside the polygon, the minimum distance from the origin to the facet is straightforwardly computed by

$$|p_0| = |d|/\sqrt{a^2 + b^2 + c^2}. \tag{4.3.6}$$

If the point is outside the polygon, we can calculate the minimum distance from the point to the polygon. The polygon is composed of line segments so that the minimum distance from the point to the polygon can be obtained by comparing the minimum distances from the point to the line segments. The two vertices of a line segment are expressed by points $p_1 = (x_1, y_1, z_1)$ and $p_2 = (x_2, y_2, z_2)$. Any point on the line can be parameterized as

$$p = p_1 + t(p_2 - p_1), \tag{4.3.7}$$

where $t$ is an unknown variable. The point on the line including points $p_1$ and $p_2$ that has the minimum distance from point $p_0$ can be described as

$$(p - p_0) \cdot (p_2 - p_1) = 0 \tag{4.3.8a}$$

$$t_m = \frac{(p_0 - p_1) \cdot (p_2 - p_1)}{(p_2 - p_1) \cdot (p_2 - p_1)}. \tag{4.3.8b}$$

Accordingly, the point that has the minimum distance from point $p_0$ is

$$p_m = \begin{cases} p_1, & t_m \leq 0; \\ p_1 + t_m(p_2 - p_1), & 0 < t_m < 1; \\ p_2, & 1 \leq t_m. \end{cases} \tag{4.3.9}$$

Afterward, we can exhaust all the line segments to find the minimum distance from the point $p_0$ to the polygon so that the minimum distance from the origin to the polygon can be easily obtained. Then, we can exhaust all the polygons to find the minimum distance from the origin to the polygons. The right minimum distance is the radius of the largest inscribed sphere.

#### *4.3.2.2 Intersection between spherical shell at certain radius and scattering particle*

Gaussian quadrature is used for the angular integrations in matrix **U**. The zenith integration ($\mu = \cos\theta$) is replaced by Gaussian quadrature, so the spherical shell at a certain radius is decomposed into a series of circles according to the cosine of the zenith angle. For one of the circles with certain $\mu$, the next step is to find the azimuthal intervals that can be obtained by locating the intersections between this circle and the facets. Then, the integrations for the zenith and azimuthal angles can be worked out so that matrix **U** can be given. The algorithm is also described by Bi et al. (2013).

Without loss of generality, the circle at imbedding radius $r_0$ and zenith angle $\theta_0$ is studied to find the azimuthal intervals. Any arbitrary point on the circle in Cartesian coordinate can be described as $(r_0 \sin\theta_0 \cos\varphi, r_0 \sin\theta_0 \sin\varphi, r_0 \cos\theta_0)$. The equation of the plane containing a facet is given in Eq. (4.3.4). The intersection between the circle and the facet satisfies the following equation:

$$\cos(\varphi - \varphi_0) = \frac{C}{\sqrt{A^2 + B^2}}, \tag{4.3.10a}$$

$$A = ar_0 \sin\theta_0, \quad B = br_0 \sin\theta_0, \quad C = d - cr_0 \cos\theta_0. \tag{4.3.10b}$$

$$\sin\varphi_0 = \frac{B}{\sqrt{A^2+B^2}}, \quad \cos\varphi_0 = \frac{A}{\sqrt{A^2+B^2}}. \tag{4.3.10c}$$

The intersection points can be described as

$$\varphi = \begin{cases} \text{no intersection}, & |C| > \sqrt{A^2+B^2}; \\ \varphi_0, & |C| = \sqrt{A^2+B^2}; \\ \varphi_0 \pm \arccos\left(\dfrac{C}{\sqrt{A^2+B^2}}\right), & |C| < \sqrt{A^2+B^2}. \end{cases} \tag{4.3.11}$$

Since no intersection and only one intersection point do not contribute to the azimuthal integration, we shall deal with the one intersection case as no intersection. If there are two intersection points, the ray-casting algorithm (Roth, 1982) can be used to determine if the two intersection points are inside the facet or not. We can check all the facets to find all the intersection points. There could be no intersection or a finite number of intersection points.

If there is no intersection between the circle and all facets, the circle could be inside or outside the scattering particle. An arbitrary point on the circle can be picked to determine if the circle is inside the particle or not. For convenience, the chosen point is $p = (r_0 \sin\theta_0, 0, r_0 \cos\theta_0)$ at 0-degree azimuthal angle, and the center of the circle is $o = (0, 0, r_0 \cos\theta_0)$. We can find all the intersection points between ray $op$ and all the facets. If there is no intersection, point $p$ and the circle must be outside the particle. If there are $N$ intersections, the distances from point $o$ to the intersection points can be sorted as

$$d_1 < d_2 \cdots < d_N. \tag{4.3.12}$$

The outward normal to the facets corresponding to the order of Eq. (4.3.12) can be given as

$$\hat{n}_1 \to \hat{n}_2 \cdots \to \hat{n}_N. \tag{4.3.13}$$

The sign function is defined as $\text{sign}(x) = x/|x|$ (the $x = 0$ case does not occur here). Accordingly, we can have the following sign function between vector $op$ and the normal to the facets as

$$\text{sign}(op \cdot \hat{n}_1) \to \text{sign}(op \cdot \hat{n}_2) \cdots \to \text{sign}(op \cdot \hat{n}_N). \tag{4.3.14}$$

The sign series in Eq. (4.3.14) alternates between 1 and $-1$. Then, we can conclude that the circle inside or outside the particle by

$$\begin{cases} |op| < d_1, & \begin{cases} \text{sign}(op \cdot \hat{n}_1) = 1, & \text{outside}, \\ \text{sign}(op \cdot \hat{n}_1) = -1, & \text{inside}, \end{cases} \\ d_N < |op|, & \text{outside}, \\ d_i < |op| < d_{i+1}, & \begin{cases} \text{sign}(op \cdot \hat{n}_i) = 1, & \text{outside}, \\ \text{sign}(op \cdot \hat{n}) = -1, & \text{inside}. \end{cases} \end{cases} \tag{4.3.15}$$

Whether the circle is inside or outside the particle is determined if there is no intersection between the circle and the facets. If there are $M$ intersection points, the points can be sorted based on the azimuthal angles as follows:

$$\varphi_1 \rightarrow \varphi_2 \cdots \rightarrow \varphi_M \rightarrow \varphi_1. \tag{4.3.16}$$

All the azimuthal angles are arranged from 0 to $2\pi$. The corresponding tangential vectors of the circle at the intersection points and outward normal to the facets stated in the order of Eq. (4.3.16) can be given as

$$\begin{aligned} &\vec{\nu}_1 \rightarrow \vec{\nu}_2 \cdots \rightarrow \vec{\nu}_M, \\ &\vec{\nu}_i = (-\sin\theta_0 \sin\varphi_i,\ \sin\theta_0 \cos\varphi_i,\ \cos\theta_0(\cos\varphi_i - \sin\varphi_i)),\quad i = 1,2,\ldots,M, \end{aligned} \tag{4.3.17a}$$

$$\hat{n}_1 \rightarrow \hat{n}_2 \cdots \rightarrow \hat{n}_M. \tag{4.3.17b}$$

Accordingly, we can have the following sign functions corresponding to Eq. (4.3.16) as follows:

$$\operatorname{sign}\left(\vec{\nu}_1 \cdot \hat{n}_1\right) \rightarrow \operatorname{sign}\left(\vec{\nu}_2 \cdot \hat{n}_2\right) \cdots \rightarrow \operatorname{sign}\left(\vec{\nu}_N \cdot \hat{n}_N\right) \rightarrow \operatorname{sign}\left(\vec{\nu}_1 \cdot \hat{n}_1\right). \tag{4.3.18}$$

The sign series in Eq. (4.3.18) alternates between 1 and $-1$. The signs equal to $-1$ and 1 mean the circle enters the particle and exits the particle, respectively. Accordingly, the corresponding azimuthal range $[\varphi_i, \varphi_{i+1}]$ can be determined to be inside or outside the particle as

$$[\varphi_i, \varphi_{i+1}] \begin{cases} \text{Inside, if } \operatorname{sign}\left(\vec{\nu}_i \cdot \hat{n}_i\right) = -1, \\ \text{Outside, if } \operatorname{sign}\left(\vec{\nu}_i \cdot \hat{n}_i\right) = 1. \end{cases} \tag{4.3.19}$$

Of course, only the range inside the particle is considered in the integration.

The same steps are used to exhaust all the quadrature zenith points. Now, the radii of the largest inscribed sphere and the smallest circumscribed sphere of the particle and the intersected area between the spherical shell and the particle are solved in this section.

### 4.3.3 Discussion about the truncation term and imbedding step size

The T-matrix method is a semianalytical method to obtain the light scattering properties of a complex shape. The term "analytical" here refers to the expansion using the vector spherical wave functions. Numerically, the computation of the T-matrix must truncate the VSWF from an infinite to a finite number of terms based on the particle size, as described in Eq. (4.3.3). The IITM is an imbedding method to obtain the

T-matrix of a scattering particle. The trapezoidal rule is used in the integration computation since the integrand is discontinuous. Accordingly, the step size can affect the computational accuracy. The truncation term $N$ and the step size $\Delta\rho = k\Delta r$ are discussed simply in this section.

The truncation term $N$ for spherical particles is given by Wiscombe (1980), and the expression usually used for spheres is

$$N = x + 4.05x^{1/3} + 2, \tag{4.3.20}$$

where $x = kr$, $r$ is the radius of the sphere, and Eq. (4.3.20) is obtained by fitting formula $N \sim x + cx^{1/3}$ suggested by Khare (1976). The physical interpretation is clear although Eq. (4.3.20) is an estimated formula. The first term $x$ in Eq. (4.3.20) corresponds to a geometric term according to the localization principle described by van de Hulst (1957, p. 208), the second term $x^{1/3}$ accounts for the edge effect (Khare, 1976; Wiscombe, 1980), and the remaining parameters are given by fitting a large amount of data. For nonspherical particles, the semianalytical T-matrix method is usually employed in light scattering. Fitting a formula for nonspherical particles is not realistic because not only the nonspherical particles have different morphology but also the memory and time requirements restrict the feasible amount of computation. When EBCM is used for a particle with axial symmetry, the truncation term is tested by determining the convergence of the T-matrix at component $m = 0$ (Mishchenko, 1993). The convergence criterion is established for particles with axial symmetry. The IITM uses the imbedding method to recursively obtain the T-matrix of the particle. The same convergence criterion for the IITM can be used to determine the truncation term at every imbedding step when the particle has axial symmetry. Moreover, the convergence criterion for a particle with finite-fold rotational symmetry or complex shape still needs to be explored, since the IITM has excellent stability and can be applied to arbitrary morphologies. However, because of the enormous number of imbedding steps, the application of the convergence criterion on every imbedding step is not computationally efficient. Due to the outstanding stability of the IITM, the truncation term at an imbedding step is given as an input parameter described as

$$N_i = \text{int}\left[x_i + 4.05x_i^{1/3}\right] + N_0, \tag{4.3.21}$$

where variable $N_0$ is an input parameter, $x_i = kr_c$, and $r_c$ is the radius of the smallest circumscribed sphere of the imbedded shape. Empirically, for small particles ($kr_c < 100$) and simple morphologies such as a spheroid, a cylinder, a hexagonal column, or a hexahedron, $N_0 = 8$ can be used to get convergent results, and $N_0 = 20$ is an adequate number for small particles. For large particles or complex morphologies, $N_0$ should be increased to 30 or larger depending on the particle size and the complexity. The advantage for IITM is that a larger truncation term can give better convergence, but the trade-off is an increase of computational time. Another conservative formula for the truncation term can be used as

$$N_i = \text{int}\left[x_c + 4.05x_c^{1/3}\right] + N_0, \tag{4.3.22}$$

where $x_c = kr_c$ and the truncation term $N_i$ at an imbedding step is the same as the one for the whole scattering particle. This formula gives a comfortable degree of convergence even for a particle with complex morphology (Liu et al., 2015). However, it takes a longer time than the one in Eq. (4.3.21). In Chapter 5, we will show that the formula in Eq. (4.3.22) can also be applied to large size particles and still give good convergence. The value of $N_0$ is important because an insufficient truncation term can give an incorrect result.

The recurrence relation for the T-matrix in Eq. (4.2.36) is derived from the quadrature expression of the integrations in Eqs. (4.2.22), (4.2.23). Using Eqs. (4.2.15), (4.2.16), the quadrature expressions of the integrations from radius $r_{n-1}$ to radius $r_n$ can be described as

$$\mathbf{T}(r_n) = ik\int_0^{r_{n-1}} dr' \left[\mathbf{J}^T(r')\mathbf{F}(r_n|\, r')\right] + ik\omega_n \mathbf{J}^T(r_n)\mathbf{F}(r_n|\, r_n), \tag{4.3.23}$$

$$\mathbf{F}(r_n|\, r) = \mathbf{U}(r)\mathbf{J}(r) + \mathbf{U}(r)\int_0^{r_{n-1}} dr' \mathrm{G}(r,r')\mathbf{F}(r_n|\, r') + \mathbf{U}(r)\mathrm{G}(r,r_n)\mathbf{F}(r_n|\, r_n). \tag{4.3.24}$$

The recurrence relation between the T-matrix $\mathbf{T}(r_n)$ at radius $r_n$ and the T-matrix $\mathbf{T}(r_{n-1})$ at radius $r_{n-1}$ is actually obtained from the quadrature expression of the integration from radius $r_{n-1}$ to radius $r_n$. Accordingly, the error originates from the quadrature expression. For the trapezoidal rule, the error can be usually expressed as follows (Atkinson, 1989):

$$\text{error} = \frac{(kr_c - kr_i)^3}{12n_0^2}\frac{d^2\mathcal{F}(kr_\xi)}{d(kr)^2}, \tag{4.3.25}$$

where $\mathcal{F}(r_\xi)$ represents the integrand in Eqs. (4.3.23), (4.3.24) and $r_\xi \in [r_i, r_c]$. Although Eq. (4.3.25) shows that more steps $n_0$ can give more accurate results, the computational time is linearly increased with the number of steps. Moreover, the step number $n_0$ should be increased or the step size $\Delta\rho$ should be decreased when the $k(r_c - r_i)$ is large. In Section 5.1, we will use the Lorenz-Mie coefficients as benchmarks to discuss the step size.

# Application examples of optical properties of small-to-moderate size particles

# 5

**(With contributions by Jiachen Ding)**

The basic principles of the invariant imbedding T-matrix method are discussed in Chapter 4. We present the applications of the IITM to particles with various morphologies in this chapter. The comparisons between the IITM and the Lorenz-Mie theory for spheres are given in Section 5.1. Applications to particles with axial symmetry, spheroids, and cylinders are given in Section 5.2, and an application to particles with finite-fold rotational symmetry, hexagonal prisms, is given in Section 5.3. Section 5.4 presents the applications to asymmetric particles, predefined aggregates and irregular hexahedra. Section 5.5 presents an application to inhomogeneous particles, nested hexahedra.

## 5.1 Spherical and multilayered spherical particles

For a particle with spherical symmetry, the T-matrix is simplified to be a diagonal matrix. Moreover, in the difference form of the IITM, the recurrence formula is also completely decoupled as in Eq. (4.2.63) and is restated here:

$$\begin{aligned}\left[\mathbf{T}(r_p)\right]_n = & \left[\mathbf{Q}_{jj}(r_p)\right]_n + \left(\mathbf{I} + \left[\mathbf{Q}_{jh}(r_p)\right]_n\right)\left[\mathbf{T}(r_{p-1})\right]_n \\ & \left(\mathbf{I} - \left[\mathbf{Q}_{hh}(r_p)\right]_n\left[\mathbf{T}(r_{p-1})\right]_n\right)^{-1}\left(\mathbf{I} + \left[\mathbf{Q}_{hj}(r_p)\right]_n\right),\end{aligned} \tag{5.1.1}$$

where all the matrices indicated with bold faced symbols are $2 \times 2$ and diagonal. The T-matrix can be formally expressed as

$$\left[\mathbf{T}(r_p)\right]_n = \begin{pmatrix} \mathrm{T}_n^{11}(r_p) & 0 \\ 0 & \mathrm{T}_n^{22}(r_p) \end{pmatrix}. \tag{5.1.2}$$

The $\mathbf{Q}$ matrix can be explicitly given by

$$[\mathbf{Q}_{\mathcal{AB}}]_n = \begin{pmatrix} \mathcal{A}_a\mathcal{C}_n^{11}\mathcal{B}_a & 0 \\ 0 & \mathcal{A}_b\mathcal{C}_n^{22}\mathcal{B}_b + \mathcal{A}_b\mathcal{C}_n^{23}\mathcal{B}_c + \mathcal{A}_c\mathcal{C}_n^{32}\mathcal{B}_b + \mathcal{A}_c\mathcal{C}_n^{33}\mathcal{B}_c \end{pmatrix}, \tag{5.1.3}$$

**Invariant Imbedding T-matrix Method for Light Scattering by Nonspherical and Inhomogeneous Particles**
**https://doi.org/10.1016/B978-0-12-818090-7.00005-X**

where the notations $\mathcal{A}$ and $\mathcal{B}$ denote the corresponding matrices of the subscripts of **Q** matrices in Eq. (5.1.1). For instance, if the subscripts of the **Q** matrix are "*jh*," then $\mathcal{A}=j$ and $\mathcal{B}=h$, and the components are given in Eq. (4.1.74), and matrix $\mathcal{C}$ is given by

$$\mathcal{C}_n=\begin{pmatrix} \dfrac{s_n}{1-s_n g_{11}} & 0 & 0 \\ 0 & \dfrac{(1-g_{33}s_n/m^2)s_n}{d_n} & \dfrac{g_{23}s_n^2/m^2}{d_n} \\ 0 & \dfrac{g_{32}s_n^2/m^2}{d_n} & \dfrac{(1-g_{22}s_n)s_n/m^2}{d_n} \end{pmatrix}, \tag{5.1.4}$$

where

$$g=\begin{pmatrix} h_a j_a & 0 & 0 \\ 0 & h_b j_b & (h_b j_c+h_c j_b)/2 \\ 0 & (h_b j_c+h_c j_b)/2 & h_c j_c \end{pmatrix}, \tag{5.1.5}$$

$$s_n=ik\omega_n(kr_p)^2(m^2-1), \tag{5.1.6}$$

$$d_n=(1-s_n g_{22})(1-s_n g_{33}/m^2)-g_{23}g_{32}s_n^2/m^2, \tag{5.1.7}$$

where $m$ is the refractive index relative to the surrounding medium and is a function of the radius $r_p$, $k$ is the wave number in the surrounding medium, and the functions $j_{a,\,b,\,c}$ and $h_{a,\,b,\,c}$ are defined in Eq. (4.1.74). Matrix $g$ in Eq. (5.1.5) is symmetric, and so is the matrix $\mathcal{C}$ in Eq. (5.1.4).

Eqs. (5.1.1)–(5.1.7) are the necessary equations to compute the T-matrix of a spherical particle. The T-matrix in this situation is actually the Lorenz-Mie coefficients $a_n$ and $b_n$ , with a difference of a minus sign, which are given in Eq. (3.3.41). Since the Lorenz-Mie coefficients can be computed more accurately, we next use these coefficients to discuss the precision of the IITM method by using different quadrature rules and different step sizes in discretizing the particle into a multilayered particle. We define the errors as follows:

$$db_n=\sqrt{\left|(b_n)_{\mathrm{IITM}}-(b_n)_{\mathrm{LM}}\right|^2}, \tag{5.1.8a}$$

$$da_n=\sqrt{\left|(a_n)_{\mathrm{IITM}}-(a_n)_{\mathrm{LM}}\right|^2}, \tag{5.1.8b}$$

where the subscript "IITM" denotes the results computed using Eqs. (5.1.1)–(5.1.7) and the subscript "LM" denotes the results computed using Eq. (3.3.41).

Fig. 5.1 gives the IITM errors of Lorenz-Mie coefficients of a sphere with $x=60$ and $m=1.311+i1.64\times10^{-8}$. The step size is 0.1 for the trapezoidal rule, so

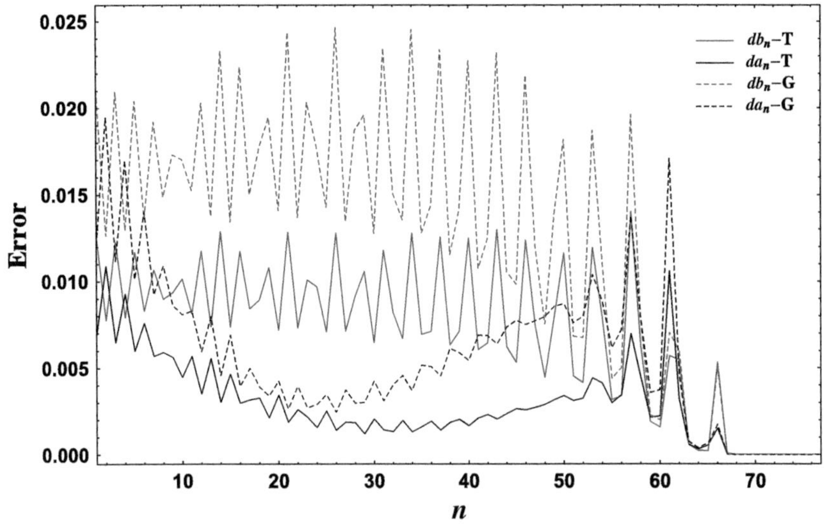


**Fig. 5.1** The errors of Lorenz-Mie coefficients of a sphere calculated by the IITM with $x=60$ and $m=1.311+i1.64\times10^{-8}$. The starting point is $x=kr_i=0$. The solid lines are the results using the trapezoidal rule, while the dashed lines are the results based on the Gaussian quadrature rule.

the step number is 600. Accordingly, the Gaussian quadrature rule uses 600 quadrature points to do the comparison. The starting point for Eq. (5.1.1) is $x=kr_i=0$. Although Gaussian quadrature is a high-order quadrature method, the computed results are worse than the ones computed by the low-order quadrature method with the trapezoidal rule. Fig. 5.2 gives the IITM errors of the Lorenz-Mie coefficients of a sphere with $x=60$ and $m=1.311+i1.64\times10^{-8}$ for different step sizes. The starting point is also $x=kr_i=0$. The step sizes $\Delta\rho$ of 0.2, 0.1, and 0.05 are used and denoted in the legends. According to Eq. (4.3.25), with the decrease of step sizes $\Delta\rho$, the errors significantly decrease. Fig. 5.3 uses the same parameters as Fig. 5.2 except that $x=30$. With the same step sizes, the larger sizes generate larger errors as evident from comparing Fig. 5.2 and Fig. 5.3.

## 5.2 Axially symmetric particles: Spheroids and cylinders

For a particle with axial symmetry, the T-matrix is decoupled into block diagonal matrices in terms of index $m$. The procedure to compute the T-matrix can be simplified by obtaining each block of the T-matrix separately. The scattering phase matrix in random orientation only has six independent elements, denoted as $P_{11}$, $P_{22}$, $P_{33}$, $P_{44}$, $P_{12}$, $P_{34}$ as shown in Eq. (2.2.78), and can be analytically obtained using the

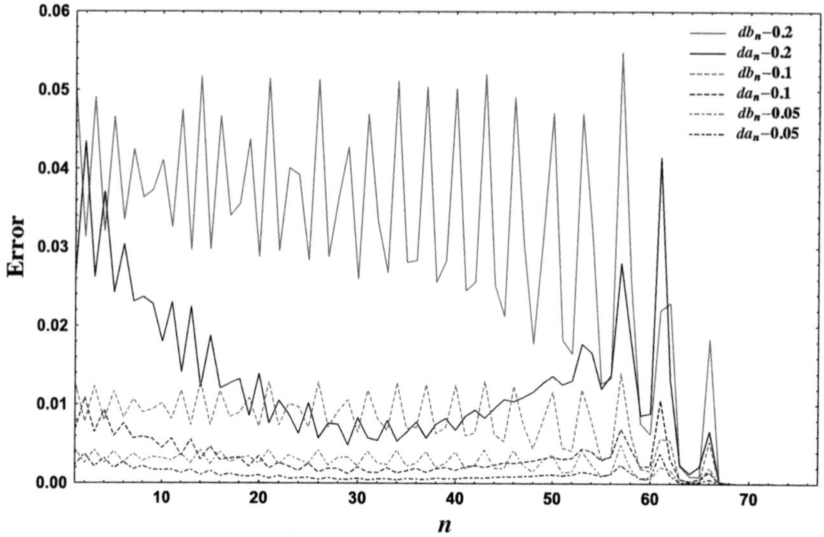


**Fig. 5.2** The IITM errors of the Lorenz-Mie coefficients of a sphere with $x=60$ and $m=1.311+i1.64\times10^{-8}$ for different step sizes. The starting point is $x=kr_i=0$. All IITM results are computed using the trapezoidal rule. The solid, dashed, and dot-dashed lines are the results for step size $\Delta\rho$ of 0.2, 0.1, and 0.05, respectively. The step sizes are also denoted in the legend.

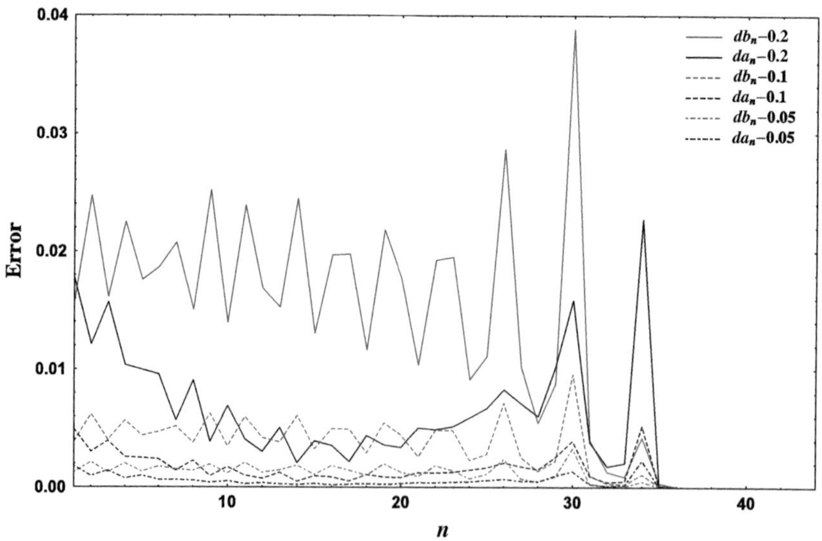


**Fig. 5.3** The same as Fig. 5.2 except that $x=30$.

T-matrix (Mishchenko et al., 2002). Specifically, the six independent elements are expanded by using the Wigner-d functions as shown in Eq. (3.4.75), and the expansion coefficients denoted as $\{\alpha_1^s\}$, $\{\alpha_2^s\}$, $\{\alpha_3^s\}$, $\{\alpha_4^s\}$, $\{\beta_1^s\}$, $\{\beta_2^s\}$ are expressed in terms of the T-matrix elements. The EBCM method introduced in Section 3.3.2 is able to efficiently compute the single-scattering properties of axially symmetric particles. Consequently, the comparisons between the EBCM and the IITM are carried out in this section to illustrate the numerical accuracy of the IITM. Moreover, the scattering properties for particles with large sizes and extreme aspect ratios can be given by using the IITM.

Fig. 5.4A shows a homogeneous spheroid with semiminor axis $a$ and semimajor axis $c$, and the aspect ratio (AR) for a spheroid is defined as $\mathrm{AR}=a/c$; Fig. 5.4B shows a homogeneous cylinder with diameter $D$ and height $H$, and the aspect ratio for a cylinder is defined as $\mathrm{AR}=D/H$. If $\mathrm{AR}>1$, the particle is oblate; if $\mathrm{AR}<1$, the particle is prolate.

### 5.2.1 Comparisons between IITM and EBCM

Suppose a prolate spheroid has parameters $2\pi c/\lambda=20$, $\mathrm{AR}=0.5$, and refractive index $1.311+i0.0$. In Table 5.1, the expansion coefficients with respect to the six independent elements computed by the EBCM and IITM are compared in a term-by-term form, and the differences are also given. In Table 5.2, the scattering phase matrix elements at special scattering angles are compared for the EBCM, IITM, and the

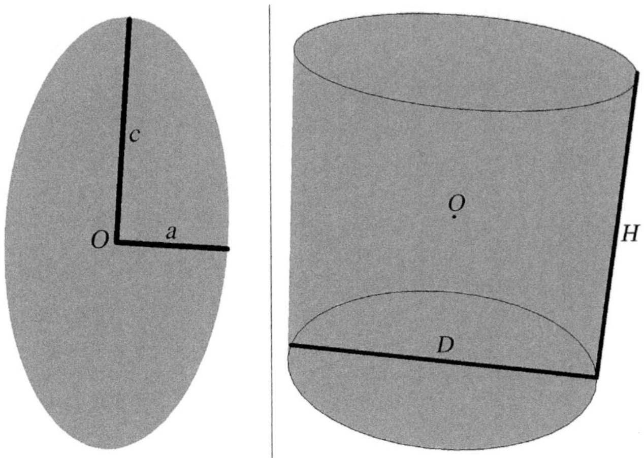


(A) (B)

**Fig. 5.4** (A) Spheroid with a semiminor axis $a$ and semimajor axis $c$; (B) cylinder with diameter $D$ and height $H$.

**Table 5.1** Expansion coefficient comparisons between the EBCM and the IITM for a spheroid in random orientation.

| $s$ | $\alpha_1^s$ | $\alpha_2^s$ | $\alpha_3^s$ | $\alpha_4^s$ | $\beta_1^s$ | $\beta_2^s$ |
|---|---|---|---|---|---|---|
| 0 | 1.0000 | 0.0000 | 0.0000 | 0.9433 | 0.0000 | 0.0000 |
| | 1.0000 | 0.0000 | 0.0000 | 0.9430 | 0.0000 | 0.0000 |
| | 0.0000 | 0.0000 | 0.0000 | −0.0003 | 0.0000 | 0.0000 |
| 1 | 2.2604 | 0.0000 | 0.0000 | 2.2884 | 0.0000 | 0.0000 |
| | 2.2608 | 0.0000 | 0.0000 | 2.2891 | 0.0000 | 0.0000 |
| | 0.0004 | 0.0000 | 0.0000 | 0.0007 | 0.0000 | 0.0000 |
| 2 | 3.0956 | 4.0384 | 3.9285 | 3.0927 | −0.0203 | 0.1173 |
| | 3.0961 | 4.0387 | 3.9291 | 3.0943 | −0.0209 | 0.1164 |
| | 0.0006 | 0.0003 | 0.0006 | 0.0016 | −0.0006 | −0.0009 |
| 3 | 3.6916 | 4.0783 | 4.0243 | 3.6649 | −0.0824 | 0.0182 |
| | 3.6925 | 4.0802 | 4.0252 | 3.6655 | −0.0839 | 0.0181 |
| | 0.0009 | 0.0019 | 0.0009 | 0.0005 | −0.0014 | −0.0001 |
| 4 | 3.7586 | 4.1420 | 4.1315 | 3.7575 | −0.0476 | 0.0114 |
| | 3.7601 | 4.1431 | 4.1326 | 3.7585 | −0.0493 | 0.0107 |
| | 0.0014 | 0.0012 | 0.0011 | 0.0010 | −0.0017 | −0.0007 |
| 5 | 3.6707 | 4.0667 | 4.0462 | 3.6616 | −0.0576 | 0.0538 |
| | 3.6720 | 4.0676 | 4.0480 | 3.6638 | −0.0601 | 0.0545 |
| | 0.0013 | 0.0009 | 0.0018 | 0.0022 | −0.0025 | 0.0006 |
| 6 | 3.7067 | 3.8595 | 3.8504 | 3.7138 | −0.0320 | 0.0736 |
| | 3.7080 | 3.8615 | 3.8520 | 3.7150 | −0.0353 | 0.0746 |
| | 0.0014 | 0.0021 | 0.0017 | 0.0012 | −0.0033 | 0.0010 |
| 7 | 3.7767 | 3.8661 | 3.8396 | 3.7724 | −0.0416 | 0.0248 |
| | 3.7780 | 3.8675 | 3.8408 | 3.7736 | −0.0451 | 0.0253 |
| | 0.0013 | 0.0014 | 0.0012 | 0.0012 | −0.0035 | 0.0005 |
| 8 | 3.7961 | 3.9519 | 3.9291 | 3.7884 | −0.0158 | −0.0150 |
| | 3.7972 | 3.9529 | 3.9302 | 3.7897 | −0.0190 | −0.0143 |
| | 0.0012 | 0.0010 | 0.0011 | 0.0013 | −0.0032 | 0.0007 |
| 9 | 3.8358 | 3.9132 | 3.9036 | 3.8412 | 0.0054 | 0.0104 |
| | 3.8369 | 3.9145 | 3.9048 | 3.8422 | 0.0027 | 0.0111 |
| | 0.0011 | 0.0013 | 0.0011 | 0.0009 | −0.0026 | 0.0007 |
| 10 | 3.9218 | 3.9664 | 3.9448 | 3.9180 | 0.0105 | −0.0068 |
| | 3.9227 | 3.9674 | 3.9459 | 3.9189 | 0.0088 | −0.0066 |
| | 0.0009 | 0.0010 | 0.0010 | 0.0009 | −0.0017 | 0.0002 |
| 11 | 4.0063 | 4.0767 | 4.0690 | 4.0157 | −0.0015 | −0.0301 |
| | 4.0073 | 4.0775 | 4.0696 | 4.0163 | −0.0026 | −0.0301 |
| | 0.0010 | 0.0008 | 0.0006 | 0.0006 | −0.0011 | −0.0000 |
| 12 | 4.0986 | 4.1449 | 4.1240 | 4.0969 | 0.0104 | −0.0494 |
| | 4.0994 | 4.1456 | 4.1247 | 4.0974 | 0.0108 | −0.0496 |
| | 0.0007 | 0.0007 | 0.0007 | 0.0005 | 0.0003 | −0.0002 |
| 13 | 4.1766 | 4.2152 | 4.2027 | 4.1836 | 0.0096 | −0.0486 |
| | 4.1771 | 4.2156 | 4.2033 | 4.1840 | 0.0112 | −0.0492 |
| | 0.0005 | 0.0004 | 0.0006 | 0.0004 | 0.0016 | −0.0006 |
| 14 | 4.2517 | 4.3001 | 4.2913 | 4.2655 | −0.0168 | −0.0625 |
| | 4.2522 | 4.3005 | 4.2916 | 4.2655 | −0.0139 | −0.0633 |
| | 0.0005 | 0.0003 | 0.0002 | 0.0000 | 0.0029 | −0.0007 |

**Table 5.1** Continued

| $s$ | $\alpha_1^s$ | $\alpha_2^s$ | $\alpha_3^s$ | $\alpha_4^s$ | $\beta_1^s$ | $\beta_2^s$ |
|---|---|---|---|---|---|---|
| 15 | 4.3187 | 4.3568 | 4.3287 | 4.3116 | −0.0333 | −0.0881 |
| | 4.3188 | 4.3566 | 4.3291 | 4.3117 | −0.0291 | −0.0890 |
| | 0.0001 | −0.0001 | 0.0004 | 0.0001 | 0.0042 | −0.0009 |
| 16 | 4.3366 | 4.3779 | 4.3713 | 4.3511 | −0.0335 | −0.0858 |
| | 4.3365 | 4.3775 | 4.3710 | 4.3506 | −0.0283 | −0.0868 |
| | −0.0002 | −0.0003 | −0.0003 | −0.0005 | 0.0052 | −0.0010 |
| 17 | 4.3231 | 4.3696 | 4.3582 | 4.3348 | −0.0821 | −0.1087 |
| | 4.3225 | 4.3689 | 4.3579 | 4.3344 | −0.0757 | −0.1097 |
| | −0.0006 | −0.0008 | −0.0002 | −0.0004 | 0.0064 | −0.0010 |
| 18 | 4.2283 | 4.2792 | 4.2638 | 4.2378 | −0.1097 | −0.1512 |
| | 4.2274 | 4.2784 | 4.2637 | 4.2378 | −0.1030 | −0.1520 |
| | −0.0008 | −0.0009 | −0.0001 | −0.0000 | 0.0067 | −0.0008 |
| 19 | 4.0227 | 4.0833 | 4.0595 | 4.0209 | −0.1079 | −0.1912 |
| | 4.0224 | 4.0831 | 4.0596 | 4.0211 | −0.1022 | −0.1921 |
| | −0.0002 | −0.0002 | 0.0001 | 0.0002 | 0.0057 | −0.0009 |
| 20 | 3.7553 | 3.8134 | 3.7744 | 3.7321 | −0.0736 | −0.2001 |
| | 3.7553 | 3.8135 | 3.7751 | 3.7331 | −0.0695 | −0.2011 |
| | −0.0001 | 0.0001 | 0.0007 | 0.0011 | 0.0042 | −0.0010 |
| 21 | 3.4192 | 3.4649 | 3.4502 | 3.4139 | −0.0373 | −0.1324 |
| | 3.4195 | 3.4654 | 3.4507 | 3.4146 | −0.0351 | −0.1343 |
| | 0.0003 | 0.0006 | 0.0005 | 0.0007 | 0.0023 | −0.0019 |
| 22 | 3.1489 | 3.1900 | 3.1802 | 3.1468 | −0.0481 | −0.1034 |
| | 3.1494 | 3.1906 | 3.1808 | 3.1475 | −0.0472 | −0.1055 |
| | 0.0005 | 0.0007 | 0.0006 | 0.0007 | 0.0010 | −0.0021 |
| 23 | 2.8927 | 2.9270 | 2.9140 | 2.8850 | −0.0446 | −0.0972 |
| | 2.8932 | 2.9277 | 2.9146 | 2.8858 | −0.0447 | −0.0995 |
| | 0.0006 | 0.0007 | 0.0006 | 0.0008 | −0.0002 | −0.0023 |
| 24 | 2.6315 | 2.6643 | 2.6548 | 2.6246 | −0.0347 | −0.0792 |
| | 2.6319 | 2.6650 | 2.6553 | 2.6254 | −0.0357 | −0.0816 |
| | 0.0004 | 0.0007 | 0.0006 | 0.0008 | −0.0010 | −0.0024 |
| 25 | 2.3784 | 2.4051 | 2.4021 | 2.3769 | −0.0325 | −0.0634 |
| | 2.3787 | 2.4056 | 2.4025 | 2.3775 | −0.0343 | −0.0659 |
| | 0.0003 | 0.0006 | 0.0004 | 0.0006 | −0.0018 | −0.0025 |
| 26 | 2.1347 | 2.1587 | 2.1562 | 2.1334 | −0.0317 | −0.0592 |
| | 2.1349 | 2.1592 | 2.1565 | 2.1339 | −0.0342 | −0.0617 |
| | 0.0002 | 0.0005 | 0.0003 | 0.0005 | −0.0025 | −0.0026 |
| 27 | 1.8909 | 1.9162 | 1.9136 | 1.8893 | −0.0302 | −0.0583 |
| | 1.8910 | 1.9166 | 1.9138 | 1.8898 | −0.0332 | −0.0609 |
| | 0.0002 | 0.0004 | 0.0002 | 0.0004 | −0.0030 | −0.0026 |
| 28 | 1.6498 | 1.6705 | 1.6687 | 1.6487 | −0.0260 | −0.0578 |
| | 1.6499 | 1.6709 | 1.6687 | 1.6491 | −0.0294 | −0.0605 |
| | 0.0001 | 0.0004 | 0.0001 | 0.0003 | −0.0034 | −0.0027 |
| 29 | 1.4083 | 1.4278 | 1.4270 | 1.4083 | −0.0224 | −0.0560 |
| | 1.4084 | 1.4281 | 1.4271 | 1.4086 | −0.0260 | −0.0586 |
| | 0.0001 | 0.0003 | 0.0000 | 0.0003 | −0.0035 | −0.0027 |

*Continued*

**Table 5.1** Continued

| $s$ | $\alpha_1^s$ | $\alpha_2^s$ | $\alpha_3^s$ | $\alpha_4^s$ | $\beta_1^s$ | $\beta_2^s$ |
|---|---|---|---|---|---|---|
| 30 | 1.1753 | 1.1937 | 1.1924 | 1.1750 | −0.0187 | −0.0559 |
| | 1.1753 | 1.1939 | 1.1924 | 1.1751 | −0.0222 | −0.0584 |
| | 0.0000 | 0.0003 | −0.0000 | 0.0001 | −0.0035 | −0.0025 |
| 31 | 0.9517 | 0.9681 | 0.9677 | 0.9524 | −0.0153 | −0.0533 |
| | 0.9517 | 0.9682 | 0.9675 | 0.9524 | −0.0186 | −0.0556 |
| | −0.0000 | 0.0001 | −0.0002 | −0.0000 | −0.0033 | −0.0023 |
| 32 | 0.7462 | 0.7604 | 0.7593 | 0.7462 | −0.0117 | −0.0534 |
| | 0.7461 | 0.7604 | 0.7590 | 0.7461 | −0.0147 | −0.0554 |
| | −0.0001 | 0.0001 | −0.0003 | −0.0001 | −0.0030 | −0.0020 |
| 33 | 0.5564 | 0.5686 | 0.5680 | 0.5569 | −0.0063 | −0.0502 |
| | 0.5563 | 0.5686 | 0.5676 | 0.5566 | −0.0088 | −0.0519 |
| | −0.0001 | −0.0000 | −0.0003 | −0.0002 | −0.0025 | −0.0017 |
| 34 | 0.3925 | 0.4025 | 0.4011 | 0.3921 | −0.0003 | −0.0454 |
| | 0.3922 | 0.4024 | 0.4007 | 0.3918 | −0.0022 | −0.0467 |
| | −0.0002 | −0.0001 | −0.0004 | −0.0003 | −0.0019 | −0.0012 |
| 35 | 0.2572 | 0.2649 | 0.2635 | 0.2567 | 0.0048 | −0.0361 |
| | 0.2570 | 0.2646 | 0.2631 | 0.2564 | 0.0034 | −0.0369 |
| | −0.0003 | −0.0002 | −0.0004 | −0.0003 | −0.0014 | −0.0009 |
| 36 | 0.1568 | 0.1619 | 0.1597 | 0.1552 | 0.0075 | −0.0261 |
| | 0.1565 | 0.1617 | 0.1594 | 0.1549 | 0.0066 | −0.0266 |
| | −0.0003 | −0.0002 | −0.0003 | −0.0003 | −0.0009 | −0.0005 |
| 37 | 0.0858 | 0.0891 | 0.0883 | 0.0855 | 0.0070 | −0.0143 |
| | 0.0857 | 0.0889 | 0.0882 | 0.0853 | 0.0065 | −0.0147 |
| | −0.0001 | −0.0001 | −0.0002 | −0.0001 | −0.0005 | −0.0003 |
| 38 | 0.0454 | 0.0471 | 0.0459 | 0.0443 | 0.0048 | −0.0090 |
| | 0.0452 | 0.0470 | 0.0458 | 0.0442 | 0.0046 | −0.0091 |
| | −0.0001 | −0.0001 | −0.0001 | −0.0001 | −0.0003 | −0.0001 |
| 39 | 0.0204 | 0.0213 | 0.0210 | 0.0202 | 0.0032 | −0.0036 |
| | 0.0203 | 0.0213 | 0.0210 | 0.0202 | 0.0031 | −0.0036 |
| | −0.0001 | −0.0001 | −0.0001 | −0.0001 | −0.0001 | −0.0001 |
| 40 | 0.0093 | 0.0098 | 0.0094 | 0.0090 | 0.0015 | −0.0019 |
| | 0.0093 | 0.0097 | 0.0094 | 0.0090 | 0.0015 | −0.0019 |
| | −0.0000 | −0.0000 | −0.0000 | −0.0000 | −0.0001 | −0.0000 |
| 41 | 0.0035 | 0.0037 | 0.0036 | 0.0034 | 0.0008 | −0.0006 |
| | 0.0035 | 0.0037 | 0.0036 | 0.0034 | 0.0008 | −0.0006 |
| | −0.0000 | −0.0000 | −0.0000 | −0.0000 | −0.0000 | −0.0000 |
| 42 | 0.0014 | 0.0015 | 0.0014 | 0.0013 | 0.0003 | −0.0003 |
| | 0.0014 | 0.0014 | 0.0014 | 0.0013 | 0.0003 | −0.0003 |
| | −0.0000 | −0.0000 | −0.0000 | −0.0000 | −0.0000 | −0.0000 |
| 43 | 0.0004 | 0.0005 | 0.0004 | 0.0004 | 0.0001 | −0.0001 |
| | 0.0004 | 0.0005 | 0.0004 | 0.0004 | 0.0001 | −0.0001 |
| | 0.0000 | 0.0000 | 0.0000 | 0.0000 | 0.0000 | −0.0000 |
| 44 | 0.0001 | 0.0001 | 0.0001 | 0.0001 | 0.0000 | −0.0000 |
| | 0.0001 | 0.0002 | 0.0001 | 0.0001 | 0.0000 | −0.0000 |
| | 0.0000 | 0.0000 | 0.0000 | 0.0000 | 0.0000 | 0.0000 |

The size parameters of the semimajor and semiminor axes of the spheroid are $2\pi c/\lambda = 20$ and $2\pi a/\lambda = 10$, respectively, and the refractive index is $1.311 + i0.0$. The expansion coefficients are defined in Eq. (3.4.76). For each expansion order, the three rows denote the results from the EBCM, the IITM, and the IITM minus the EBCM. All data are rounded to four digits after the decimal point from the original computed data.

**Table 5.2** Scattering phase matrix element comparisons.

| $\Theta(^\circ)$ | $P_{11}$ | $P_{22}$ | $P_{33}$ | $P_{44}$ | $P_{12}$ | $P_{34}$ |
|---|---|---|---|---|---|---|
| 0.0 | 103.9885 | 103.9534 | 103.9534 | 103.9183 | 0.0000 | 0.0000 |
| | 104.0012 | 103.9682 | 103.9682 | 103.9351 | 0.0000 | 0.0000 |
| | 0.0127 | 0.0148 | 0.0148 | 0.0168 | 0.0000 | 0.0000 |
| 10.0 | 12.6586 | 12.6509 | 12.6060 | 12.6052 | 0.3132 | 0.5662 |
| | 12.6654 | 12.6576 | 12.6119 | 12.6109 | 0.2971 | 0.5715 |
| | 0.0069 | 0.0067 | 0.0059 | 0.0057 | −0.0161 | 0.0053 |
| 20.0 | 6.3850 | 6.3773 | 6.3203 | 6.3224 | −0.0219 | −0.2387 |
| | 6.3851 | 6.3777 | 6.3216 | 6.3240 | −0.0013 | −0.2389 |
| | 0.0001 | 0.0004 | 0.0013 | 0.0016 | 0.0206 | −0.0002 |
| 30.0 | 2.7215 | 2.7130 | 2.6808 | 2.6861 | 0.1863 | −0.0312 |
| | 2.7218 | 2.7134 | 2.6806 | 2.6860 | 0.1891 | −0.0321 |
| | 0.0003 | 0.0004 | −0.0002 | −0.0001 | 0.0028 | −0.0009 |
| 40.0 | 1.4510 | 1.4406 | 1.4096 | 1.4162 | 0.0595 | −0.1064 |
| | 1.4501 | 1.4399 | 1.4092 | 1.4159 | 0.0621 | −0.1052 |
| | −0.0009 | −0.0007 | −0.0004 | −0.0003 | 0.0026 | 0.0012 |
| 50.0 | 0.6175 | 0.6066 | 0.5760 | 0.5844 | 0.0808 | −0.0287 |
| | 0.6175 | 0.6065 | 0.5758 | 0.5842 | 0.0806 | −0.0280 |
| | −0.0000 | −0.0001 | −0.0002 | −0.0002 | −0.0002 | 0.0007 |
| 60.0 | 0.3729 | 0.3603 | 0.3364 | 0.3459 | 0.0195 | −0.0415 |
| | 0.3726 | 0.3599 | 0.3363 | 0.3457 | 0.0201 | −0.0405 |
| | −0.0003 | −0.0004 | −0.0001 | −0.0002 | 0.0006 | 0.0010 |
| 70.0 | 0.1937 | 0.1770 | 0.1417 | 0.1555 | 0.0412 | −0.0350 |
| | 0.1937 | 0.1769 | 0.1416 | 0.1553 | 0.0413 | −0.0344 |
| | −0.0000 | −0.0001 | −0.0001 | −0.0002 | 0.0001 | 0.0006 |
| 80.0 | 0.2073 | 0.1776 | 0.1397 | 0.1665 | 0.0068 | −0.0627 |
| | 0.2072 | 0.1774 | 0.1394 | 0.1661 | 0.0074 | −0.0623 |
| | −0.0001 | −0.0002 | −0.0003 | −0.0004 | 0.0006 | 0.0004 |
| 90.0 | 0.2468 | 0.1804 | 0.1178 | 0.1806 | 0.0088 | −0.1061 |
| | 0.2468 | 0.1803 | 0.1171 | 0.1799 | 0.0089 | −0.1060 |
| | 0.0000 | −0.0001 | −0.0007 | −0.0007 | 0.0001 | 0.0001 |
| 100.0 | 0.2737 | 0.1400 | 0.0813 | 0.2113 | −0.0226 | −0.0778 |
| | 0.2737 | 0.1399 | 0.0803 | 0.2100 | −0.0225 | −0.0779 |
| | −0.0000 | −0.0001 | −0.0010 | −0.0013 | 0.0001 | −0.0001 |
| 110.0 | 0.2563 | 0.0578 | −0.0096 | 0.1840 | −0.0080 | −0.0526 |
| | 0.2562 | 0.0579 | −0.0102 | 0.1827 | −0.0081 | −0.0523 |
| | −0.0001 | 0.0001 | −0.0006 | −0.0013 | −0.0001 | 0.0003 |
| 120.0 | 0.2007 | 0.0085 | −0.0645 | 0.1202 | −0.0134 | −0.0335 |
| | 0.2004 | 0.0086 | −0.0644 | 0.1194 | −0.0135 | −0.0327 |
| | −0.0003 | 0.0001 | 0.0001 | −0.0008 | −0.0001 | 0.0008 |
| 130.0 | 0.1482 | 0.0372 | −0.0317 | 0.0717 | −0.0248 | −0.0280 |
| | 0.1480 | 0.0371 | −0.0314 | 0.0713 | −0.0252 | −0.0272 |
| | −0.0002 | −0.0001 | 0.0003 | −0.0004 | −0.0004 | 0.0008 |
| 140.0 | 0.1260 | 0.0518 | −0.0412 | 0.0277 | −0.0228 | −0.0318 |
| | 0.1255 | 0.0511 | −0.0405 | 0.0280 | −0.0228 | −0.0312 |
| | −0.0005 | −0.0007 | 0.0007 | 0.0003 | −0.0000 | 0.0006 |

*Continued*

**Table 5.2** Continued

| $\Theta(^\circ)$ | $P_{11}$ | $P_{22}$ | $P_{33}$ | $P_{44}$ | $P_{12}$ | $P_{34}$ |
|---|---|---|---|---|---|---|
| 150.0 | 0.0871 | 0.0361 | −0.0325 | 0.0138 | −0.0257 | −0.0092 |
| | 0.0870 | 0.0358 | −0.0322 | 0.0137 | −0.0257 | −0.0089 |
| | −0.0001 | −0.0003 | 0.0003 | −0.0001 | −0.0000 | 0.0003 |
| 160.0 | 0.0785 | 0.0324 | −0.0236 | 0.0186 | −0.0195 | −0.0004 |
| | 0.0785 | 0.0321 | −0.0233 | 0.0187 | −0.0196 | −0.0001 |
| | −0.0000 | −0.0003 | 0.0003 | 0.0001 | −0.0001 | 0.0003 |
| 170.0 | 0.0553 | 0.0344 | −0.0196 | −0.0028 | −0.0110 | 0.0060 |
| | 0.0554 | 0.0343 | −0.0195 | −0.0031 | −0.0112 | 0.0058 |
| | 0.0001 | −0.0001 | 0.0001 | −0.0003 | −0.0002 | −0.0002 |
| 180.0 | 0.0936 | 0.0320 | −0.0320 | 0.0296 | 0.0000 | 0.0000 |
| | 0.0932 | 0.0319 | −0.0319 | 0.0295 | 0.0000 | 0.0000 |
| | −0.0004 | −0.0001 | 0.0001 | −0.0001 | 0.0000 | 0.0000 |

Other parameters are the same as for Table 5.1.

**Table 5.3** Scattering and extinction cross sections (the wavelength is assumed to be $2\pi$ in units of the particle radius) and asymmetry factor comparisons.

| $(\lambda = 2\pi)$ | EBCM | IITM | IITM-EBCM |
|---|---|---|---|
| $C_{sca} = C_{ext}$ | 1144.33 | 1145.06 | 0.73 |
| $g$ | 0.7535 | 0.7536 | 0.0001 |

Other parameters are the same as for Table 5.1.

differences. In Table 5.3, the scattering and extinction cross sections are given when the incident wavelength is $2\pi$ in units of the particle radius, and the asymmetry factor can be directly given by the expansion coefficient as $g = \alpha_1^1/3$. Fig. 5.5 compares the scattering phase matrix elements at all scattering angles between the EBCM and the IITM. The differences are so small that they cannot be identified in Fig. 5.5.

We consider an oblate spheroid with $2\pi c/\lambda = 10$, AR $= 1.5$, and refractive index $1.53 + i0.008$. The comparisons of the expansion coefficients are given in Table 5.4. The comparisons of the scattering phase matrix elements are given in Table 5.5. The comparisons of the scattering and extinction cross sections and asymmetry factor are given in Table 5.6. The scattering phase matrix comparisons between the EBCM and IITM are given in Fig. 5.6.

For a cylinder with an aspect ratio of unity, $2\pi D/\lambda = 20$, and a refractive index of $1.311 + i0.0$, the expansion coefficients, scattering phase matrix elements, scattering and extinction cross sections, and asymmetry factor are compared between the EBCM and the IITM in Tables 5.7–5.9 and Fig. 5.7.

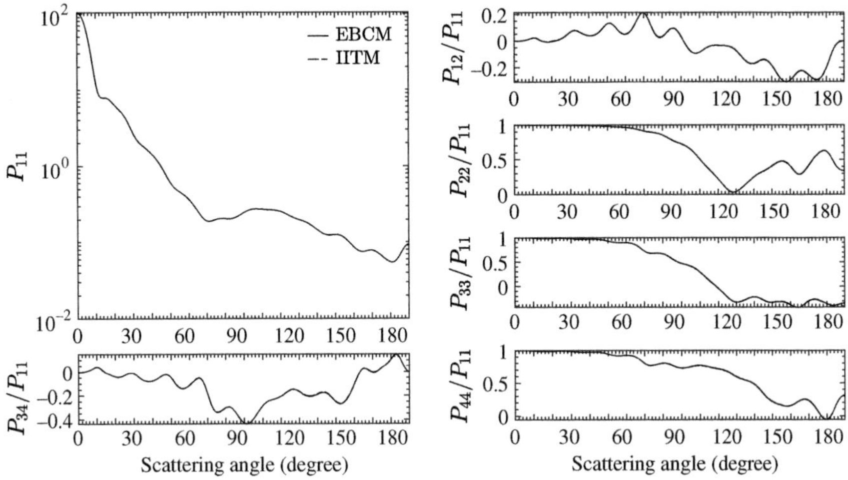


**Fig 5.5** Scattering phase matrix elements between the EBCM and the IITM. The parameters are the same as for Table 5.1.

**Table 5.4** Expansion coefficient comparisons between the EBCM and the IITM for a spheroid in random orientation.

| $s$ | $\alpha_1^s$ | $\alpha_2^s$ | $\alpha_3^s$ | $\alpha_4^s$ | $\beta_1^s$ | $\beta_2^s$ |
|---|---|---|---|---|---|---|
| 0 | 1.0000 | 0.0000 | 0.0000 | 0.8797 | 0.0000 | 0.0000 |
| | 1.0000 | 0.0000 | 0.0000 | 0.8794 | 0.0000 | 0.0000 |
| | 0.0000 | 0.0000 | 0.0000 | −0.0003 | 0.0000 | 0.0000 |
| 1 | 2.2700 | 0.0000 | 0.0000 | 2.3421 | 0.0000 | 0.0000 |
| | 2.2690 | 0.0000 | 0.0000 | 2.3417 | 0.0000 | 0.0000 |
| | −0.0010 | 0.0000 | 0.0000 | −0.0004 | 0.0000 | 0.0000 |
| 2 | 3.5433 | 4.1796 | 3.9340 | 3.4584 | 0.0075 | 0.0832 |
| | 3.5424 | 4.1795 | 3.9324 | 3.4572 | 0.0067 | 0.0851 |
| | −0.0009 | −0.0001 | −0.0016 | −0.0012 | −0.0008 | 0.0019 |
| 3 | 4.0822 | 4.5007 | 4.4880 | 4.1063 | −0.0184 | −0.0888 |
| | 4.0804 | 4.4989 | 4.4864 | 4.1049 | −0.0182 | −0.0870 |
| | −0.0017 | −0.0018 | −0.0016 | −0.0013 | 0.0002 | 0.0018 |
| 4 | 4.7842 | 5.0813 | 4.9922 | 4.7493 | 0.0053 | 0.0926 |
| | 4.7822 | 5.0798 | 4.9901 | 4.7469 | 0.0034 | 0.0957 |
| | −0.0020 | −0.0015 | −0.0020 | −0.0023 | −0.0019 | 0.0031 |
| 5 | 5.3349 | 5.4331 | 5.4021 | 5.3475 | 0.0687 | −0.1174 |
| | 5.3332 | 5.4307 | 5.3998 | 5.3463 | 0.0686 | −0.1149 |
| | −0.0017 | −0.0024 | −0.0023 | −0.0013 | −0.0001 | 0.0024 |
| 6 | 5.8520 | 6.1136 | 6.0223 | 5.7862 | 0.0134 | −0.0490 |
| | 5.8506 | 6.1127 | 6.0209 | 5.7845 | 0.0128 | −0.0467 |
| | −0.0014 | −0.0009 | −0.0014 | −0.0018 | −0.0006 | 0.0023 |
| 7 | 6.1981 | 6.2433 | 6.2694 | 6.2441 | 0.1158 | −0.0891 |
| | 6.1971 | 6.2415 | 6.2677 | 6.2428 | 0.1155 | −0.0870 |
| | −0.0010 | −0.0018 | −0.0017 | −0.0012 | −0.0003 | 0.0022 |

*Continued*

**Table 5.4** Continued

| $s$ | $\alpha_1^s$ | $\alpha_2^s$ | $\alpha_3^s$ | $\alpha_4^s$ | $\beta_1^s$ | $\beta_2^s$ |
|---|---|---|---|---|---|---|
| 8 | 6.6078 | 6.8041 | 6.7329 | 6.5721 | 0.0166 | −0.1182 |
| | 6.6068 | 6.8033 | 6.7323 | 6.5714 | 0.0162 | −0.1164 |
| | −0.0010 | −0.0008 | −0.0005 | −0.0007 | −0.0004 | 0.0018 |
| 9 | 6.8092 | 6.8851 | 6.8583 | 6.7993 | 0.1192 | −0.1757 |
| | 6.8087 | 6.8843 | 6.8574 | 6.7985 | 0.1191 | −0.1740 |
| | −0.0006 | −0.0009 | −0.0009 | −0.0008 | −0.0001 | 0.0016 |
| 10 | 6.9676 | 7.1208 | 7.1091 | 6.9810 | 0.0467 | −0.1504 |
| | 6.9669 | 7.1202 | 7.1087 | 6.9805 | 0.0462 | −0.1491 |
| | −0.0007 | −0.0007 | −0.0003 | −0.0005 | −0.0005 | 0.0013 |
| 11 | 7.0531 | 7.1375 | 7.1078 | 7.0530 | 0.0934 | −0.2385 |
| | 7.0525 | 7.1368 | 7.1072 | 7.0525 | 0.0930 | −0.2374 |
| | −0.0006 | −0.0008 | −0.0006 | −0.0006 | −0.0003 | 0.0010 |
| 12 | 7.0515 | 7.1929 | 7.1554 | 7.0430 | 0.0703 | −0.2325 |
| | 7.0506 | 7.1922 | 7.1549 | 7.0423 | 0.0698 | −0.2318 |
| | −0.0009 | −0.0008 | −0.0005 | −0.0007 | −0.0005 | 0.0007 |
| 13 | 6.9611 | 7.0447 | 7.0382 | 6.9862 | 0.0892 | −0.2822 |
| | 6.9602 | 7.0437 | 7.0372 | 6.9853 | 0.0887 | −0.2816 |
| | −0.0009 | −0.0010 | −0.0010 | −0.0009 | −0.0004 | 0.0006 |
| 14 | 6.8430 | 6.9707 | 6.9274 | 6.8360 | 0.0965 | −0.2783 |
| | 6.8417 | 6.9694 | 6.9264 | 6.8350 | 0.0958 | −0.2782 |
| | −0.0014 | −0.0013 | −0.0010 | −0.0010 | −0.0007 | 0.0001 |
| 15 | 6.6514 | 6.7321 | 6.7199 | 6.6756 | 0.0764 | −0.3037 |
| | 6.6497 | 6.7305 | 6.7183 | 6.6741 | 0.0758 | −0.3039 |
| | −0.0017 | −0.0016 | −0.0016 | −0.0016 | −0.0006 | −0.0002 |
| 16 | 6.4559 | 6.5664 | 6.5107 | 6.4339 | 0.1095 | −0.3054 |
| | 6.4538 | 6.5644 | 6.5088 | 6.4320 | 0.1088 | −0.3059 |
| | −0.0021 | −0.0020 | −0.0019 | −0.0019 | −0.0008 | −0.0006 |
| 17 | 6.1867 | 6.2570 | 6.2784 | 6.2465 | 0.0758 | −0.2793 |
| | 6.1844 | 6.2547 | 6.2760 | 6.2441 | 0.0752 | −0.2803 |
| | −0.0024 | −0.0024 | −0.0024 | −0.0024 | −0.0007 | −0.0010 |
| 18 | 5.9358 | 6.0438 | 5.9976 | 5.9354 | 0.0254 | −0.3190 |
| | 5.9333 | 6.0413 | 5.9952 | 5.9330 | 0.0250 | −0.3203 |
| | −0.0026 | −0.0026 | −0.0024 | −0.0024 | −0.0005 | −0.0014 |
| 19 | 5.5424 | 5.6241 | 5.6222 | 5.5910 | −0.0284 | −0.3729 |
| | 5.5401 | 5.6217 | 5.6198 | 5.5887 | −0.0285 | −0.3742 |
| | −0.0024 | −0.0024 | −0.0024 | −0.0023 | −0.0001 | −0.0013 |
| 20 | 5.0159 | 5.1472 | 5.1146 | 5.0380 | −0.0498 | −0.4785 |
| | 5.0139 | 5.1452 | 5.1129 | 5.0364 | −0.0501 | −0.4799 |
| | −0.0020 | −0.0020 | −0.0017 | −0.0016 | −0.0003 | −0.0014 |
| 21 | 4.3344 | 4.4293 | 4.3878 | 4.3486 | 0.0010 | −0.5197 |
| | 4.3327 | 4.4276 | 4.3863 | 4.3472 | 0.0001 | −0.5208 |
| | −0.0017 | −0.0017 | −0.0014 | −0.0014 | −0.0009 | −0.0011 |
| 22 | 3.5848 | 3.7133 | 3.6510 | 3.5667 | 0.0410 | −0.5054 |
| | 3.5836 | 3.7121 | 3.6497 | 3.5654 | 0.0394 | −0.5063 |
| | −0.0012 | −0.0012 | −0.0013 | −0.0013 | −0.0016 | −0.0009 |

**Table 5.4** Continued

| $s$ | $\alpha_1^s$ | $\alpha_2^s$ | $\alpha_3^s$ | $\alpha_4^s$ | $\beta_1^s$ | $\beta_2^s$ |
|---|---|---|---|---|---|---|
| 23 | 2.8237 | 2.8895 | 2.8784 | 2.8564 | 0.0688 | −0.4444 |
| | 2.8230 | 2.8887 | 2.8769 | 2.8549 | 0.0667 | −0.4443 |
| | −0.0008 | −0.0008 | −0.0015 | −0.0015 | −0.0021 | 0.0000 |
| 24 | 2.1770 | 2.2881 | 2.2201 | 2.1471 | 0.1053 | −0.3945 |
| | 2.1767 | 2.2875 | 2.2191 | 2.1460 | 0.1029 | −0.3936 |
| | −0.0004 | −0.0006 | −0.0010 | −0.0011 | −0.0025 | 0.0009 |
| 25 | 1.6007 | 1.6200 | 1.6157 | 1.6266 | 0.0626 | −0.2959 |
| | 1.6003 | 1.6196 | 1.6156 | 1.6265 | 0.0604 | −0.2940 |
| | −0.0003 | −0.0004 | −0.0001 | −0.0001 | −0.0023 | 0.0019 |
| 26 | 1.2329 | 1.3230 | 1.2735 | 1.2245 | 0.0926 | −0.3019 |
| | 1.2333 | 1.3231 | 1.2743 | 1.2251 | 0.0912 | −0.3014 |
| | 0.0004 | 0.0001 | 0.0007 | 0.0007 | −0.0014 | 0.0005 |
| 27 | 0.8144 | 0.8196 | 0.8447 | 0.8705 | −0.0098 | −0.2029 |
| | 0.8145 | 0.8199 | 0.8459 | 0.8718 | −0.0102 | −0.2026 |
| | 0.0002 | 0.0003 | 0.0012 | 0.0013 | −0.0004 | 0.0002 |
| 28 | 0.6944 | 0.7644 | 0.6299 | 0.5951 | 0.0915 | −0.3198 |
| | 0.6949 | 0.7648 | 0.6297 | 0.5951 | 0.0917 | −0.3204 |
| | 0.0005 | 0.0004 | −0.0002 | 0.0000 | 0.0002 | −0.0006 |
| 29 | 0.2959 | 0.2997 | 0.3247 | 0.3391 | −0.0024 | −0.0820 |
| | 0.2959 | 0.2998 | 0.3251 | 0.3394 | −0.0024 | −0.0822 |
| | 0.0000 | 0.0001 | 0.0004 | 0.0004 | 0.0000 | −0.0002 |
| 30 | 0.2856 | 0.3134 | 0.2052 | 0.1933 | 0.0500 | −0.1767 |
| | 0.2854 | 0.3131 | 0.2048 | 0.1930 | 0.0497 | −0.1768 |
| | −0.0003 | −0.0004 | −0.0005 | −0.0003 | −0.0003 | −0.0001 |
| 31 | 0.0621 | 0.0644 | 0.0850 | 0.0874 | 0.0016 | −0.0099 |
| | 0.0621 | 0.0644 | 0.0848 | 0.0872 | 0.0015 | −0.0101 |
| | −0.0000 | −0.0000 | −0.0002 | −0.0002 | −0.0001 | −0.0002 |
| 32 | 0.0721 | 0.0796 | 0.0446 | 0.0413 | 0.0112 | −0.0534 |
| | 0.0719 | 0.0793 | 0.0443 | 0.0411 | 0.0109 | −0.0534 |
| | −0.0003 | −0.0003 | −0.0002 | −0.0002 | −0.0002 | 0.0000 |
| 33 | 0.0117 | 0.0124 | 0.0152 | 0.0150 | 0.0053 | −0.0004 |
| | 0.0117 | 0.0124 | 0.0152 | 0.0150 | 0.0053 | −0.0005 |
| | 0.0000 | −0.0000 | −0.0000 | −0.0000 | −0.0000 | −0.0001 |
| 34 | 0.0087 | 0.0095 | 0.0051 | 0.0050 | 0.0006 | −0.0060 |
| | 0.0087 | 0.0094 | 0.0051 | 0.0049 | 0.0006 | −0.0059 |
| | −0.0000 | −0.0000 | −0.0000 | −0.0000 | −0.0000 | 0.0000 |
| 35 | 0.0006 | 0.0008 | 0.0009 | 0.0009 | 0.0001 | −0.0005 |
| | 0.0006 | 0.0008 | 0.0009 | 0.0009 | 0.0001 | −0.0005 |
| | −0.0000 | −0.0000 | −0.0000 | −0.0000 | 0.0000 | −0.0000 |
| 36 | 0.0011 | 0.0012 | 0.0003 | 0.0003 | 0.0006 | −0.0007 |
| | 0.0011 | 0.0012 | 0.0003 | 0.0003 | 0.0006 | −0.0007 |
| | −0.0000 | −0.0000 | −0.0000 | −0.0000 | −0.0000 | 0.0000 |
| 37 | 0.0001 | 0.0001 | 0.0002 | 0.0001 | 0.0001 | 0.0000 |
| | 0.0001 | 0.0001 | 0.0002 | 0.0002 | 0.0001 | −0.0000 |
| | 0.0000 | −0.0000 | −0.0000 | 0.0000 | 0.0000 | −0.0000 |

The size parameters of the semimajor and semiminor axes of the spheroid are $2\pi c/\lambda = 10$ and $2\pi a/\lambda = 15$, and the refractive index is $1.53 + i0.008$.

**Table 5.5** Scattering phase matrix element comparisons. Other parameters are the same as for Table 5.4.

| $\Theta(°)$ | $\mathbf{P_{11}}$ | $\mathbf{P_{22}}$ | $\mathbf{P_{33}}$ | $\mathbf{P_{44}}$ | $\mathbf{P_{12}}$ | $\mathbf{P_{34}}$ |
|---|---|---|---|---|---|---|
| 0.0 | 135.1466 | 135.0845 | 135.0845 | 135.0224 | 0.0000 | 0.0000 |
| | 135.1138 | 135.0526 | 135.0526 | 134.9914 | 0.0000 | 0.0000 |
| | −0.0328 | −0.0319 | −0.0319 | −0.0310 | 0.0000 | 0.0000 |
| 10.0 | 20.8697 | 20.8518 | 20.6699 | 20.6605 | −0.4231 | 2.3646 |
| | 20.8658 | 20.8479 | 20.6663 | 20.6566 | −0.4172 | 2.3638 |
| | −0.0039 | −0.0039 | −0.0036 | −0.0039 | 0.0059 | −0.0008 |
| 20.0 | 4.7080 | 4.6828 | 4.6463 | 4.6360 | −0.2554 | 0.0581 |
| | 4.7021 | 4.6770 | 4.6408 | 4.6301 | −0.2542 | 0.0491 |
| | −0.0059 | −0.0058 | −0.0055 | −0.0059 | 0.0012 | −0.0090 |
| 30.0 | 1.2035 | 1.1849 | 1.1228 | 1.1164 | 0.0492 | −0.0492 |
| | 1.2042 | 1.1857 | 1.1237 | 1.1173 | 0.0488 | −0.0522 |
| | 0.0007 | 0.0008 | 0.0009 | 0.0009 | −0.0004 | −0.0030 |
| 40.0 | 1.2603 | 1.2475 | 1.1924 | 1.1919 | −0.0460 | 0.0141 |
| | 1.2607 | 1.2480 | 1.1932 | 1.1928 | −0.0437 | 0.0101 |
| | 0.0004 | 0.0005 | 0.0008 | 0.0009 | 0.0023 | −0.0040 |
| 50.0 | 0.5970 | 0.5849 | 0.4877 | 0.4907 | 0.0392 | −0.0988 |
| | 0.5976 | 0.5856 | 0.4878 | 0.4910 | 0.0402 | −0.1014 |
| | 0.0006 | 0.0007 | 0.0001 | 0.0003 | 0.0010 | −0.0026 |
| 60.0 | 0.3855 | 0.3718 | 0.3032 | 0.3075 | 0.0263 | 0.0129 |
| | 0.3856 | 0.3722 | 0.3028 | 0.3072 | 0.0264 | 0.0119 |
| | 0.0001 | 0.0004 | −0.0004 | −0.0003 | 0.0001 | −0.0010 |
| 70.0 | 0.2542 | 0.2391 | 0.1682 | 0.1732 | 0.0053 | −0.0383 |
| | 0.2543 | 0.2394 | 0.1686 | 0.1736 | 0.0054 | −0.0390 |
| | 0.0001 | 0.0003 | 0.0004 | 0.0004 | 0.0001 | −0.0007 |
| 80.0 | 0.1705 | 0.1537 | 0.0878 | 0.0928 | 0.0046 | −0.0149 |
| | 0.1705 | 0.1539 | 0.0877 | 0.0928 | 0.0047 | −0.0157 |
| | 0.0000 | 0.0002 | −0.0001 | 0.0000 | 0.0001 | −0.0008 |
| 90.0 | 0.1223 | 0.1029 | 0.0256 | 0.0321 | −0.0063 | −0.0344 |
| | 0.1223 | 0.1031 | 0.0254 | 0.0318 | −0.0063 | −0.0347 |
| | −0.0000 | 0.0002 | −0.0002 | −0.0003 | −0.0000 | −0.0003 |
| 100.0 | 0.0818 | 0.0598 | −0.0130 | −0.0034 | −0.0096 | −0.0047 |
| | 0.0818 | 0.0599 | −0.0131 | −0.0036 | −0.0097 | −0.0050 |
| | −0.0000 | 0.0001 | −0.0001 | −0.0002 | −0.0001 | −0.0003 |
| 110.0 | 0.0801 | 0.0575 | −0.0181 | −0.0075 | −0.0166 | −0.0270 |
| | 0.0804 | 0.0579 | −0.0182 | −0.0077 | −0.0163 | −0.0275 |
| | 0.0003 | 0.0004 | −0.0001 | −0.0002 | 0.0003 | −0.0005 |
| 120.0 | 0.1508 | 0.1284 | −0.0258 | −0.0184 | −0.0304 | −0.0988 |
| | 0.1511 | 0.1290 | −0.0258 | −0.0184 | −0.0297 | −0.0995 |
| | 0.0003 | 0.0006 | −0.0000 | −0.0000 | 0.0007 | −0.0007 |
| 130.0 | 0.2278 | 0.2070 | 0.0284 | 0.0307 | −0.0240 | −0.1550 |
| | 0.2283 | 0.2078 | 0.0285 | 0.0310 | −0.0228 | −0.1557 |
| | 0.0005 | 0.0008 | 0.0001 | 0.0003 | 0.0012 | −0.0007 |
| 140.0 | 0.2640 | 0.2373 | 0.0848 | 0.0940 | −0.0060 | −0.1621 |
| | 0.2646 | 0.2385 | 0.0849 | 0.0941 | −0.0046 | −0.1630 |
| | 0.0006 | 0.0012 | 0.0001 | 0.0001 | 0.0014 | −0.0009 |

**Table 5.5** Continued

| $\Theta(^\circ)$ | $P_{11}$ | $P_{22}$ | $P_{33}$ | $P_{44}$ | $P_{12}$ | $P_{34}$ |
|---|---|---|---|---|---|---|
| 150.0 | 0.2663 | 0.2002 | 0.0430 | 0.0881 | 0.0683 | −0.0801 |
| | 0.2672 | 0.2016 | 0.0430 | 0.0878 | 0.0695 | −0.0808 |
| | 0.0009 | 0.0014 | −0.0000 | −0.0003 | 0.0012 | −0.0007 |
| 160.0 | 0.2900 | 0.1166 | −0.0850 | 0.0532 | 0.0682 | −0.0406 |
| | 0.2908 | 0.1179 | −0.0854 | 0.0526 | 0.0692 | −0.0406 |
| | 0.0008 | 0.0013 | −0.0004 | −0.0006 | 0.0010 | 0.0000 |
| 170.0 | 0.4036 | 0.3475 | 0.0793 | 0.1068 | 0.0286 | −0.0345 |
| | 0.4037 | 0.3483 | 0.0779 | 0.1055 | 0.0291 | −0.0340 |
| | 0.0001 | 0.0008 | −0.0014 | −0.0013 | 0.0005 | 0.0005 |
| 180.0 | 1.0809 | 0.5655 | −0.5655 | −0.0500 | 0.0000 | 0.0000 |
| | 1.0812 | 0.5668 | −0.5668 | −0.0523 | 0.0000 | 0.0000 |
| | 0.0003 | 0.0013 | −0.0013 | −0.0023 | 0.0000 | 0.0000 |

**Table 5.6** Scattering and extinction cross section and asymmetry factor comparisons.

| $(\lambda = 2\pi)$ | EBCM | IITM | IITM-EBCM |
|---|---|---|---|
| $C_{sca}$ | 1079.64 | 1079.69 | 0.05 |
| $C_{ext}$ | 1316.31 | 1316.31 | 0.00 |
| $g$ | 0.7567 | 0.7563 | −0.0004 |

Other parameters are the same as for Table 5.4.

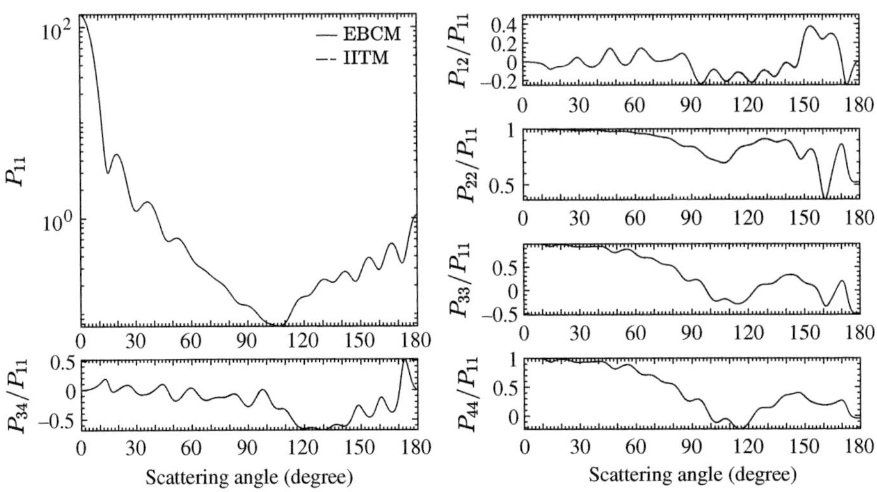


**Fig 5.6** Scattering phase matrix elements between the EBCM and the IITM. The parameters are the same as for Table 5.4.

**Table 5.7** Expansion coefficient comparisons for a cylinder with aspect ratio 1 and $2\pi D/\lambda = 20$.

| $s$ | $\alpha_1^s$ | $\alpha_2^s$ | $\alpha_3^s$ | $\alpha_4^s$ | $\beta_1^s$ | $\beta_2^s$ |
|---|---|---|---|---|---|---|
| 0 | 1.0000 | 0.0000 | 0.0000 | 0.9416 | 0.0000 | 0.0000 |
| | 1.0000 | 0.0000 | 0.0000 | 0.9414 | 0.0000 | 0.0000 |
| | 0.0000 | 0.0000 | 0.0000 | −0.0002 | 0.0000 | 0.0000 |
| 1 | 2.1183 | 0.0000 | 0.0000 | 2.1395 | 0.0000 | 0.0000 |
| | 2.1180 | 0.0000 | 0.0000 | 2.1393 | 0.0000 | 0.0000 |
| | −0.0004 | 0.0000 | 0.0000 | −0.0002 | 0.0000 | 0.0000 |
| 2 | 2.6322 | 3.8614 | 3.7541 | 2.5841 | −0.0300 | 0.1435 |
| | 2.6316 | 3.8612 | 3.7534 | 2.5838 | −0.0305 | 0.1455 |
| | −0.0006 | −0.0003 | −0.0007 | −0.0003 | −0.0005 | 0.0021 |
| 3 | 2.6354 | 3.1214 | 3.1178 | 2.6563 | −0.1239 | 0.0754 |
| | 2.6347 | 3.1211 | 3.1171 | 2.6559 | −0.1242 | 0.0769 |
| | −0.0007 | −0.0002 | −0.0008 | −0.0004 | −0.0002 | 0.0015 |
| 4 | 2.6110 | 3.0023 | 2.9477 | 2.5937 | −0.1073 | 0.0874 |
| | 2.6104 | 3.0019 | 2.9468 | 2.5933 | −0.1077 | 0.0887 |
| | −0.0006 | −0.0004 | −0.0009 | −0.0005 | −0.0004 | 0.0013 |
| 5 | 2.5517 | 2.7423 | 2.7193 | 2.5599 | −0.1371 | 0.0480 |
| | 2.5514 | 2.7421 | 2.7188 | 2.5599 | −0.1371 | 0.0489 |
| | −0.0003 | −0.0002 | −0.0005 | −0.0000 | −0.0000 | 0.0009 |
| 6 | 2.5576 | 2.7046 | 2.6667 | 2.5472 | −0.0912 | 0.0440 |
| | 2.5578 | 2.7048 | 2.6666 | 2.5473 | −0.0913 | 0.0446 |
| | 0.0002 | 0.0002 | −0.0001 | 0.0002 | −0.0001 | 0.0006 |
| 7 | 2.6388 | 2.7069 | 2.6898 | 2.6477 | −0.0758 | 0.0307 |
| | 2.6390 | 2.7073 | 2.6901 | 2.6481 | −0.0756 | 0.0308 |
| | 0.0002 | 0.0005 | 0.0003 | 0.0004 | 0.0002 | 0.0001 |
| 8 | 2.7648 | 2.8348 | 2.7992 | 2.7534 | −0.0445 | 0.0167 |
| | 2.7652 | 2.8352 | 2.7996 | 2.7540 | −0.0444 | 0.0165 |
| | 0.0004 | 0.0004 | 0.0004 | 0.0006 | 0.0001 | −0.0002 |
| 9 | 2.9200 | 2.9489 | 2.9468 | 2.9418 | −0.0094 | 0.0052 |
| | 2.9204 | 2.9495 | 2.9469 | 2.9419 | −0.0093 | 0.0046 |
| | 0.0004 | 0.0006 | 0.0002 | 0.0001 | 0.0001 | −0.0005 |
| 10 | 3.0902 | 3.1399 | 3.0926 | 3.0654 | −0.0112 | −0.0072 |
| | 3.0902 | 3.1399 | 3.0929 | 3.0658 | −0.0114 | −0.0078 |
| | 0.0000 | 0.0000 | 0.0003 | 0.0004 | −0.0002 | −0.0007 |
| 11 | 3.2373 | 3.2580 | 3.2632 | 3.2578 | 0.0376 | −0.0175 |
| | 3.2373 | 3.2582 | 3.2629 | 3.2574 | 0.0374 | −0.0182 |
| | 0.0000 | 0.0002 | −0.0003 | −0.0004 | −0.0002 | −0.0007 |
| 12 | 3.3965 | 3.4343 | 3.4059 | 3.3887 | 0.0141 | −0.0032 |
| | 3.3960 | 3.4338 | 3.4055 | 3.3884 | 0.0137 | −0.0039 |
| | −0.0004 | −0.0005 | −0.0004 | −0.0003 | −0.0004 | −0.0007 |
| 13 | 3.5515 | 3.5680 | 3.5457 | 3.5394 | 0.0368 | −0.0405 |
| | 3.5509 | 3.5675 | 3.5451 | 3.5386 | 0.0364 | −0.0409 |
| | −0.0007 | −0.0005 | −0.0006 | −0.0008 | −0.0003 | −0.0004 |

**Table 5.7** Continued

| $s$ | $\alpha_1^s$ | $\alpha_2^s$ | $\alpha_3^s$ | $\alpha_4^s$ | $\beta_1^s$ | $\beta_2^s$ |
|---|---|---|---|---|---|---|
| 14 | 3.6882 | 3.7164 | 3.7242 | 3.7117 | 0.0609 | −0.0039 |
| | 3.6873 | 3.7154 | 3.7231 | 3.7107 | 0.0603 | −0.0044 |
| | −0.0009 | −0.0010 | −0.0011 | −0.0010 | −0.0006 | −0.0005 |
| 15 | 3.8782 | 3.8865 | 3.8752 | 3.8873 | −0.0067 | −0.0204 |
| | 3.8769 | 3.8854 | 3.8741 | 3.8860 | −0.0072 | −0.0207 |
| | −0.0013 | −0.0012 | −0.0011 | −0.0013 | −0.0005 | −0.0002 |
| 16 | 4.0199 | 4.0620 | 4.0195 | 3.9910 | −0.0148 | −0.0782 |
| | 4.0186 | 4.0606 | 4.0181 | 3.9897 | −0.0155 | −0.0781 |
| | −0.0013 | −0.0014 | −0.0014 | −0.0013 | −0.0007 | 0.0002 |
| 17 | 4.0943 | 4.1012 | 4.1054 | 4.1046 | 0.0515 | −0.0900 |
| | 4.0931 | 4.1000 | 4.1041 | 4.1032 | 0.0504 | −0.0894 |
| | −0.0012 | −0.0013 | −0.0013 | −0.0015 | −0.0011 | 0.0006 |
| 18 | 4.1353 | 4.1782 | 4.2134 | 4.1908 | −0.0119 | −0.0426 |
| | 4.1333 | 4.1763 | 4.2120 | 4.1896 | −0.0133 | −0.0424 |
| | −0.0020 | −0.0019 | −0.0014 | −0.0012 | −0.0013 | 0.0003 |
| 19 | 4.1061 | 4.1352 | 4.1485 | 4.1557 | −0.1473 | −0.1927 |
| | 4.1038 | 4.1332 | 4.1468 | 4.1539 | −0.1491 | −0.1925 |
| | −0.0023 | −0.0020 | −0.0017 | −0.0018 | −0.0019 | 0.0002 |
| 20 | 3.7745 | 3.8917 | 3.8491 | 3.7765 | −0.1962 | −0.3823 |
| | 3.7725 | 3.8897 | 3.8468 | 3.7745 | −0.1988 | −0.3825 |
| | −0.0020 | −0.0021 | −0.0023 | −0.0021 | −0.0026 | −0.0002 |
| 21 | 3.1101 | 3.1949 | 3.1628 | 3.1097 | −0.1108 | −0.4249 |
| | 3.1077 | 3.1929 | 3.1604 | 3.1072 | −0.1146 | −0.4244 |
| | −0.0024 | −0.0020 | −0.0024 | −0.0025 | −0.0038 | 0.0005 |
| 22 | 2.3541 | 2.4602 | 2.4105 | 2.3317 | −0.0613 | −0.4244 |
| | 2.3518 | 2.4578 | 2.4076 | 2.3289 | −0.0657 | −0.4239 |
| | −0.0023 | −0.0025 | −0.0030 | −0.0028 | −0.0043 | 0.0005 |
| 23 | 1.5322 | 1.6080 | 1.5729 | 1.5122 | 0.0261 | −0.3269 |
| | 1.5299 | 1.6057 | 1.5705 | 1.5098 | 0.0216 | −0.3259 |
| | −0.0023 | −0.0023 | −0.0024 | −0.0025 | −0.0045 | 0.0010 |
| 24 | 0.9150 | 0.9658 | 0.9400 | 0.8996 | 0.0656 | −0.2132 |
| | 0.9133 | 0.9641 | 0.9381 | 0.8977 | 0.0617 | −0.2127 |
| | −0.0017 | −0.0018 | −0.0019 | −0.0019 | −0.0039 | 0.0005 |
| 25 | 0.4746 | 0.5053 | 0.4928 | 0.4669 | 0.0654 | −0.1118 |
| | 0.4734 | 0.5041 | 0.4916 | 0.4659 | 0.0624 | −0.1111 |
| | −0.0012 | −0.0013 | −0.0011 | −0.0011 | −0.0030 | 0.0007 |
| 26 | 0.2366 | 0.2527 | 0.2444 | 0.2308 | 0.0425 | −0.0567 |
| | 0.2359 | 0.2519 | 0.2438 | 0.2301 | 0.0407 | −0.0566 |
| | −0.0007 | −0.0008 | −0.0007 | −0.0006 | −0.0019 | 0.0001 |
| 27 | 0.0930 | 0.0998 | 0.0952 | 0.0895 | 0.0258 | −0.0227 |
| | 0.0926 | 0.0994 | 0.0950 | 0.0893 | 0.0248 | −0.0224 |
| | −0.0004 | −0.0004 | −0.0003 | −0.0002 | −0.0009 | 0.0002 |
| 28 | 0.0376 | 0.0406 | 0.0382 | 0.0358 | 0.0120 | −0.0079 |
| | 0.0374 | 0.0404 | 0.0381 | 0.0357 | 0.0116 | −0.0079 |
| | −0.0002 | −0.0002 | −0.0001 | −0.0001 | −0.0005 | 0.0000 |

*Continued*

**Table 5.7** Continued

| $s$ | $\alpha_1^s$ | $\alpha_2^s$ | $\alpha_3^s$ | $\alpha_4^s$ | $\beta_1^s$ | $\beta_2^s$ |
|---|---|---|---|---|---|---|
| 29 | 0.0128 | 0.0138 | 0.0128 | 0.0120 | 0.0049 | −0.0024 |
| | 0.0127 | 0.0138 | 0.0128 | 0.0119 | 0.0047 | −0.0024 |
| | −0.0001 | −0.0001 | −0.0000 | −0.0000 | −0.0002 | 0.0000 |
| 30 | 0.0040 | 0.0044 | 0.0040 | 0.0037 | 0.0018 | −0.0007 |
| | 0.0040 | 0.0044 | 0.0040 | 0.0037 | 0.0017 | −0.0007 |
| | −0.0000 | −0.0000 | −0.0000 | −0.0000 | −0.0001 | −0.0000 |
| 31 | 0.0011 | 0.0012 | 0.0011 | 0.0010 | 0.0006 | −0.0002 |
| | 0.0011 | 0.0012 | 0.0011 | 0.0010 | 0.0006 | −0.0002 |
| | −0.0000 | −0.0000 | −0.0000 | −0.0000 | −0.0000 | −0.0000 |
| 32 | 0.0003 | 0.0003 | 0.0003 | 0.0003 | 0.0002 | −0.0000 |
| | 0.0003 | 0.0003 | 0.0003 | 0.0003 | 0.0002 | −0.0000 |
| | −0.0000 | −0.0000 | −0.0000 | −0.0000 | −0.0000 | 0.0000 |
| 33 | 0.0001 | 0.0001 | 0.0001 | 0.0001 | 0.0000 | −0.0000 |
| | 0.0001 | 0.0001 | 0.0001 | 0.0001 | 0.0000 | −0.0000 |
| | 0.0000 | −0.0000 | 0.0000 | −0.0000 | 0.0000 | 0.0000 |

The refractive index is $1.311 + i0.0$.

**Table 5.8** Scattering phase matrix element comparisons.

| $\Theta$(°) | $\mathbf{P_{11}}$ | $\mathbf{P_{22}}$ | $\mathbf{P_{33}}$ | $\mathbf{P_{44}}$ | $\mathbf{P_{12}}$ | $\mathbf{P_{34}}$ |
|---|---|---|---|---|---|---|
| 0.0 | 74.1734 | 74.1505 | 74.1505 | 74.1275 | 0.0000 | 0.0000 |
| | 74.1488 | 74.1265 | 74.1265 | 74.1043 | 0.0000 | 0.0000 |
| | −0.0246 | −0.0240 | −0.0240 | −0.0232 | 0.0000 | 0.0000 |
| 10.0 | 10.0005 | 9.9819 | 9.9212 | 9.9159 | 0.1724 | 0.9238 |
| | 10.0060 | 9.9880 | 9.9272 | 9.9224 | 0.1840 | 0.9232 |
| | 0.0054 | 0.0061 | 0.0060 | 0.0065 | 0.0116 | −0.0006 |
| 20.0 | 7.5148 | 7.4953 | 7.4709 | 7.4660 | 0.0097 | −0.4539 |
| | 7.5102 | 7.4912 | 7.4667 | 7.4620 | 0.0073 | −0.4537 |
| | −0.0046 | −0.0041 | −0.0042 | −0.0040 | −0.0024 | 0.0002 |
| 30.0 | 2.1336 | 2.1206 | 2.0377 | 2.0377 | 0.3955 | −0.1145 |
| | 2.1326 | 2.1197 | 2.0362 | 2.0364 | 0.3945 | −0.1170 |
| | −0.0010 | −0.0009 | −0.0015 | −0.0013 | −0.0010 | −0.0025 |
| 40.0 | 2.2863 | 2.2711 | 2.2554 | 2.2596 | 0.1208 | −0.0490 |
| | 2.2866 | 2.2716 | 2.2559 | 2.2603 | 0.1226 | −0.0518 |
| | 0.0003 | 0.0005 | 0.0005 | 0.0007 | 0.0018 | −0.0028 |
| 50.0 | 1.0789 | 1.0667 | 1.0030 | 1.0059 | 0.0757 | −0.2047 |
| | 1.0791 | 1.0670 | 1.0025 | 1.0056 | 0.0759 | −0.2067 |
| | 0.0002 | 0.0003 | −0.0005 | −0.0003 | 0.0002 | −0.0020 |
| 60.0 | 0.6486 | 0.6327 | 0.6107 | 0.6176 | 0.0869 | −0.0374 |
| | 0.6488 | 0.6331 | 0.6107 | 0.6178 | 0.0872 | −0.0388 |
| | 0.0002 | 0.0004 | 0.0000 | 0.0002 | 0.0003 | −0.0014 |

**Table 5.8** Continued

| $\Theta(°)$ | $P_{11}$ | $P_{22}$ | $P_{33}$ | $P_{44}$ | $P_{12}$ | $P_{34}$ |
|---|---|---|---|---|---|---|
| 70.0 | 0.5018 | 0.4896 | 0.4384 | 0.4436 | −0.0209 | −0.1288 |
| | 0.5022 | 0.4900 | 0.4382 | 0.4434 | −0.0206 | −0.1302 |
| | 0.0004 | 0.0004 | −0.0002 | −0.0002 | 0.0003 | −0.0014 |
| 80.0 | 0.2990 | 0.2809 | 0.2627 | 0.2736 | 0.0370 | −0.0454 |
| | 0.2989 | 0.2809 | 0.2625 | 0.2734 | 0.0372 | −0.0464 |
| | −0.0001 | 0.0000 | −0.0002 | −0.0002 | 0.0002 | −0.0010 |
| 90.0 | 0.2604 | 0.2451 | 0.1956 | 0.2055 | −0.0335 | −0.0911 |
| | 0.2605 | 0.2453 | 0.1953 | 0.2053 | −0.0332 | −0.0920 |
| | 0.0001 | 0.0002 | −0.0003 | −0.0002 | 0.0003 | −0.0009 |
| 100.0 | 0.1725 | 0.1484 | 0.1347 | 0.1523 | 0.0169 | −0.0230 |
| | 0.1725 | 0.1485 | 0.1344 | 0.1521 | 0.0169 | −0.0237 |
| | 0.0000 | 0.0001 | −0.0003 | −0.0002 | 0.0000 | −0.0007 |
| 110.0 | 0.1765 | 0.1468 | 0.0859 | 0.1094 | −0.0358 | −0.0814 |
| | 0.1766 | 0.1469 | 0.0857 | 0.1092 | −0.0356 | −0.0820 |
| | 0.0001 | 0.0001 | −0.0002 | −0.0002 | 0.0002 | −0.0006 |
| 120.0 | 0.1283 | 0.0771 | 0.0494 | 0.0936 | −0.0096 | −0.0168 |
| | 0.1284 | 0.0773 | 0.0493 | 0.0935 | −0.0095 | −0.0173 |
| | 0.0001 | 0.0002 | −0.0001 | −0.0001 | 0.0001 | −0.0005 |
| 130.0 | 0.1499 | 0.0840 | 0.0268 | 0.0851 | −0.0192 | −0.0603 |
| | 0.1502 | 0.0841 | 0.0265 | 0.0849 | −0.0191 | −0.0606 |
| | 0.0003 | 0.0001 | −0.0003 | −0.0002 | 0.0001 | −0.0003 |
| 140.0 | 0.1543 | 0.0735 | −0.0187 | 0.0533 | −0.0480 | −0.0320 |
| | 0.1543 | 0.0734 | −0.0185 | 0.0535 | −0.0477 | −0.0322 |
| | −0.0000 | −0.0001 | 0.0002 | 0.0002 | 0.0003 | −0.0002 |
| 150.0 | 0.1749 | 0.0587 | −0.0204 | 0.0827 | −0.0304 | −0.0192 |
| | 0.1749 | 0.0587 | −0.0206 | 0.0825 | −0.0302 | −0.0195 |
| | 0.0000 | −0.0000 | −0.0002 | −0.0002 | 0.0002 | −0.0003 |
| 160.0 | 0.1941 | 0.0781 | −0.0290 | 0.0702 | −0.0112 | −0.0197 |
| | 0.1944 | 0.0783 | −0.0295 | 0.0697 | −0.0108 | −0.0200 |
| | 0.0003 | 0.0002 | −0.0005 | −0.0005 | 0.0004 | −0.0003 |
| 170.0 | 0.1771 | 0.1277 | −0.0412 | −0.0159 | 0.0255 | −0.0192 |
| | 0.1771 | 0.1274 | −0.0413 | −0.0160 | 0.0256 | −0.0185 |
| | −0.0000 | −0.0003 | −0.0001 | −0.0001 | 0.0001 | 0.0007 |
| 180.0 | 0.2623 | 0.1489 | −0.1489 | −0.0355 | 0.0000 | 0.0000 |
| | 0.2627 | 0.1486 | -0.1486 | -0.0344 | 0.0000 | 0.0000 |
| | 0.0004 | -0.0003 | 0.0003 | 0.0011 | 0.0000 | 0.0000 |

Other parameters are the same as for Table 5.7.

For a cylinder with aspect ratio unity, $2\pi D/\lambda = 20$, and refractive index $1.53 + i0.008$, the expansion coefficient comparisons are given in Table 5.10. The scattering phase matrix elements and the scattering and extinction cross sections and asymmetry factor are given in Tables 5.11 and 5.12. Fig. 5.8 compares the scattering phase matrix elements.

**Table 5.9** Scattering and extinction cross sections and asymmetry factor comparisons.

| $(\lambda=2\pi)$ | EBCM | IITM | IITM-EBCM |
|---|---|---|---|
| $C_{sca}=C_{ext}$ | 837.77 | 837.46 | 0.31 |
| g | 0.7061 | 0.7060 | −0.0001 |

Other parameters are the same as for Table 5.7.

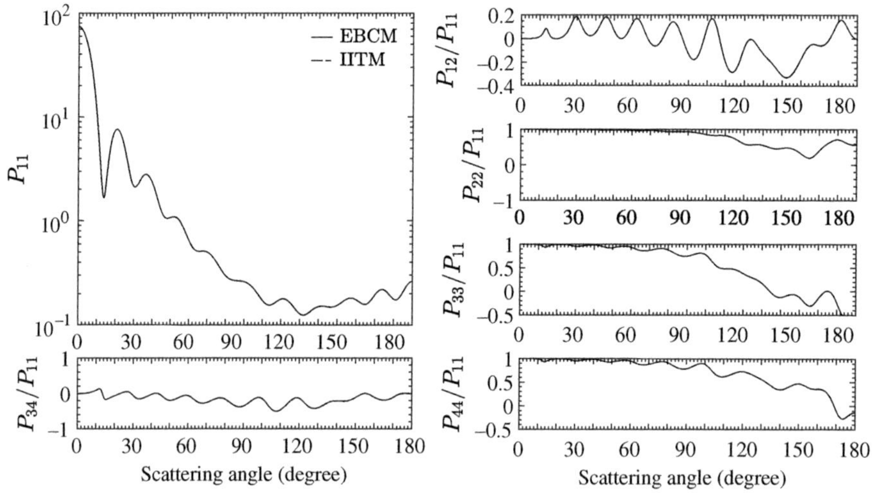


**Fig. 5.7** Scattering phase matrix element comparisons between the EBCM and the IITM. All parameters are the same as for Table 5.7.

**Table 5.10** Expansion coefficient comparisons for a cylinder with aspect ratio 1 and $2\pi D/\lambda=20$.

| $s$ | $\alpha_1^s$ | $\alpha_2^s$ | $\alpha_3^s$ | $\alpha_4^s$ | $\beta_1^s$ | $\beta_2^s$ |
|---|---|---|---|---|---|---|
| 0 | 1.0000 | 0.0000 | 0.0000 | 0.8930 | 0.0000 | 0.0000 |
| | 1.0000 | 0.0000 | 0.0000 | 0.8913 | 0.0000 | 0.0000 |
| | 0.0000 | 0.0000 | 0.0000 | −0.0017 | 0.0000 | 0.0000 |
| 1 | 2.2589 | 0.0000 | 0.0000 | 2.3175 | 0.0000 | 0.0000 |
| | 2.2589 | 0.0000 | 0.0000 | 2.3172 | 0.0000 | 0.0000 |
| | 0.0000 | 0.0000 | 0.0000 | −0.0003 | 0.0000 | 0.0000 |
| 2 | 3.4389 | 4.1114 | 3.9071 | 3.3707 | −0.0014 | 0.1104 |
| | 3.4397 | 4.1116 | 3.9063 | 3.3722 | −0.0015 | 0.1137 |
| | 0.0008 | 0.0002 | −0.0008 | 0.0015 | −0.0001 | 0.0032 |
| 3 | 4.3001 | 4.4925 | 4.4931 | 4.3467 | 0.0326 | −0.0160 |
| | 4.3006 | 4.4947 | 4.4936 | 4.3467 | 0.0317 | −0.0145 |
| | 0.0005 | 0.0022 | 0.0005 | 0.0000 | −0.0008 | 0.0015 |

**Table 5.10** Continued

| $s$ | $\alpha_1^s$ | $\alpha_2^s$ | $\alpha_3^s$ | $\alpha_4^s$ | $\beta_1^s$ | $\beta_2^s$ |
|---|---|---|---|---|---|---|
| 4 | 5.1489 | 5.4120 | 5.2975 | 5.0483 | 0.0150 | −0.0359 |
| | 5.1489 | 5.4120 | 5.2979 | 5.0492 | 0.0131 | −0.0360 |
| | −0.0001 | 0.0000 | 0.0004 | 0.0009 | −0.0020 | −0.0001 |
| 5 | 5.6110 | 5.8194 | 5.8668 | 5.6560 | −0.0043 | −0.0493 |
| | 5.6112 | 5.8202 | 5.8663 | 5.6556 | −0.0078 | −0.0489 |
| | 0.0002 | 0.0008 | −0.0005 | −0.0004 | −0.0035 | 0.0004 |
| 6 | 6.0739 | 6.2666 | 6.1969 | 6.0126 | −0.0405 | −0.0438 |
| | 6.0725 | 6.2659 | 6.1965 | 6.0123 | −0.0443 | −0.0442 |
| | −0.0014 | −0.0007 | −0.0004 | −0.0002 | −0.0038 | −0.0004 |
| 7 | 6.3020 | 6.4755 | 6.5025 | 6.3421 | −0.0531 | −0.0680 |
| | 6.3009 | 6.4750 | 6.5014 | 6.3415 | −0.0578 | −0.0685 |
| | −0.0011 | −0.0006 | −0.0011 | −0.0005 | −0.0047 | −0.0005 |
| 8 | 6.5438 | 6.7004 | 6.6415 | 6.4998 | −0.1075 | −0.0697 |
| | 6.5427 | 6.6998 | 6.6398 | 6.4987 | −0.1118 | −0.0707 |
| | −0.0011 | −0.0006 | −0.0016 | −0.0011 | −0.0043 | −0.0009 |
| 9 | 6.6308 | 6.7658 | 6.7778 | 6.6644 | −0.0760 | −0.0947 |
| | 6.6296 | 6.7651 | 6.7767 | 6.6634 | −0.0807 | −0.0958 |
| | −0.0012 | −0.0007 | −0.0011 | −0.0010 | −0.0047 | −0.0011 |
| 10 | 6.7238 | 6.8673 | 6.8277 | 6.7153 | −0.1389 | −0.0789 |
| | 6.7224 | 6.8662 | 6.8261 | 6.7144 | −0.1430 | −0.0808 |
| | −0.0014 | −0.0011 | −0.0016 | −0.0009 | −0.0042 | −0.0019 |
| 11 | 6.7377 | 6.8383 | 6.8213 | 6.7469 | −0.0933 | −0.1513 |
| | 6.7363 | 6.8374 | 6.8198 | 6.7454 | −0.0982 | −0.1528 |
| | −0.0014 | −0.0010 | −0.0015 | −0.0016 | −0.0049 | −0.0016 |
| 12 | 6.6590 | 6.8066 | 6.7774 | 6.6665 | −0.1558 | −0.1245 |
| | 6.6578 | 6.8052 | 6.7753 | 6.6649 | −0.1603 | −0.1267 |
| | −0.0012 | −0.0014 | −0.0021 | −0.0016 | −0.0044 | −0.0021 |
| 13 | 6.5386 | 6.6310 | 6.6114 | 6.5517 | −0.1139 | −0.2105 |
| | 6.5367 | 6.6297 | 6.6106 | 6.5506 | −0.1193 | −0.2122 |
| | −0.0019 | −0.0013 | −0.0008 | −0.0011 | −0.0054 | −0.0017 |
| 14 | 6.2901 | 6.4318 | 6.4141 | 6.3173 | −0.1619 | −0.2074 |
| | 6.2889 | 6.4306 | 6.4121 | 6.3165 | −0.1671 | −0.2105 |
| | −0.0011 | −0.0012 | −0.0019 | −0.0008 | −0.0051 | −0.0030 |
| 15 | 5.9733 | 6.0765 | 6.0507 | 5.9884 | −0.1512 | −0.3047 |
| | 5.9724 | 6.0764 | 6.0495 | 5.9874 | −0.1572 | −0.3069 |
| | −0.0009 | −0.0001 | −0.0012 | −0.0010 | −0.0060 | −0.0022 |
| 16 | 5.5067 | 5.6574 | 5.6221 | 5.5170 | −0.1292 | −0.3704 |
| | 5.5059 | 5.6568 | 5.6220 | 5.5176 | −0.1360 | −0.3738 |
| | −0.0008 | −0.0005 | −0.0001 | 0.0006 | −0.0068 | −0.0034 |
| 17 | 4.9626 | 5.0467 | 5.0578 | 5.0213 | −0.0862 | −0.3607 |
| | 4.9621 | 5.0471 | 5.0582 | 5.0223 | −0.0950 | −0.3654 |
| | −0.0005 | 0.0004 | 0.0004 | 0.0010 | −0.0087 | −0.0047 |
| 18 | 4.3618 | 4.5269 | 4.4921 | 4.3867 | −0.1610 | −0.4511 |
| | 4.3624 | 4.5280 | 4.4915 | 4.3873 | −0.1699 | −0.4576 |
| | 0.0007 | 0.0011 | −0.0006 | 0.0006 | −0.0089 | −0.0065 |

*Continued*

**Table 5.10** Continued

| $s$ | $\alpha_1^s$ | $\alpha_2^s$ | $\alpha_3^s$ | $\alpha_4^s$ | $\beta_1^s$ | $\beta_2^s$ |
|---|---|---|---|---|---|---|
| 19 | 3.6638 | 3.7521 | 3.7134 | 3.6743 | −0.1055 | −0.5108 |
| | 3.6660 | 3.7550 | 3.7138 | 3.6754 | −0.1162 | −0.5158 |
| | 0.0022 | 0.0028 | 0.0004 | 0.0011 | −0.0108 | −0.0051 |
| 20 | 2.8850 | 3.0271 | 2.9571 | 2.8643 | −0.0319 | −0.5063 |
| | 2.8869 | 3.0292 | 2.9577 | 2.8657 | −0.0425 | −0.5116 |
| | 0.0019 | 0.0021 | 0.0006 | 0.0014 | −0.0106 | −0.0052 |
| 21 | 2.1427 | 2.1986 | 2.1781 | 2.1533 | 0.0372 | −0.3618 |
| | 2.1434 | 2.1999 | 2.1793 | 2.1549 | 0.0264 | −0.3668 |
| | 0.0007 | 0.0013 | 0.0012 | 0.0015 | −0.0108 | −0.0050 |
| 22 | 1.5939 | 1.6972 | 1.6553 | 1.5801 | 0.0369 | −0.3267 |
| | 1.5939 | 1.6974 | 1.6533 | 1.5790 | 0.0283 | −0.3344 |
| | 0.0000 | 0.0002 | −0.0020 | −0.0010 | −0.0085 | −0.0076 |
| 23 | 1.1271 | 1.1615 | 1.1341 | 1.1191 | 0.0378 | −0.2618 |
| | 1.1265 | 1.1621 | 1.1328 | 1.1180 | 0.0336 | −0.2674 |
| | −0.0006 | 0.0005 | −0.0013 | −0.0010 | −0.0042 | −0.0056 |
| 24 | 0.7481 | 0.7894 | 0.7273 | 0.7053 | 0.0700 | −0.1999 |
| | 0.7481 | 0.7894 | 0.7265 | 0.7050 | 0.0665 | −0.2025 |
| | −0.0000 | −0.0000 | −0.0008 | −0.0003 | −0.0036 | −0.0026 |
| 25 | 0.3934 | 0.4199 | 0.4309 | 0.4039 | 0.0249 | −0.0752 |
| | 0.3928 | 0.4195 | 0.4309 | 0.4040 | 0.0213 | −0.0765 |
| | −0.0006 | −0.0004 | −0.0000 | 0.0001 | −0.0036 | −0.0013 |
| 26 | 0.2880 | 0.2993 | 0.2801 | 0.2720 | 0.0194 | −0.0991 |
| | 0.2870 | 0.2985 | 0.2789 | 0.2714 | 0.0167 | −0.1002 |
| | −0.0011 | −0.0008 | −0.0012 | −0.0006 | −0.0028 | −0.0011 |
| 27 | 0.1443 | 0.1568 | 0.1416 | 0.1312 | 0.0204 | −0.0640 |
| | 0.1439 | 0.1566 | 0.1406 | 0.1304 | 0.0187 | −0.0649 |
| | −0.0005 | −0.0002 | −0.0010 | −0.0008 | −0.0017 | −0.0008 |
| 28 | 0.0749 | 0.0779 | 0.0681 | 0.0669 | 0.0218 | −0.0293 |
| | 0.0750 | 0.0780 | 0.0678 | 0.0668 | 0.0210 | −0.0303 |
| | 0.0001 | 0.0002 | −0.0002 | −0.0000 | −0.0008 | −0.0011 |
| 29 | 0.0195 | 0.0222 | 0.0197 | 0.0176 | 0.0070 | −0.0077 |
| | 0.0195 | 0.0223 | 0.0198 | 0.0177 | 0.0068 | −0.0076 |
| | 0.0000 | 0.0001 | 0.0001 | 0.0001 | −0.0002 | 0.0001 |
| 30 | 0.0099 | 0.0103 | 0.0086 | 0.0084 | 0.0042 | −0.0035 |
| | 0.0099 | 0.0103 | 0.0086 | 0.0085 | 0.0041 | −0.0036 |
| | 0.0000 | 0.0000 | 0.0000 | 0.0000 | −0.0002 | −0.0001 |
| 31 | 0.0020 | 0.0022 | 0.0019 | 0.0017 | 0.0009 | −0.0005 |
| | 0.0020 | 0.0022 | 0.0019 | 0.0017 | 0.0009 | −0.0005 |
| | −0.0000 | 0.0000 | 0.0000 | 0.0000 | −0.0000 | 0.0000 |
| 32 | 0.0007 | 0.0008 | 0.0006 | 0.0006 | 0.0003 | −0.0002 |
| | 0.0007 | 0.0007 | 0.0006 | 0.0006 | 0.0003 | −0.0002 |
| | −0.0000 | −0.0000 | 0.0000 | 0.0000 | −0.0000 | −0.0000 |
| 33 | 0.0002 | 0.0002 | 0.0001 | 0.0001 | 0.0001 | −0.0000 |
| | 0.0002 | 0.0002 | 0.0001 | 0.0001 | 0.0001 | −0.0000 |
| | −0.0000 | −0.0000 | -0.0000 | −0.0000 | −0.0000 | −0.0000 |

The refractive index is $1.53 + i0.008$.

**Table 5.11** Scattering phase matrix element comparisons.

| $\Theta(^\circ)$ | $P_{11}$ | $P_{22}$ | $P_{33}$ | $P_{44}$ | $P_{12}$ | $P_{34}$ |
|---|---|---|---|---|---|---|
| 0.0 | 114.1554 | 114.1082 | 114.1082 | 114.0610 | 0.0000 | 0.0000 |
| | 114.1456 | 114.0998 | 114.0998 | 114.0541 | 0.0000 | 0.0000 |
| | −0.0098 | −0.0084 | −0.0084 | −0.0069 | 0.0000 | 0.0000 |
| 10.0 | 27.9760 | 27.9341 | 27.8361 | 27.8263 | 0.5586 | 2.0291 |
| | 27.9729 | 27.9336 | 27.8325 | 27.8231 | 0.6085 | 2.0548 |
| | −0.0031 | −0.0005 | −0.0036 | −0.0032 | 0.0499 | 0.0257 |
| 20.0 | 3.1961 | 3.1522 | 3.0897 | 3.0824 | 0.1262 | −0.4160 |
| | 3.2018 | 3.1580 | 3.0941 | 3.0861 | 0.1262 | −0.4217 |
| | 0.0057 | 0.0058 | 0.0044 | 0.0037 | −0.0000 | −0.0057 |
| 30.0 | 1.3826 | 1.3507 | 1.2991 | 1.2945 | 0.0103 | 0.2066 |
| | 1.3816 | 1.3499 | 1.2974 | 1.2916 | 0.0188 | 0.2069 |
| | −0.0010 | −0.0008 | −0.0017 | −0.0029 | 0.0085 | 0.0003 |
| 40.0 | 0.6807 | 0.6433 | 0.6187 | 0.6251 | −0.0873 | −0.0583 |
| | 0.6820 | 0.6445 | 0.6191 | 0.6242 | −0.0874 | −0.0603 |
| | 0.0013 | 0.0012 | 0.0004 | −0.0009 | −0.0001 | −0.0020 |
| 50.0 | 0.3834 | 0.3490 | 0.3071 | 0.3181 | −0.0150 | −0.0167 |
| | 0.3826 | 0.3484 | 0.3059 | 0.3156 | −0.0147 | −0.0191 |
| | −0.0008 | −0.0006 | −0.0012 | −0.0025 | 0.0003 | −0.0024 |
| 60.0 | 0.3702 | 0.3312 | 0.2873 | 0.3043 | −0.0369 | −0.0484 |
| | 0.3698 | 0.3308 | 0.2863 | 0.3019 | −0.0369 | −0.0505 |
| | −0.0004 | −0.0004 | −0.0010 | −0.0024 | −0.0000 | −0.0021 |
| 70.0 | 0.2674 | 0.2288 | 0.1701 | 0.1856 | 0.0095 | −0.0641 |
| | 0.2668 | 0.2283 | 0.1693 | 0.1836 | 0.0083 | −0.0655 |
| | −0.0006 | −0.0005 | −0.0008 | −0.0020 | −0.0012 | −0.0014 |
| 80.0 | 0.2938 | 0.2556 | 0.1952 | 0.2115 | −0.0123 | −0.0715 |
| | 0.2934 | 0.2553 | 0.1935 | 0.2086 | −0.0122 | −0.0750 |
| | −0.0004 | −0.0003 | −0.0017 | −0.0029 | 0.0001 | −0.0035 |
| 90.0 | 0.2310 | 0.1819 | 0.0859 | 0.1162 | 0.0268 | −0.0964 |
| | 0.2312 | 0.1822 | 0.0854 | 0.1143 | 0.0268 | −0.0974 |
| | 0.0002 | 0.0003 | −0.0005 | −0.0019 | 0.0000 | −0.0010 |
| 100.0 | 0.2045 | 0.1626 | 0.0957 | 0.1156 | −0.0040 | −0.0404 |
| | 0.2041 | 0.1621 | 0.0945 | 0.1135 | −0.0044 | −0.0425 |
| | −0.0004 | −0.0005 | −0.0012 | −0.0021 | −0.0004 | −0.0021 |
| 110.0 | 0.1994 | 0.1395 | 0.0551 | 0.0976 | 0.0191 | −0.0817 |
| | 0.1987 | 0.1393 | 0.0544 | 0.0957 | 0.0193 | −0.0829 |
| | −0.0007 | −0.0002 | −0.0007 | −0.0019 | 0.0002 | −0.0012 |
| 120.0 | 0.1697 | 0.1077 | 0.0198 | 0.0678 | 0.0102 | −0.0403 |
| | 0.1696 | 0.1081 | 0.0188 | 0.0656 | 0.0104 | −0.0409 |
| | −0.0001 | 0.0004 | −0.0010 | −0.0022 | 0.0002 | −0.0006 |
| 130.0 | 0.1597 | 0.0762 | −0.0119 | 0.0521 | 0.0210 | −0.0399 |
| | 0.1602 | 0.0766 | −0.0130 | 0.0502 | 0.0206 | −0.0405 |
| | 0.0005 | 0.0004 | −0.0011 | −0.0019 | −0.0004 | −0.0006 |
| 140.0 | 0.1703 | 0.0740 | −0.0292 | 0.0488 | −0.0180 | −0.0332 |
| | 0.1708 | 0.0746 | −0.0297 | 0.0474 | −0.0176 | −0.0332 |
| | 0.0005 | 0.0006 | −0.0005 | −0.0014 | 0.0004 | −0.0000 |

*Continued*

**Table 5.11** Continued

| $\Theta(°)$ | $P_{11}$ | $P_{22}$ | $P_{33}$ | $P_{44}$ | $P_{12}$ | $P_{34}$ |
|---|---|---|---|---|---|---|
| 150.0 | 0.1893 | 0.0783 | −0.0530 | 0.0429 | 0.0138 | 0.0038 |
| | 0.1903 | 0.0781 | −0.0535 | 0.0426 | 0.0133 | 0.0041 |
| | 0.0010 | −0.0002 | −0.0005 | −0.0003 | −0.0005 | 0.0003 |
| 160.0 | 0.2738 | 0.1822 | −0.0893 | −0.0250 | 0.0424 | −0.0707 |
| | 0.2737 | 0.1809 | −0.0893 | −0.0252 | 0.0419 | −0.0705 |
| | −0.0001 | −0.0013 | 0.0000 | −0.0002 | −0.0005 | 0.0002 |
| 170.0 | 0.3497 | 0.2735 | −0.1256 | −0.1060 | 0.0852 | 0.0041 |
| | 0.3491 | 0.2715 | −0.1252 | −0.1046 | 0.0837 | 0.0065 |
| | −0.0006 | −0.0020 | 0.0004 | 0.0014 | −0.0015 | 0.0024 |
| 180.0 | 0.5396 | 0.3756 | −0.3756 | −0.2115 | 0.0000 | 0.0000 |
| | 0.5401 | 0.3755 | -0.3755 | -0.2108 | 0.0000 | 0.0000 |
| | 0.0005 | -0.0001 | 0.0001 | 0.0007 | 0.0000 | 0.0000 |

Other parameters are the same as for Table 5.10.

**Table 5.12** Scattering and extinction cross section and asymmetry factor comparisons.

| $(\lambda = 2\pi)$ | EBCM | IITM | IITM-EBCM |
|---|---|---|---|
| $C_{sca}$ | 983.38 | 983.30 | −0.08 |
| $C_{ext}$ | 1158.56 | 1158.59 | 0.03 |
| g | 0.7530 | 0.7530 | 0.0000 |

Other parameters are the same as for Table 5.10.

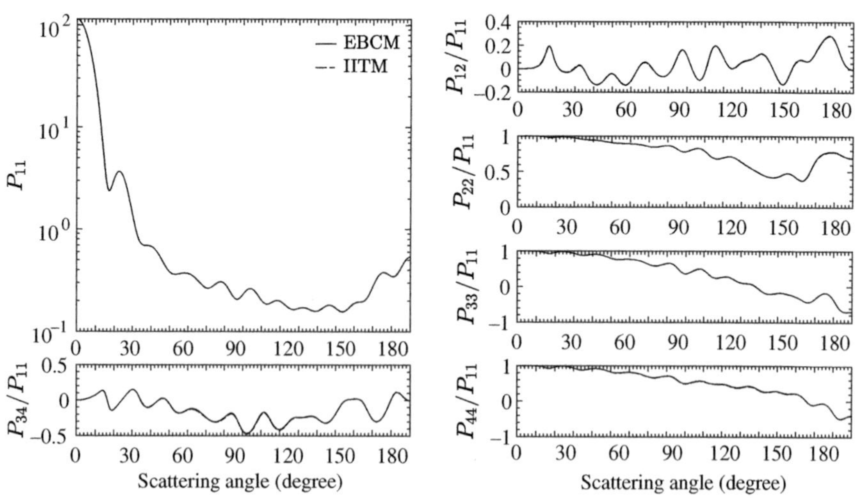


**Fig. 5.8** Scattering phase matrix element comparisons between the EBCM and the IITM. All parameters are the same as for Table 5.10.

### 5.2.2 Spheroids and cylinders with extreme aspect ratios and sizes

In this section, we present several IITM computational results for spheroids with large size parameters and for cylinders with aspect ratios far from unity and large size parameters.

We choose a prolate spheroid with size parameter $2\pi c/\lambda = 150$, AR $= 0.1$, and refractive index $1.311 + i0.0$. In Table 5.13, the scattering phase matrix elements at 19 scattering angles are given. Fig. 5.9 shows the scattering phase matrix elements versus the scattering angle.

We choose an oblate spheroid with size parameter $2\pi c/\lambda = 15$, AR $= 10$, and refractive index $1.53 + i0.008$. In Table 5.14, the scattering phase matrix elements at 19 special scattering angles are given. Fig. 5.10 shows the scattering phase matrix elements versus scattering angle.

For a prolate cylinder, we choose a size parameter $2\pi H/\lambda = 300$, AR $= 0.1$, and refractive index $1.311 + i0.0$. The scattering phase matrix elements at 19 scattering angles are given in Table 5.15. Fig. 5.11 shows the scattering phase matrix elements at 721 scattering angles.

For an oblate cylinder, we choose size parameter $2\pi D/\lambda = 300$, AR $= 10$, and refractive index $1.53 + i0.008$. The scattering phase matrix elements at 19 scattering angles are given in Table 5.16. Fig. 5.12 shows the scattering phase matrix elements versus scattering angle.

**Table 5.13** Scattering phase matrix elements of a prolate spheroid with size parameter $2\pi c/\lambda = 150$, AR $= 0.1$, and refractive index $1.311 + i0.0$.

| $\Theta(°)$ | $P_{11}$ | $P_{22}$ | $P_{33}$ | $P_{44}$ | $P_{12}$ | $P_{34}$ |
|---|---|---|---|---|---|---|
| 0.0 | 1173.5280 | 1172.8800 | 1172.8800 | 1172.2320 | 0.0000 | 0.0000 |
| 10.0 | 13.9453 | 13.7148 | 13.6798 | 13.8377 | −0.2430 | 0.5025 |
| 20.0 | 3.4234 | 3.3018 | 3.2769 | 3.3686 | −0.1310 | −0.0331 |
| 30.0 | 1.4453 | 1.3606 | 1.3406 | 1.4092 | 0.0138 | −0.0145 |
| 40.0 | 0.7921 | 0.7161 | 0.6956 | 0.7623 | 0.0098 | 0.0051 |
| 50.0 | 0.4338 | 0.3786 | 0.3599 | 0.4081 | −0.0011 | −0.0098 |
| 60.0 | 0.2404 | 0.1923 | 0.1607 | 0.2043 | −0.0021 | −0.0255 |
| 70.0 | 0.1415 | 0.0935 | 0.0664 | 0.1109 | 0.0002 | −0.0127 |
| 80.0 | 0.1015 | 0.0529 | 0.0218 | 0.0685 | −0.0048 | −0.0041 |
| 90.0 | 0.0787 | 0.0439 | 0.0151 | 0.0484 | −0.0109 | −0.0035 |
| 100.0 | 0.0771 | 0.0253 | −0.0107 | 0.0394 | −0.0118 | −0.0081 |
| 110.0 | 0.0690 | 0.0231 | −0.0202 | 0.0241 | −0.0088 | −0.0140 |
| 120.0 | 0.0721 | 0.0224 | −0.0252 | 0.0229 | −0.0104 | −0.0185 |
| 130.0 | 0.0758 | 0.0267 | −0.0243 | 0.0234 | −0.0167 | −0.0210 |
| 140.0 | 0.0758 | 0.0322 | −0.0254 | 0.0167 | −0.0231 | −0.0172 |
| 150.0 | 0.0603 | 0.0259 | −0.0214 | 0.0115 | −0.0225 | 0.0021 |
| 160.0 | 0.0468 | 0.0207 | −0.0201 | 0.0047 | −0.0071 | 0.0109 |
| 170.0 | 0.0448 | 0.0298 | −0.0129 | 0.0011 | 0.0109 | 0.0076 |
| 180.0 | 0.0765 | 0.0271 | −0.0271 | 0.0222 | 0.0000 | 0.0000 |

All data are rounded to four digits after the decimal point from the original computed results.

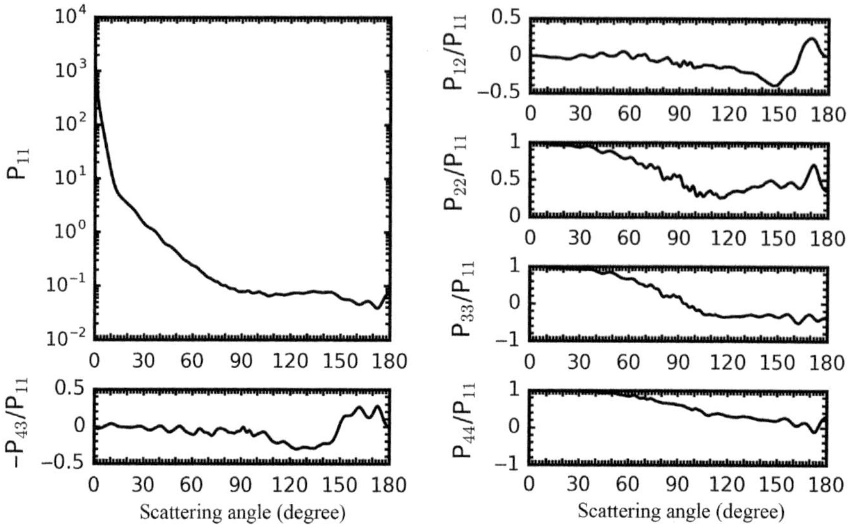


**Fig. 5.9** Scattering phase matrix elements of a prolate spheroid computed by the IITM. All parameters are the same as for Table 5.13.

**Table 5.14** Scattering phase matrix elements of an oblate spheroid with size parameter $2\pi c/\lambda = 15$, AR = 10, and refractive index $1.53 + i0.008$.

| $\Theta(°)$ | $P_{11}$ | $P_{22}$ | $P_{33}$ | $P_{44}$ | $P_{12}$ | $P_{34}$ |
|---|---|---|---|---|---|---|
| 0.0 | 9863.2860 | 9862.8410 | 9862.8410 | 9862.3960 | 0.0000 | 0.0000 |
| 10.0 | 9.9516 | 9.8317 | 9.4321 | 9.3305 | 0.2104 | −2.2496 |
| 20.0 | 2.4454 | 2.3804 | 2.1656 | 2.1250 | 0.1694 | −0.8466 |
| 30.0 | 0.5717 | 0.5419 | 0.4799 | 0.4646 | −0.0573 | −0.1759 |
| 40.0 | 0.2084 | 0.1972 | 0.1537 | 0.1481 | −0.0767 | −0.0512 |
| 50.0 | 0.1079 | 0.1019 | 0.0544 | 0.0521 | −0.0698 | −0.0281 |
| 60.0 | 0.0739 | 0.0698 | 0.0178 | 0.0168 | −0.0593 | −0.0191 |
| 70.0 | 0.0569 | 0.0535 | −0.0018 | −0.0020 | −0.0476 | −0.0142 |
| 80.0 | 0.0484 | 0.0456 | −0.0143 | −0.0139 | −0.0394 | −0.0107 |
| 90.0 | 0.0434 | 0.0405 | −0.0217 | −0.0211 | −0.0312 | −0.0081 |
| 100.0 | 0.0399 | 0.0373 | −0.0262 | −0.0249 | −0.0235 | −0.0069 |
| 110.0 | 0.0393 | 0.0365 | −0.0299 | −0.0283 | −0.0182 | −0.0049 |
| 120.0 | 0.0388 | 0.0360 | −0.0327 | −0.0307 | −0.0134 | −0.0034 |
| 130.0 | 0.0374 | 0.0346 | −0.0331 | −0.0308 | −0.0090 | −0.0025 |
| 140.0 | 0.0367 | 0.0338 | −0.0333 | −0.0308 | −0.0056 | −0.0014 |
| 150.0 | 0.0369 | 0.0340 | −0.0337 | −0.0311 | −0.0032 | −0.0008 |
| 160.0 | 0.0376 | 0.0346 | −0.0343 | −0.0316 | −0.0013 | −0.0004 |
| 170.0 | 0.0383 | 0.0354 | −0.0349 | −0.0322 | −0.0003 | −0.0003 |
| 180.0 | 0.0387 | 0.0356 | −0.0356 | −0.0324 | 0.0000 | 0.0000 |

All data are rounded to four decimal points.

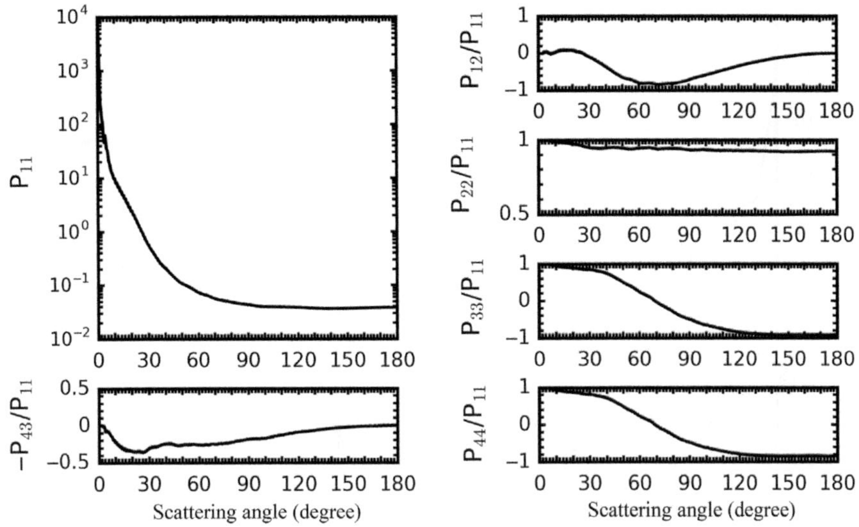


**Fig. 5.10** Scattering phase matrix elements of an oblate spheroid computed by the IITM. All parameters are the same as for Table 5.14.

**Table 5.15** Scattering phase matrix elements of a prolate cylinder with size parameter $2\pi H/\lambda = 300$, AR $= 0.1$, and refractive index $1.311 + i0.0$.

| $\Theta(^\circ)$ | $P_{11}$ | $P_{22}$ | $P_{33}$ | $P_{44}$ | $P_{12}$ | $P_{34}$ |
|---|---|---|---|---|---|---|
| 0.0 | 1832.3780 | 1831.7510 | 1831.7510 | 1831.1230 | 0.0000 | 0.0000 |
| 10.0 | 10.4827 | 10.3844 | 10.3049 | 10.3684 | −0.2365 | 0.7675 |
| 20.0 | 2.9472 | 2.8812 | 2.8557 | 2.9077 | −0.1829 | 0.0232 |
| 30.0 | 0.9053 | 0.8495 | 0.8237 | 0.8711 | 0.0199 | −0.0125 |
| 40.0 | 0.5937 | 0.5391 | 0.5119 | 0.5595 | 0.0172 | 0.0469 |
| 50.0 | 0.4246 | 0.3715 | 0.3518 | 0.3993 | −0.0154 | 0.0212 |
| 60.0 | 0.2553 | 0.2075 | 0.1917 | 0.2360 | −0.0129 | −0.0250 |
| 70.0 | 0.1424 | 0.0978 | 0.0696 | 0.1117 | 0.0015 | −0.0286 |
| 80.0 | 0.0946 | 0.0524 | 0.0195 | 0.0594 | 0.0066 | −0.0087 |
| 90.0 | 0.0717 | 0.0330 | 0.0087 | 0.0451 | −0.0003 | 0.0050 |
| 100.0 | 0.0650 | 0.0297 | 0.0013 | 0.0343 | −0.0169 | −0.0022 |
| 110.0 | 0.0659 | 0.0157 | −0.0188 | 0.0288 | −0.0127 | −0.0160 |
| 120.0 | 0.0624 | 0.0230 | −0.0219 | 0.0151 | −0.0073 | −0.0172 |
| 130.0 | 0.0680 | 0.0200 | −0.0197 | 0.0260 | −0.0129 | −0.0094 |
| 140.0 | 0.0736 | 0.0341 | −0.0257 | 0.0111 | −0.0236 | −0.0144 |
| 150.0 | 0.0513 | 0.0177 | −0.0087 | 0.0212 | −0.0163 | 0.0077 |
| 160.0 | 0.0431 | 0.0130 | −0.0010 | 0.0238 | −0.0070 | 0.0032 |
| 170.0 | 0.0622 | 0.0324 | −0.0071 | 0.0159 | 0.0192 | 0.0049 |
| 180.0 | 0.1047 | 0.0419 | −0.0419 | 0.0210 | 0.0000 | 0.0000 |

All data are rounded to four decimal points.

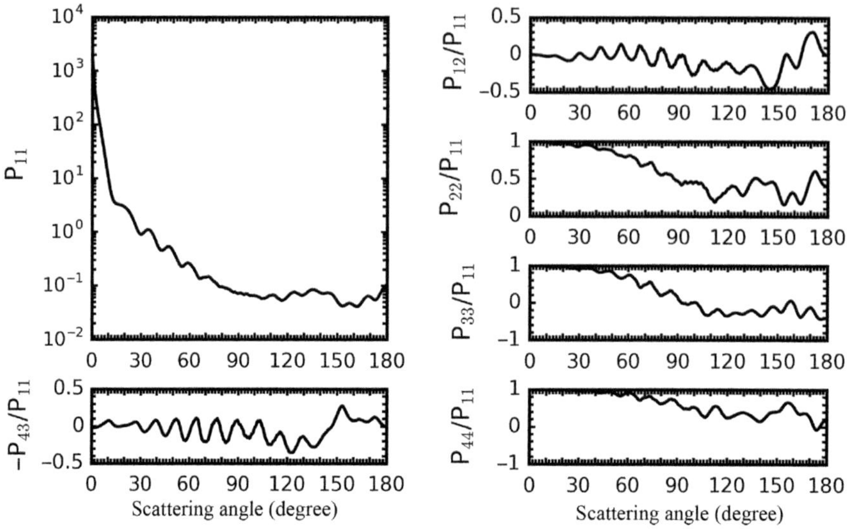


**Fig. 5.11** Scattering phase matrix elements of a prolate cylinder computed by the IITM. All parameters are the same as for Table 5.15.

**Table 5.16** Scattering phase matrix elements of an oblate cylinder with size parameter $2\pi D/\lambda = 300$, AR = 10, and refractive index $1.53 + i0.008$.

| $\Theta(°)$ | $P_{11}$ | $P_{22}$ | $P_{33}$ | $P_{44}$ | $P_{12}$ | $P_{34}$ |
|---|---|---|---|---|---|---|
| 0.0 | 14239.1700 | 14216.0500 | 14216.0500 | 14192.9300 | 0.0000 | 0.0000 |
| 10.0 | 2.8690 | 2.8098 | 2.5242 | 2.4740 | 0.1016 | −0.7675 |
| 20.0 | 0.5571 | 0.5386 | 0.4372 | 0.4253 | −0.0630 | −0.1738 |
| 30.0 | 0.2825 | 0.2733 | 0.2273 | 0.2224 | −0.0799 | −0.0641 |
| 40.0 | 0.2114 | 0.2053 | 0.1442 | 0.1413 | −0.1101 | −0.0348 |
| 50.0 | 0.1319 | 0.1272 | 0.0446 | 0.0434 | −0.0926 | −0.0302 |
| 60.0 | 0.0604 | 0.0552 | 0.0127 | 0.0138 | −0.0301 | −0.0287 |
| 70.0 | 0.0719 | 0.0668 | 0.0097 | 0.0116 | −0.0514 | −0.0204 |
| 80.0 | 0.0832 | 0.0786 | −0.0198 | −0.0177 | −0.0605 | −0.0203 |
| 90.0 | 0.0458 | 0.0392 | −0.0154 | −0.0108 | −0.0174 | −0.0175 |
| 100.0 | 0.0403 | 0.0330 | −0.0128 | −0.0070 | −0.0183 | −0.0107 |
| 110.0 | 0.0655 | 0.0589 | −0.0437 | −0.0384 | −0.0252 | −0.0126 |
| 120.0 | 0.0522 | 0.0456 | −0.0354 | −0.0298 | −0.0135 | −0.0065 |
| 130.0 | 0.0279 | 0.0212 | −0.0162 | −0.0102 | −0.0020 | −0.0032 |
| 140.0 | 0.0308 | 0.0243 | −0.0208 | −0.0147 | −0.0033 | −0.0041 |
| 150.0 | 0.0467 | 0.0408 | −0.0377 | −0.0322 | −0.0025 | −0.0032 |
| 160.0 | 0.0571 | 0.0509 | −0.0473 | −0.0414 | −0.0011 | −0.0030 |
| 170.0 | 0.0645 | 0.0545 | −0.0507 | −0.0409 | 0.0012 | −0.0022 |
| 180.0 | 0.0884 | 0.0621 | −0.0621 | −0.0358 | 0.0000 | 0.0000 |

All data are rounded to four decimal points.

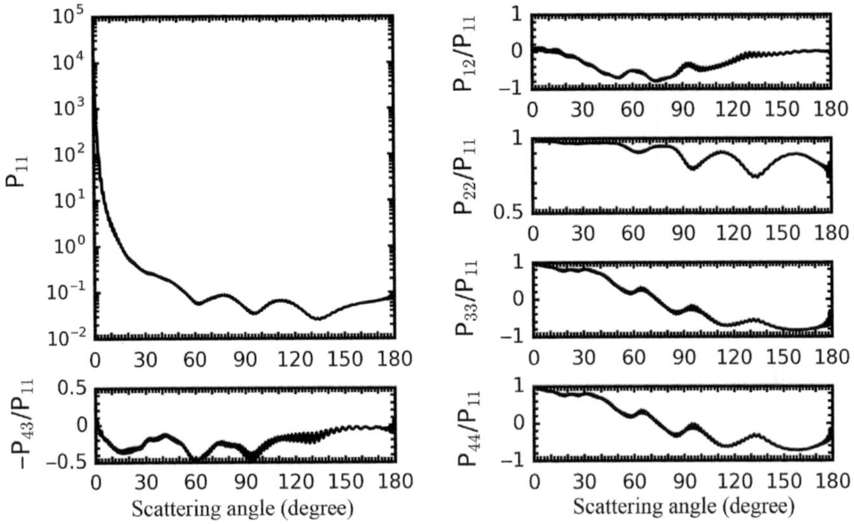


**Fig. 5.12** Scattering phase matrix elements of an oblate cylinder computed by the IITM. All parameters are the same as for Table 5.16.

Another prolate spheroid is chosen with size parameter $2\pi c/\lambda = 400$, AR $= 0.5$, and refractive index $1.53 + i0.008$. In Table 5.17, the scattering phase matrix elements at 19 special scattering angles are given. Fig. 5.13 shows the scattering phase matrix elements over the scattering angle range.

Another cylinder is chosen with size parameter $2\pi H/\lambda = 500$, AR $= 1.0$, and refractive index $1.311 + i0.0$. The scattering phase matrix elements at 19 special scattering angles are given in Table 5.18. Fig. 5.14 shows the scattering phase matrix elements over the scattering angle range.

The scattering and extinction cross sections and asymmetry factors of the particles in this section are summarized in Table 5.19.

Similar verification and applications for spheroids and cylinders are also given by Bi et al. (2013).

## 5.3 Finite-fold rotationally symmetric particles—Hexagonal prisms

For a particle with $N$-fold rotational symmetry, the T-matrix is decoupled into $N$ submatrices and only $N/2 + 1$ submatrices for even $N$ and $(N+1)/2$ submatrices for odd $N$ are independent. Since a particle with a finite-fold rotational symmetry always has mirror symmetry, the scattering phase matrix of the particle in random orientation also has six independent elements denoted as $P_{11}$, $P_{22}$, $P_{33}$, $P_{44}$, $P_{12}$, $P_{34}$, which can be analytically expressed with respect to the T-matrix. In this section, we use hexagonal prisms to show the potential applications of the IITM for finite-fold rotationally

**Table 5.17** Scattering phase matrix elements of a prolate spheroid with size parameter $2\pi c/\lambda = 400$, AR = 0.5, and refractive index $1.53 + i0.008$.

| $\Theta(°)$ | $P_{11}$ | $P_{22}$ | $P_{33}$ | $P_{44}$ | $P_{12}$ | $P_{34}$ |
|---|---|---|---|---|---|---|
| 0.0 | 65432.5100 | 65432.2800 | 65432.2800 | 65432.0600 | 0.0000 | 0.0000 |
| 10.0 | 1.4961 | 1.4935 | 1.4638 | 1.4613 | −0.2605 | −0.0274 |
| 20.0 | 0.4621 | 0.4593 | 0.4245 | 0.4219 | −0.1716 | −0.0152 |
| 30.0 | 0.2575 | 0.2550 | 0.2087 | 0.2063 | −0.1414 | −0.0136 |
| 40.0 | 0.1674 | 0.1655 | 0.1102 | 0.1085 | −0.1190 | −0.0117 |
| 50.0 | 0.1178 | 0.1165 | 0.0539 | 0.0527 | −0.0994 | −0.0092 |
| 60.0 | 0.0899 | 0.0888 | 0.0191 | 0.0182 | −0.0798 | −0.0099 |
| 70.0 | 0.0716 | 0.0706 | −0.0032 | −0.0041 | −0.0628 | −0.0087 |
| 80.0 | 0.0577 | 0.0570 | −0.0177 | −0.0183 | −0.0500 | −0.0053 |
| 90.0 | 0.0491 | 0.0487 | −0.0267 | −0.0271 | −0.0389 | −0.0031 |
| 100.0 | 0.0444 | 0.0441 | −0.0320 | −0.0322 | −0.0296 | −0.0020 |
| 110.0 | 0.0418 | 0.0416 | −0.0352 | −0.0353 | −0.0219 | −0.0013 |
| 120.0 | 0.0406 | 0.0404 | −0.0372 | −0.0373 | −0.0156 | −0.0009 |
| 130.0 | 0.0404 | 0.0401 | −0.0387 | −0.0388 | −0.0106 | −0.0005 |
| 140.0 | 0.0408 | 0.0406 | −0.0400 | −0.0400 | −0.0068 | −0.0002 |
| 150.0 | 0.0401 | 0.0399 | −0.0398 | −0.0398 | −0.0037 | −0.0001 |
| 160.0 | 0.0396 | 0.0393 | −0.0393 | −0.0393 | −0.0016 | −0.0001 |
| 170.0 | 0.0390 | 0.0388 | −0.0388 | −0.0388 | −0.0004 | 0.0000 |
| 180.0 | 0.0401 | 0.0399 | −0.0399 | −0.0397 | 0.0000 | 0.0000 |

All data are rounded to four digits after the decimal point from the original computed results.

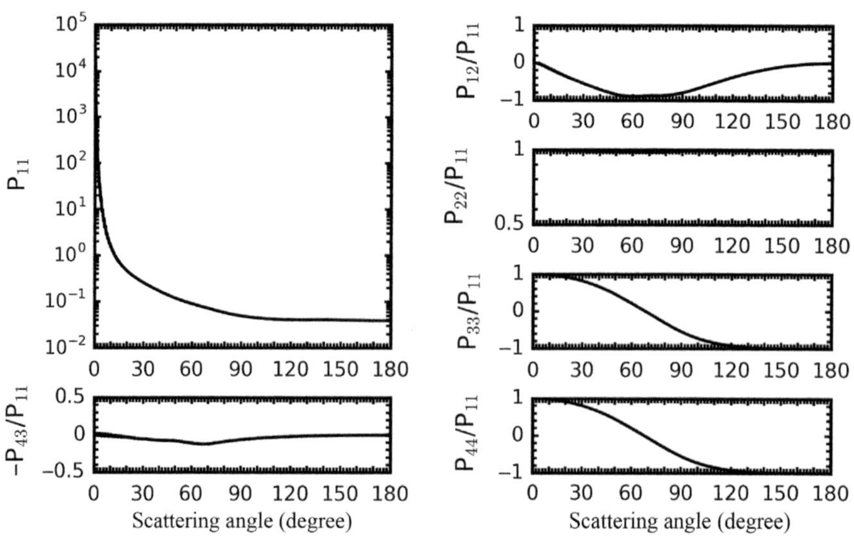


**Fig. 5.13** Scattering phase matrix elements of a prolate spheroid computed by the IITM. All parameters are the same as for Table 5.17.

**Table 5.18** Scattering phase matrix elements of a cylinder with size parameter $2\pi H/\lambda = 500$, AR = 1.0, and refractive index $1.311 + i0.0$.

| $\Theta(°)$ | $P_{11}$ | $P_{22}$ | $P_{33}$ | $P_{44}$ | $P_{12}$ | $P_{34}$ |
|---|---|---|---|---|---|---|
| 0.0 | 49178.5300 | 49178.1300 | 49178.1300 | 49177.7300 | 0.0000 | 0.0000 |
| 10.0 | 5.0200 | 4.9926 | 4.9816 | 4.9691 | −0.0534 | 0.0212 |
| 20.0 | 1.7934 | 1.7686 | 1.7520 | 1.7425 | 0.0117 | −0.0596 |
| 30.0 | 1.1568 | 1.1290 | 1.0706 | 1.0589 | 0.0162 | −0.1855 |
| 40.0 | 0.6916 | 0.6639 | 0.5966 | 0.5859 | −0.0012 | −0.1974 |
| 50.0 | 0.6501 | 0.6244 | 0.5465 | 0.5383 | 0.0436 | −0.1533 |
| 60.0 | 0.3359 | 0.3126 | 0.2271 | 0.2217 | −0.0016 | −0.1669 |
| 70.0 | 0.2813 | 0.2578 | 0.1646 | 0.1660 | −0.0093 | −0.1673 |
| 80.0 | 0.2509 | 0.2247 | 0.1246 | 0.1328 | −0.0050 | −0.1596 |
| 90.0 | 0.2514 | 0.2205 | 0.1165 | 0.1314 | −0.0081 | −0.1534 |
| 100.0 | 0.2322 | 0.1952 | 0.0976 | 0.1187 | −0.0017 | −0.1378 |
| 110.0 | 0.1805 | 0.1377 | 0.0614 | 0.0885 | −0.0045 | −0.0908 |
| 120.0 | 0.1224 | 0.0819 | 0.0199 | 0.0451 | −0.0079 | −0.0460 |
| 130.0 | 0.0945 | 0.0568 | −0.0053 | 0.0168 | −0.0178 | −0.0214 |
| 140.0 | 0.1014 | 0.0581 | −0.0169 | 0.0032 | −0.0145 | −0.0002 |
| 150.0 | 0.1082 | 0.0563 | −0.0349 | −0.0138 | −0.0107 | −0.0124 |
| 160.0 | 0.1202 | 0.0595 | −0.0473 | −0.0210 | 0.0033 | −0.0268 |
| 170.0 | 0.2061 | 0.1040 | −0.0951 | −0.0416 | 0.0141 | −0.0279 |
| 180.0 | 0.6755 | 0.4756 | −0.4756 | −0.2756 | 0.0000 | 0.0000 |

All data are rounded to four digits after the decimal point from the original computed results.

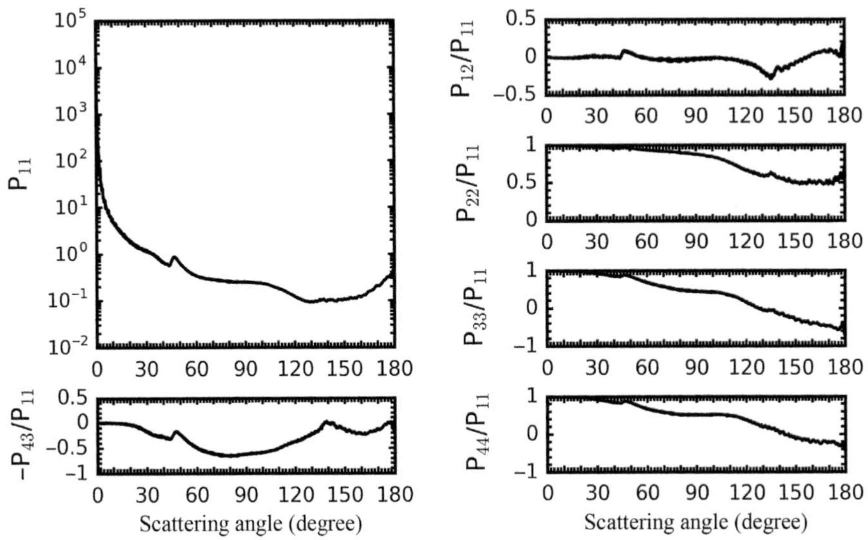


**Fig. 5.14** Scattering phase matrix elements of a cylinder computed by the IITM. All parameters are the same as for Table 5.18.

**Table 5.19** Scattering and extinction cross sections (the wavelength is assumed to be $2\pi$) and asymmetry factors of the chosen spheroids and cylinders with extreme shapes.

| Shape | $m$ | $C_{sca}$ | $C_{ext}$ | G |
|---|---|---|---|---|
| Prolate spheroid ($2\pi c/\lambda = 150$, AR = 0.1) | $1.311 + i0.0$ | 13393.0625 | 13393.0376 | 0.8813 |
| Oblate spheroid ($2\pi c/\lambda = 15$, AR = 10) | $1.53 + i0.008$ | 60587.1254 | 76558.9102 | 0.9441 |
| Prolate cylinder ($2\pi H/\lambda = 300$, AR = 0.1) | $1.311 + i0.0$ | 19786.2908 | 19786.2217 | 0.8933 |
| Oblate cylinder ($2\pi D/\lambda = 300$, AR = 10) | $1.53 + i0.008$ | 67648.0867 | 88439.1021 | 0.9412 |
| Prolate spheroid ($2\pi c/\lambda = 400$, AR = 0.5) | $1.53 + i0.008$ | 241779.9053 | 440100.5215 | 0.9476 |
| Cylinder ($2\pi H/\lambda = 500$, AR = 1.0) | $1.311 + i0.0$ | 593902.9156 | 593902.9157 | 0.7953 |

All data are rounded to four decimal points.

symmetric particles. The comparisons of the phase matrix elements of hexagonal prisms computed with the IITM, the DDA, and PSTD methods are given by Bi and Yang (2014).

The shape of a hexagonal prism is shown in Fig. 5.15. $H$ is the height of the hexagonal column, and $a$ is the semiwidth of the geometry. The aspect ratio is defined as AR $= 2a/H$.

Fig. 5.16 gives the expansion coefficients of a hexagonal prism with unity aspect ratio and $2\pi H/\lambda = 300$. The refractive index is $1.308 + i1.43 \times 10^{-8}$. Fig. 5.17 is for the same parameters as Fig. 5.16 except that the refractive index is $1.2762 + i0.4133$.

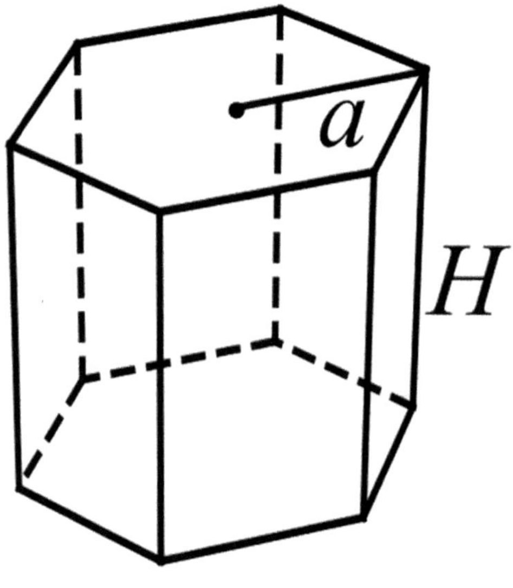


**Fig. 5.15** A hexagonal prism defined by height $H$ and semiwidth $a$.

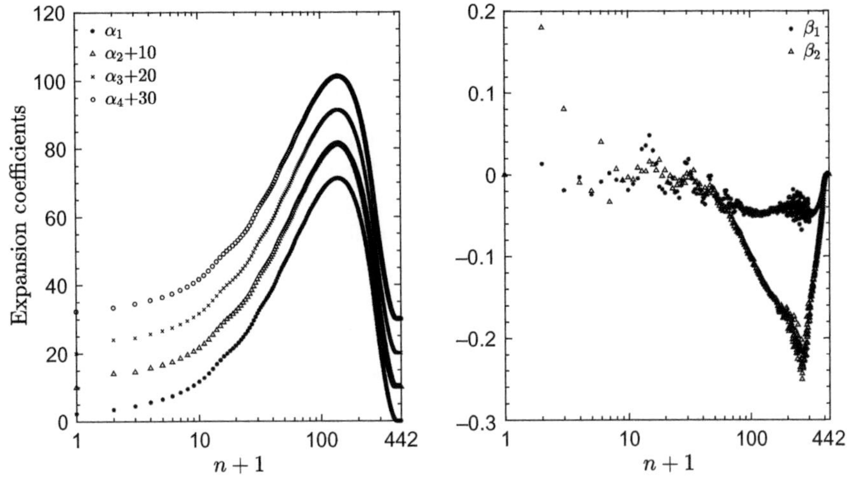


**Fig. 5.16** Expansion coefficients of a hexagonal prism with unity aspect ratio (AR = 1.0) and $2\pi H/\lambda = 300$. The refractive index for the prism is $1.308 + i1.43 \times 10^{-8}$.

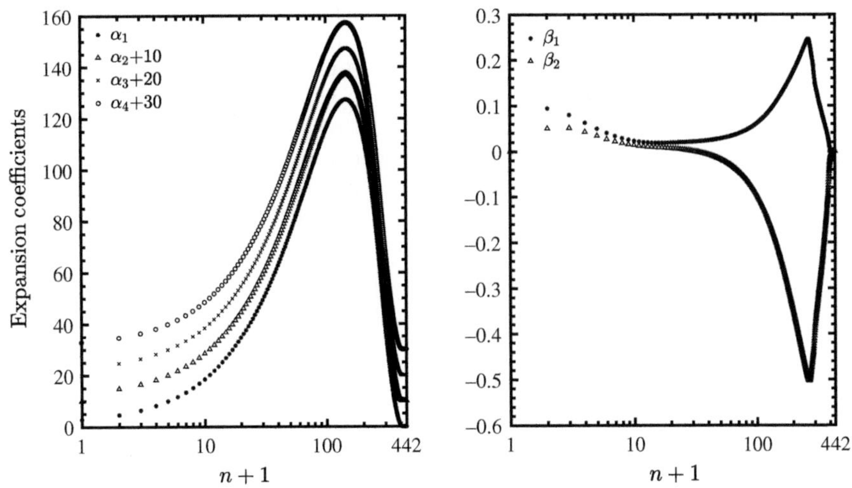


**Fig. 5.17** Expansion coefficients of a hexagonal prism with unity aspect ratio (AR = 1.0) and $2\pi H/\lambda = 300$. The refractive index for the prism is $1.2762 + i0.4133$.

We compare the IITM using the ADDA (Yurkin and Hoekstra, 2011) for a randomly oriented hexagonal prism with size parameter $2\pi H/\lambda = 30$, unity aspect ratio, and refractive index $1.308 + i1.43 \times 10^{-8}$. In the ADDA calculation, the number of dipoles per wavelength (dpl) is set to be 40, and 1025 random orientations in total are used to obtain the orientation-averaged result.

Table 5.20 gives the six independent scattering phase matrix elements by the ADDA and IITM, respectively, and the IITM-ADDA difference at 19 special scattering angles, and Fig. 5.18 shows the IITM and ADDA comparison at all the scattering

**Table 5.20** Scattering phase matrix element comparisons between ADDA and IITM for a hexagonal prism with size parameter $2\pi H/\lambda = 30$ and unity aspect ratio.

| $\Theta$(°) | $\mathbf{P_{11}}$ | $\mathbf{P_{22}}$ | $\mathbf{P_{33}}$ | $\mathbf{P_{44}}$ | $\mathbf{P_{12}}$ | $\mathbf{P_{34}}$ |
|---|---|---|---|---|---|---|
| 0.0 | 186.9337 | 186.9007 | 186.9007 | 186.8677 | 0.0000 | 0.0000 |
| | 186.9359 | 186.9055 | 186.9055 | 186.8750 | 0.0000 | 0.0000 |
| | 0.0022 | 0.0048 | 0.0048 | 0.0073 | 0.0000 | 0.0000 |
| 10.0 | 15.7611 | 15.7448 | 15.7083 | 15.7062 | −0.1414 | 0.8197 |
| | 15.8149 | 15.7998 | 15.7630 | 15.7615 | −0.0661 | 0.8559 |
| | 0.0539 | 0.0550 | 0.0547 | 0.0553 | 0.0753 | 0.0362 |
| 20.0 | 1.6983 | 1.6874 | 1.6675 | 1.6667 | −0.0852 | −0.0314 |
| | 1.6931 | 1.6825 | 1.6631 | 1.6624 | −0.0919 | −0.0272 |
| | −0.0052 | −0.0049 | −0.0044 | −0.0044 | −0.0066 | 0.0042 |
| 30.0 | 1.4665 | 1.4548 | 1.4280 | 1.4276 | 0.0462 | −0.0875 |
| | 1.4574 | 1.4459 | 1.4189 | 1.4186 | 0.0385 | −0.0859 |
| | −0.0091 | −0.0089 | −0.0091 | −0.0090 | −0.0077 | 0.0016 |
| 40.0 | 1.1026 | 1.0898 | 1.0587 | 1.0604 | 0.0879 | −0.1261 |
| | 1.0989 | 1.0861 | 1.0553 | 1.0570 | 0.0866 | −0.1262 |
| | −0.0038 | −0.0037 | −0.0034 | −0.0034 | −0.0013 | −0.0001 |
| 50.0 | 0.6624 | 0.6485 | 0.6201 | 0.6243 | 0.0843 | −0.0810 |
| | 0.6607 | 0.6469 | 0.6182 | 0.6225 | 0.0845 | −0.0832 |
| | −0.0018 | −0.0016 | −0.0019 | −0.0018 | 0.0002 | −0.0022 |
| 60.0 | 0.4807 | 0.4656 | 0.4399 | 0.4462 | 0.0335 | −0.0661 |
| | 0.4810 | 0.4662 | 0.4395 | 0.4458 | 0.0337 | −0.0694 |
| | 0.0004 | 0.0005 | −0.0004 | −0.0003 | 0.0003 | −0.0033 |
| 70.0 | 0.3638 | 0.3470 | 0.3204 | 0.3296 | 0.0086 | −0.0755 |
| | 0.3646 | 0.3479 | 0.3201 | 0.3293 | 0.0081 | −0.0788 |
| | 0.0008 | 0.0009 | −0.0003 | −0.0003 | −0.0004 | −0.0033 |
| 80.0 | 0.2741 | 0.2561 | 0.2296 | 0.2406 | −0.0010 | −0.0687 |
| | 0.2735 | 0.2556 | 0.2282 | 0.2392 | −0.0006 | −0.0717 |
| | −0.0006 | −0.0005 | −0.0014 | −0.0015 | 0.0004 | −0.0030 |
| 90.0 | 0.2128 | 0.1919 | 0.1637 | 0.1788 | −0.0085 | −0.0544 |
| | 0.2122 | 0.1913 | 0.1627 | 0.1776 | −0.0078 | −0.0564 |
| | −0.0007 | −0.0006 | −0.0010 | −0.0012 | 0.0007 | −0.0020 |
| 100.0 | 0.1674 | 0.1392 | 0.1104 | 0.1329 | −0.0037 | −0.0456 |
| | 0.1675 | 0.1394 | 0.1096 | 0.1321 | −0.0040 | −0.0475 |
| | 0.0001 | 0.0002 | −0.0007 | −0.0008 | −0.0003 | −0.0018 |
| 110.0 | 0.1427 | 0.1066 | 0.0771 | 0.1072 | −0.0233 | −0.0282 |
| | 0.1427 | 0.1064 | 0.0764 | 0.1064 | −0.0231 | −0.0304 |
| | −0.0001 | −0.0002 | −0.0008 | −0.0008 | 0.0003 | −0.0022 |
| 120.0 | 0.1317 | 0.0826 | 0.0415 | 0.0835 | −0.0154 | −0.0447 |
| | 0.1319 | 0.0824 | 0.0412 | 0.0832 | −0.0145 | −0.0458 |
| | 0.0002 | −0.0002 | −0.0002 | −0.0003 | 0.0009 | −0.0010 |
| 130.0 | 0.1272 | 0.0658 | 0.0172 | 0.0712 | −0.0331 | −0.0293 |
| | 0.1271 | 0.0655 | 0.0174 | 0.0713 | −0.0324 | −0.0296 |
| | −0.0001 | −0.0002 | 0.0002 | 0.0001 | 0.0007 | −0.0003 |
| 140.0 | 0.1160 | 0.0400 | −0.0061 | 0.0611 | −0.0178 | −0.0172 |
| | 0.1168 | 0.0409 | −0.0066 | 0.0603 | −0.0175 | −0.0179 |
| | 0.0008 | 0.0009 | −0.0005 | −0.0008 | 0.0003 | −0.0007 |

**Table 5.20** Continued

| $\Theta(°)$ | $P_{11}$ | $P_{22}$ | $P_{33}$ | $P_{44}$ | $P_{12}$ | $P_{34}$ |
|---|---|---|---|---|---|---|
| 150.0 | 0.1244 | 0.0529 | −0.0187 | 0.0409 | −0.0198 | −0.0084 |
| | 0.1237 | 0.0525 | −0.0183 | 0.0412 | −0.0196 | −0.0091 |
| | −0.0006 | −0.0004 | 0.0004 | 0.0003 | 0.0003 | −0.0007 |
| 160.0 | 0.1645 | 0.0625 | −0.0542 | 0.0345 | −0.0099 | −0.0015 |
| | 0.1650 | 0.0635 | −0.0554 | 0.0330 | −0.0092 | −0.0027 |
| | 0.0006 | 0.0009 | −0.0012 | −0.0015 | 0.0007 | −0.0011 |
| 170.0 | 0.2297 | 0.1560 | −0.0158 | 0.0347 | 0.0183 | −0.0180 |
| | 0.2282 | 0.1538 | −0.0154 | 0.0342 | 0.0184 | −0.0190 |
| | −0.0015 | −0.0021 | 0.0003 | −0.0005 | 0.0002 | −0.0010 |
| 180.0 | 0.3154 | 0.1949 | −0.1949 | −0.0744 | 0.0000 | 0.0000 |
| | 0.3017 | 0.1814 | −0.1814 | −0.0610 | 0.0000 | 0.0000 |
| | −0.0136 | −0.0135 | 0.0135 | 0.0135 | 0.0000 | 0.0000 |

The refractive index is $1.308+i1.43\times10^{-8}$ . For each scattering angle, the three rows denote the results from the ADDA, the IITM, and the IITM minus the ADDA difference. All data are rounded to four digits after the decimal point from the original computed results.

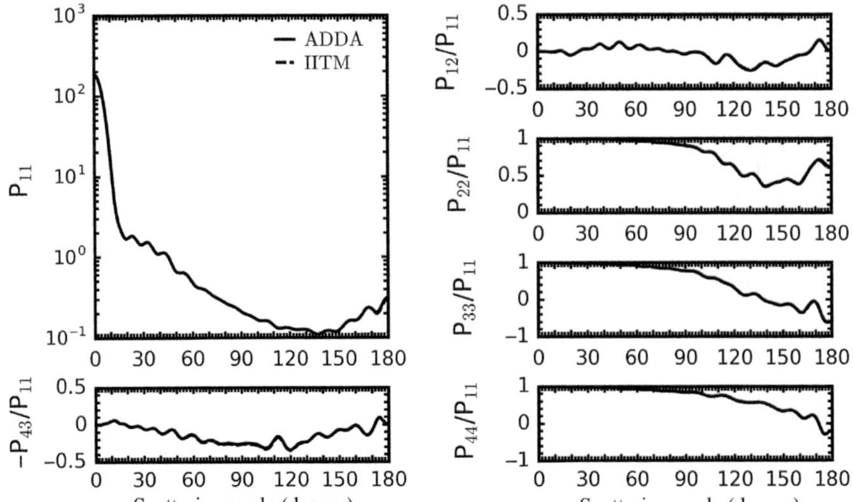


**Fig. 5.18** Scattering phase matrix element comparisons between the ADDA and the IITM. The parameters are the same as for Table 5.20.

angles used in the computations. The scattering and extinction cross sections and asymmetry factor comparisons are given in Table 5.21.

In the aforementioned cases, the refractive index has a very small imaginary part, so the particle is almost nonabsorbing. Here, we consider the refractive index $1.2762+i0.4133$ with a large imaginary part. All other parameters of the hexagonal prism remain the same. In the ADDA calculation, the dpl is set to be 20, and 4097 random orientations in total are used to obtain the orientation-averaged result. Table 5.22 gives

**Table 5.21** Scattering and extinction cross section and asymmetry factor comparisons.

| ($\lambda = 2\pi$) | ADDA | IITM | IITM-ADDA |
|---|---|---|---|
| $C_{sca}$ | 2223.6484 | 2223.9024 | 0.2540 |
| $C_{ext}$ | 2223.6492 | 2223.9032 | 0.2540 |
| $g$ | 0.7743 | 0.7744 | 0.0002 |

Other parameters are the same as for Table 5.20.

**Table 5.22** The same as Table 5.20 except that the refractive index is 1.2762+*i*0.4133.

| $\Theta$(°) | $P_{11}$ | $P_{22}$ | $P_{33}$ | $P_{44}$ | $P_{12}$ | $P_{34}$ |
|---|---|---|---|---|---|---|
| 0.0 | 312.3055 | 312.2796 | 312.2796 | 312.2538 | 0.0000 | 0.0000 |
| | 312.8384 | 312.8149 | 312.8149 | 312.7915 | 0.0000 | 0.0000 |
| | 0.5329 | 0.5353 | 0.5353 | 0.5377 | 0.0000 | 0.0000 |
| 10.0 | 12.7960 | 12.7731 | 12.6104 | 12.5917 | −1.4203 | 1.2933 |
| | 12.8051 | 12.7846 | 12.6164 | 12.5998 | −1.3318 | 1.4231 |
| | 0.0091 | 0.0115 | 0.0060 | 0.0081 | 0.0885 | 0.1298 |
| 20.0 | 2.0527 | 2.0473 | 1.9848 | 1.9811 | −0.3922 | 0.1145 |
| | 2.0529 | 2.0478 | 1.9863 | 1.9825 | −0.3810 | 0.1337 |
| | 0.0003 | 0.0004 | 0.0015 | 0.0014 | 0.0112 | 0.0192 |
| 30.0 | 0.6208 | 0.6172 | 0.5667 | 0.5647 | −0.1806 | −0.0655 |
| | 0.6212 | 0.6175 | 0.5685 | 0.5661 | −0.1782 | −0.0609 |
| | 0.0004 | 0.0003 | 0.0018 | 0.0014 | 0.0024 | 0.0046 |
| 40.0 | 0.2333 | 0.2316 | 0.1689 | 0.1682 | −0.1183 | −0.0675 |
| | 0.2327 | 0.2310 | 0.1700 | 0.1690 | −0.1170 | −0.0666 |
| | −0.0005 | −0.0006 | 0.0011 | 0.0008 | 0.0013 | 0.0009 |
| 50.0 | 0.1372 | 0.1361 | 0.0659 | 0.0654 | −0.1034 | −0.0397 |
| | 0.1361 | 0.1349 | 0.0664 | 0.0658 | −0.1019 | −0.0393 |
| | −0.0011 | −0.0012 | 0.0005 | 0.0003 | 0.0015 | 0.0004 |
| 60.0 | 0.1060 | 0.1053 | 0.0337 | 0.0335 | −0.0912 | −0.0301 |
| | 0.1047 | 0.1038 | 0.0343 | 0.0340 | −0.0897 | −0.0295 |
| | −0.0014 | −0.0015 | 0.0006 | 0.0005 | 0.0015 | 0.0006 |
| 70.0 | 0.0830 | 0.0825 | 0.0078 | 0.0078 | −0.0740 | −0.0302 |
| | 0.0816 | 0.0810 | 0.0089 | 0.0087 | −0.0727 | −0.0296 |
| | −0.0015 | −0.0015 | 0.0011 | 0.0009 | 0.0013 | 0.0006 |
| 80.0 | 0.0663 | 0.0659 | −0.0145 | −0.0146 | −0.0578 | −0.0253 |
| | 0.0648 | 0.0643 | −0.0132 | −0.0134 | −0.0567 | −0.0249 |
| | −0.0015 | −0.0015 | 0.0013 | 0.0012 | 0.0011 | 0.0004 |
| 90.0 | 0.0567 | 0.0564 | −0.0274 | −0.0274 | −0.0454 | −0.0178 |
| | 0.0552 | 0.0548 | −0.0260 | −0.0261 | −0.0444 | −0.0175 |
| | −0.0016 | −0.0016 | 0.0015 | 0.0013 | 0.0010 | 0.0003 |
| 100.0 | 0.0513 | 0.0511 | −0.0340 | −0.0340 | −0.0354 | −0.0130 |
| | 0.0497 | 0.0494 | −0.0325 | −0.0325 | −0.0346 | −0.0128 |
| | −0.0016 | −0.0017 | 0.0015 | 0.0015 | 0.0007 | 0.0002 |

**Table 5.22** Continued

| $\Theta(°)$ | $P_{11}$ | $P_{22}$ | $P_{33}$ | $P_{44}$ | $P_{12}$ | $P_{34}$ |
|---|---|---|---|---|---|---|
| 110.0 | 0.0480 | 0.0478 | −0.0387 | −0.0387 | −0.0260 | −0.0097 |
| | 0.0463 | 0.0461 | −0.0371 | −0.0371 | −0.0255 | −0.0095 |
| | −0.0017 | −0.0017 | 0.0016 | 0.0016 | 0.0006 | 0.0002 |
| 120.0 | 0.0462 | 0.0460 | −0.0416 | −0.0415 | −0.0187 | −0.0063 |
| | 0.0445 | 0.0443 | −0.0399 | −0.0399 | −0.0183 | −0.0062 |
| | −0.0017 | −0.0017 | 0.0017 | 0.0016 | 0.0004 | 0.0001 |
| 130.0 | 0.0451 | 0.0450 | −0.0429 | −0.0428 | −0.0128 | −0.0044 |
| | 0.0434 | 0.0432 | −0.0412 | −0.0411 | −0.0125 | −0.0043 |
| | −0.0018 | −0.0018 | 0.0017 | 0.0017 | 0.0003 | 0.0001 |
| 140.0 | 0.0448 | 0.0447 | −0.0439 | −0.0438 | −0.0079 | −0.0026 |
| | 0.0431 | 0.0430 | −0.0422 | −0.0422 | −0.0077 | −0.0026 |
| | −0.0017 | −0.0017 | 0.0017 | 0.0017 | 0.0002 | 0.0001 |
| 150.0 | 0.0447 | 0.0446 | −0.0443 | −0.0442 | −0.0045 | −0.0014 |
| | 0.0429 | 0.0428 | −0.0426 | −0.0425 | −0.0044 | −0.0013 |
| | −0.0018 | −0.0018 | 0.0018 | 0.0017 | 0.0001 | 0.0000 |
| 160.0 | 0.0448 | 0.0447 | −0.0447 | −0.0446 | −0.0019 | −0.0006 |
| | 0.0429 | 0.0428 | −0.0428 | −0.0427 | −0.0019 | −0.0006 |
| | −0.0020 | −0.0019 | 0.0019 | 0.0019 | 0.0000 | 0.0000 |
| 170.0 | 0.0442 | 0.0441 | −0.0441 | −0.0441 | −0.0006 | −0.0001 |
| | 0.0429 | 0.0428 | −0.0428 | −0.0428 | −0.0005 | −0.0001 |
| | −0.0013 | −0.0013 | 0.0013 | 0.0013 | 0.0001 | 0.0000 |
| 180.0 | 0.0463 | 0.0462 | −0.0462 | −0.0461 | 0.0000 | 0.0000 |
| | 0.0430 | 0.0429 | −0.0429 | −0.0429 | 0.0000 | 0.0000 |
| | −0.0033 | −0.0033 | 0.0033 | 0.0033 | 0.0000 | 0.0000 |

the six independent scattering phase matrix elements by the IITM and ADDA, respectively, at 19 scattering angles, and Fig. 5.19 shows the IITM and ADDA comparison at all the scattering angles used in the computations. The scattering and extinction cross section and asymmetry factor comparisons are given in Table 5.23.

The scattering phase matrices computed from the IITM and ADDA shown in Figs. 5.18 and 5.19 are almost fully overlapped. For the scattering phase matrix elements in Table 5.20 and 5.22, the IITM and ADDA results are consistent up to the fourth decimal place.

For a large hexagonal prism with size parameter $2\pi H/\lambda = 300$ and unity aspect ratio, we compare the IITM results with the PGOM (Sun et al., 2017). Two refractive indices $1.308 + i1.43 \times 10^{-8}$ and $1.2762 + i0.4133$ are used in the calculations.

When the refractive index is $1.308 + i1.43 \times 10^{-8}$, Table 5.24 gives the six independent scattering phase matrix elements by the IITM and PGOM at 19 special scattering angles, and Fig. 5.20 shows the IITM and PGOM comparison at all scattering angles used in the computations. The scattering and extinction cross section and asymmetry factor comparisons are given in Table 5.25.

When the refractive index is $1.2762 + i0.4133$, Table 5.26 gives the six independent scattering phase matrix elements by the IITM and PGOM at 19 special scattering angles, and Fig. 5.21 shows the IITM and PGOM comparison at all scattering angles

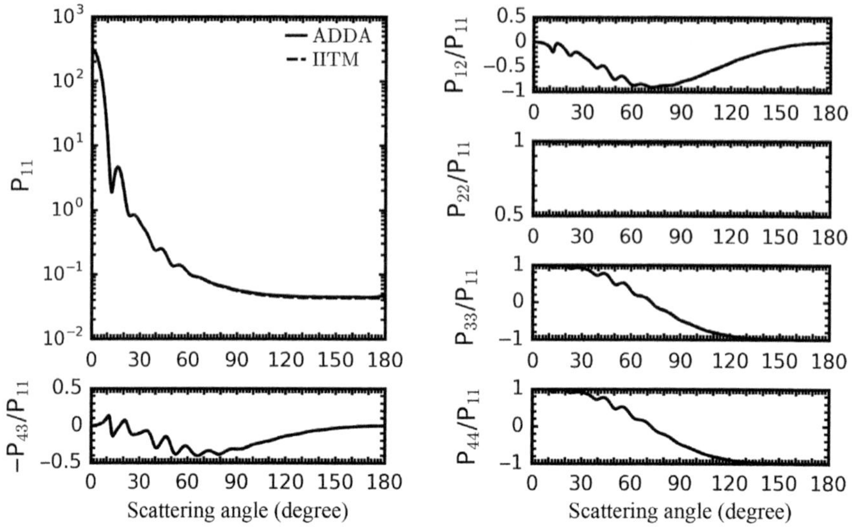


**Fig. 5.19** Scattering phase matrix element comparisons between the ADDA and the IITM. The parameters are the same as for Table 5.22.

**Table 5.23** Scattering and extinction cross sections and asymmetry factor comparisons.

| $(\lambda=2\pi)$ | ADDA | IITM | IITM-ADDA |
|---|---|---|---|
| $C_{sca}$ | 1003.5826 | 1001.4258 | −2.1568 |
| $C_{ext}$ | 1976.3567 | 1976.1594 | −0.1972 |
| $g$ | 0.9305 | 0.9322 | 0.0016 |

Other parameters are the same as for Table 5.22.

used in the computations. The scattering and extinction cross section and asymmetry factor comparisons are given in Table 5.27.

As shown in the earlier IITM and PGOM comparisons, in general, the IITM and PGOM results are consistent. The PGOM is an approximate method and is expected to be accurate at large size parameters. However, at size parameter $2\pi H/\lambda=300$, it still has some approximation errors as compared with the IITM.

## 5.4 Asymmetric particles: Aggregates and hexahedra

For an asymmetric particle, the scattering phase matrix of the particle with random orientation has 10 independent elements denoted as $P_{11}$, $P_{22}$,$P_{33}$, $P_{44}$, $P_{12}$, $P_{34}$, $P_{13}$, $P_{23}$, $P_{14}$, and $P_{24}$, which can be analytically expressed in terms of the T-matrix. For randomly oriented particles, the mirror symmetry may approximately hold; therefore, elements $P_{13}$, $P_{23}$, $P_{14}$, and $P_{24}$ are relatively small, and we only consider the six

**Table 5.24** Scattering phase matrix element comparisons between PGOM and IITM for a hexagonal prism with size parameter $2\pi H/\lambda = 300$ and unity aspect ratio.

| $\Theta$(°) | $P_{11}$ | $P_{22}$ | $P_{33}$ | $P_{44}$ | $P_{12}$ | $P_{34}$ |
|---|---|---|---|---|---|---|
| 0.0 | 16112.1865 | 16111.6415 | 16111.6415 | 16111.1896 | 0.0000 | 0.0000 |
| | 16221.3000 | 16220.7800 | 16220.7800 | 16220.2600 | 0.0000 | 0.0000 |
| | 109.1135 | 109.1385 | 109.1385 | 109.0704 | 0.0000 | 0.0000 |
| 10.0 | 1.8301 | 1.8073 | 1.8010 | 1.8070 | −0.0352 | 0.0007 |
| | 1.8155 | 1.7923 | 1.7799 | 1.7732 | −0.0088 | −0.0163 |
| | −0.0146 | −0.0151 | −0.0211 | −0.0338 | 0.0264 | −0.0169 |
| 20.0 | 1.3369 | 1.3133 | 1.3054 | 1.3085 | −0.0263 | −0.0180 |
| | 1.3181 | 1.2914 | 1.2850 | 1.2743 | −0.0212 | −0.0202 |
| | −0.0188 | −0.0219 | −0.0204 | −0.0342 | 0.0051 | −0.0022 |
| 30.0 | 1.2857 | 1.2528 | 1.2348 | 1.2424 | 0.0022 | −0.0311 |
| | 1.3110 | 1.2836 | 1.2572 | 1.2583 | 0.0161 | −0.0634 |
| | 0.0253 | 0.0307 | 0.0225 | 0.0159 | 0.0138 | −0.0323 |
| 40.0 | 0.4581 | 0.4296 | 0.3912 | 0.4035 | −0.0365 | −0.0488 |
| | 0.4768 | 0.4524 | 0.4226 | 0.4240 | −0.0244 | −0.0740 |
| | 0.0188 | 0.0229 | 0.0315 | 0.0205 | 0.0121 | −0.0252 |
| 50.0 | 0.6567 | 0.6226 | 0.5875 | 0.5990 | 0.0398 | −0.0675 |
| | 0.6734 | 0.6446 | 0.6117 | 0.6115 | 0.0424 | −0.0956 |
| | 0.0167 | 0.0219 | 0.0242 | 0.0125 | 0.0026 | −0.0281 |
| 60.0 | 0.3455 | 0.3044 | 0.2705 | 0.2850 | −0.0208 | −0.0938 |
| | 0.3512 | 0.3198 | 0.2828 | 0.2835 | −0.0130 | −0.1122 |
| | 0.0057 | 0.0154 | 0.0123 | −0.0015 | 0.0078 | −0.0184 |
| 70.0 | 0.3205 | 0.2845 | 0.2345 | 0.2492 | −0.0172 | −0.1238 |
| | 0.3119 | 0.2834 | 0.2265 | 0.2301 | −0.0167 | −0.1376 |
| | −0.0086 | −0.0011 | −0.0080 | −0.0191 | 0.0006 | −0.0138 |
| 80.0 | 0.2954 | 0.2616 | 0.2137 | 0.2309 | −0.0058 | −0.1126 |
| | 0.2991 | 0.2678 | 0.2107 | 0.2219 | −0.0091 | −0.1291 |
| | 0.0037 | 0.0062 | −0.0030 | −0.0091 | −0.0034 | −0.0165 |
| 90.0 | 0.2855 | 0.2391 | 0.1873 | 0.2224 | −0.0061 | −0.1144 |
| | 0.2787 | 0.2345 | 0.1784 | 0.2050 | −0.0079 | −0.1245 |
| | −0.0068 | −0.0047 | −0.0089 | −0.0174 | −0.0019 | −0.0101 |
| 100.0 | 0.2707 | 0.2010 | 0.1529 | 0.2134 | −0.0094 | −0.1026 |
| | 0.2619 | 0.1961 | 0.1406 | 0.1914 | −0.0102 | −0.1126 |
| | −0.0088 | −0.0049 | −0.0123 | −0.0220 | −0.0008 | −0.0100 |
| 110.0 | 0.2163 | 0.1383 | 0.1061 | 0.1735 | 0.0048 | −0.0525 |
| | 0.2211 | 0.1404 | 0.1001 | 0.1648 | −0.0043 | −0.0619 |
| | 0.0048 | 0.0021 | −0.0060 | −0.0088 | −0.0092 | −0.0095 |
| 120.0 | 0.1107 | 0.0529 | 0.0249 | 0.0732 | −0.0173 | −0.0029 |
| | 0.1226 | 0.0592 | 0.0264 | 0.0725 | −0.0170 | −0.0097 |
| | 0.0119 | 0.0063 | 0.0015 | −0.0007 | 0.0004 | −0.0068 |
| 130.0 | 0.0998 | 0.0310 | −0.0111 | 0.0433 | −0.0175 | −0.0052 |
| | 0.1056 | 0.0366 | −0.0078 | 0.0384 | −0.0169 | −0.0093 |
| | 0.0057 | 0.0056 | 0.0033 | −0.0049 | 0.0006 | −0.0041 |
| 140.0 | 0.1223 | 0.0257 | −0.0400 | 0.0437 | −0.0192 | −0.0078 |
| | 0.1217 | 0.0326 | −0.0324 | 0.0348 | −0.0162 | −0.0181 |
| | −0.0007 | 0.0069 | 0.0076 | −0.0089 | 0.0030 | −0.0103 |

*Continued*

**Table 5.24** Continued

| $\Theta$(°) | $P_{11}$ | $P_{22}$ | $P_{33}$ | $P_{44}$ | $P_{12}$ | $P_{34}$ |
|---|---|---|---|---|---|---|
| 150.0 | 0.2113 | 0.0105 | −0.0606 | 0.1125 | −0.0157 | −0.0366 |
| | 0.1981 | 0.0204 | −0.0500 | 0.0704 | −0.0146 | −0.0455 |
| | −0.0132 | 0.0099 | 0.0106 | −0.0421 | 0.0011 | −0.0089 |
| 160.0 | 0.1780 | 0.0193 | −0.0664 | 0.0614 | −0.0050 | −0.0296 |
| | 0.1779 | 0.0208 | −0.0710 | 0.0336 | −0.0021 | −0.0374 |
| | −0.0001 | 0.0014 | −0.0045 | −0.0278 | 0.0029 | −0.0078 |
| 170.0 | 0.1926 | 0.0300 | −0.0912 | 0.0381 | 0.0171 | −0.0237 |
| | 0.1989 | 0.0343 | −0.0933 | 0.0274 | 0.0188 | −0.0320 |
| | 0.0063 | 0.0043 | −0.0022 | −0.0107 | 0.0017 | −0.0083 |
| 180.0 | 1.1945 | 0.8417 | −0.8417 | −0.5665 | 0.0000 | 0.0000 |
| | 1.3267 | 0.9548 | −0.9548 | −0.5829 | 0.0000 | 0.0000 |
| | 0.1322 | 0.1131 | −0.1131 | −0.0164 | 0.0000 | 0.0000 |

For each scattering angle, the three lines denote the results from the PGOM, the IITM, and the IITM minus the PGOM. All data are rounded to four decimal points.

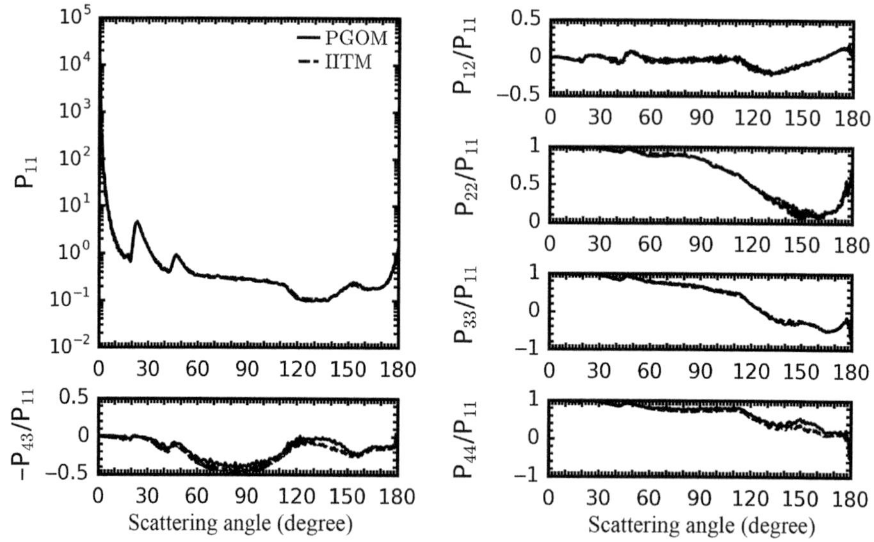


**Fig. 5.20** Scattering phase matrix element comparisons between the PGOM and the IITM. The parameters are the same as for Table 5.24.

independent scattering phase matrix elements $P_{11}$, $P_{22}$, $P_{33}$, $P_{44}$, $P_{12}$, and $P_{34}$. In this subsection, we show the application of the IITM to two kinds of asymmetric particles, aggregates and irregular hexahedra. Here, we use the maximum dimension to describe the size of an asymmetric particle. The maximum dimension of a particle is the longest distance between two arbitrary points on the particle.

**Table 5.25** Scattering and extinction cross section and asymmetry factor comparisons.

| $(\lambda = 2\pi)$ | PGOM | IITM | IITM-PGOM |
|---|---|---|---|
| $C_{sca}$ | 189081.9900 | 190478.4934 | 1396.5034 |
| $C_{ext}$ | 189082.0600 | 190479.3463 | 1397.2863 |
| $g$ | 0.7657 | 0.7661 | 0.0004 |

Other parameters are the same as for Table 5.24.

**Table 5.26** The same as Table 5.24 except that the refractive index is $1.2762+i0.4133$.

| $\Theta$(°) | $P_{11}$ | $P_{22}$ | $P_{33}$ | $P_{44}$ | $P_{12}$ | $P_{34}$ |
|---|---|---|---|---|---|---|
| 0.0 | 27891.1224 | 27891.1224 | 27891.1224 | 27891.1106 | 0.0000 | 0.0000 |
| | 28423.9900 | 28423.9700 | 28423.9700 | 28423.9500 | 0.0000 | 0.0000 |
| | 532.8676 | 532.8476 | 532.8476 | 532.8394 | 0.0000 | 0.0000 |
| 10.0 | 1.8479 | 1.8440 | 1.8295 | 1.8321 | −0.1403 | −0.0624 |
| | 1.7603 | 1.7592 | 1.7418 | 1.7410 | −0.1357 | −0.0638 |
| | −0.0876 | −0.0848 | −0.0878 | −0.0911 | 0.0046 | −0.0014 |
| 20.0 | 0.5732 | 0.5703 | 0.5421 | 0.5442 | −0.1397 | −0.0681 |
| | 0.5057 | 0.5050 | 0.4801 | 0.4796 | −0.1158 | −0.0762 |
| | −0.0675 | −0.0654 | −0.0620 | −0.0646 | 0.0239 | −0.0081 |
| 30.0 | 0.3192 | 0.3172 | 0.2728 | 0.2742 | −0.1348 | −0.0722 |
| | 0.2942 | 0.2935 | 0.2556 | 0.2551 | −0.1159 | −0.0765 |
| | −0.0250 | −0.0237 | −0.0171 | −0.0192 | 0.0189 | −0.0043 |
| 40.0 | 0.2086 | 0.2072 | 0.1498 | 0.1508 | −0.1219 | −0.0677 |
| | 0.1928 | 0.1921 | 0.1433 | 0.1427 | −0.1060 | −0.0682 |
| | −0.0158 | −0.0151 | −0.0064 | −0.0081 | 0.0159 | −0.0005 |
| 50.0 | 0.1432 | 0.1421 | 0.0752 | 0.0760 | −0.1038 | −0.0573 |
| | 0.1347 | 0.1339 | 0.0773 | 0.0767 | −0.0927 | −0.0571 |
| | −0.0086 | −0.0082 | 0.0021 | 0.0007 | 0.0111 | 0.0002 |
| 60.0 | 0.1069 | 0.1060 | 0.0325 | 0.0332 | −0.0885 | −0.0455 |
| | 0.0980 | 0.0973 | 0.0351 | 0.0346 | −0.0784 | −0.0455 |
| | −0.0089 | −0.0087 | 0.0025 | 0.0013 | 0.0101 | 0.0000 |
| 70.0 | 0.0818 | 0.0811 | 0.0037 | 0.0043 | −0.0722 | −0.0348 |
| | 0.0745 | 0.0740 | 0.0077 | 0.0073 | −0.0646 | −0.0352 |
| | −0.0073 | −0.0072 | 0.0040 | 0.0031 | 0.0076 | −0.0004 |
| 80.0 | 0.0662 | 0.0656 | −0.0145 | −0.0141 | −0.0575 | −0.0268 |
| | 0.0597 | 0.0592 | −0.0098 | −0.0102 | −0.0520 | −0.0266 |
| | −0.0065 | −0.0064 | 0.0047 | 0.0039 | 0.0055 | 0.0003 |
| 90.0 | 0.0563 | 0.0559 | −0.0258 | −0.0255 | −0.0451 | −0.0194 |
| | 0.0505 | 0.0500 | −0.0210 | −0.0213 | −0.0409 | −0.0196 |
| | −0.0059 | −0.0058 | 0.0048 | 0.0042 | 0.0042 | −0.0002 |
| 100.0 | 0.0503 | 0.0500 | −0.0330 | −0.0328 | −0.0345 | −0.0142 |
| | 0.0448 | 0.0445 | −0.0281 | −0.0283 | −0.0314 | −0.0142 |
| | −0.0055 | −0.0055 | 0.0049 | 0.0045 | 0.0030 | 0.0001 |
| 110.0 | 0.0459 | 0.0457 | −0.0367 | −0.0365 | −0.0251 | −0.0097 |
| | 0.0416 | 0.0413 | −0.0325 | −0.0326 | −0.0234 | −0.0100 |
| | −0.0044 | −0.0044 | 0.0042 | 0.0039 | 0.0017 | −0.0002 |

*Continued*

**Table 5.26** Continued

| $\Theta$(°) | $P_{11}$ | $P_{22}$ | $P_{33}$ | $P_{44}$ | $P_{12}$ | $P_{34}$ |
|---|---|---|---|---|---|---|
| 120.0 | 0.0446 | 0.0444 | −0.0399 | −0.0397 | −0.0183 | −0.0067 |
| | 0.0398 | 0.0395 | −0.0351 | −0.0352 | −0.0168 | −0.0068 |
| | −0.0048 | −0.0049 | 0.0047 | 0.0045 | 0.0015 | −0.0001 |
| 130.0 | 0.0428 | 0.0426 | −0.0406 | −0.0405 | −0.0122 | −0.0043 |
| | 0.0389 | 0.0387 | −0.0367 | −0.0367 | −0.0114 | −0.0044 |
| | −0.0039 | −0.0039 | 0.0039 | 0.0038 | 0.0007 | −0.0001 |
| 140.0 | 0.0419 | 0.0418 | −0.0411 | −0.0410 | −0.0076 | −0.0026 |
| | 0.0385 | 0.0384 | −0.0376 | −0.0376 | −0.0072 | −0.0027 |
| | −0.0034 | −0.0035 | 0.0035 | 0.0034 | 0.0004 | −0.0001 |
| 150.0 | 0.0417 | 0.0416 | −0.0414 | −0.0413 | −0.0042 | −0.0014 |
| | 0.0384 | 0.0383 | −0.0381 | −0.0380 | −0.0040 | −0.0014 |
| | −0.0033 | −0.0033 | 0.0033 | 0.0033 | 0.0002 | 0.0000 |
| 160.0 | 0.0418 | 0.0418 | −0.0417 | −0.0416 | −0.0019 | −0.0006 |
| | 0.0385 | 0.0384 | −0.0383 | −0.0383 | −0.0018 | −0.0006 |
| | −0.0034 | −0.0034 | 0.0034 | 0.0034 | 0.0001 | 0.0000 |
| 170.0 | 0.0418 | 0.0418 | −0.0418 | −0.0417 | −0.0005 | −0.0001 |
| | 0.0385 | 0.0384 | −0.0384 | −0.0384 | −0.0004 | −0.0002 |
| | −0.0033 | −0.0033 | 0.0033 | 0.0033 | 0.0000 | 0.0000 |
| 180.0 | 0.0404 | 0.0403 | −0.0403 | −0.0403 | 0.0000 | 0.0000 |
| | 0.0385 | 0.0385 | −0.0385 | −0.0384 | 0.0000 | 0.0000 |
| | −0.0018 | −0.0018 | 0.0018 | 0.0019 | 0.0000 | 0.0000 |

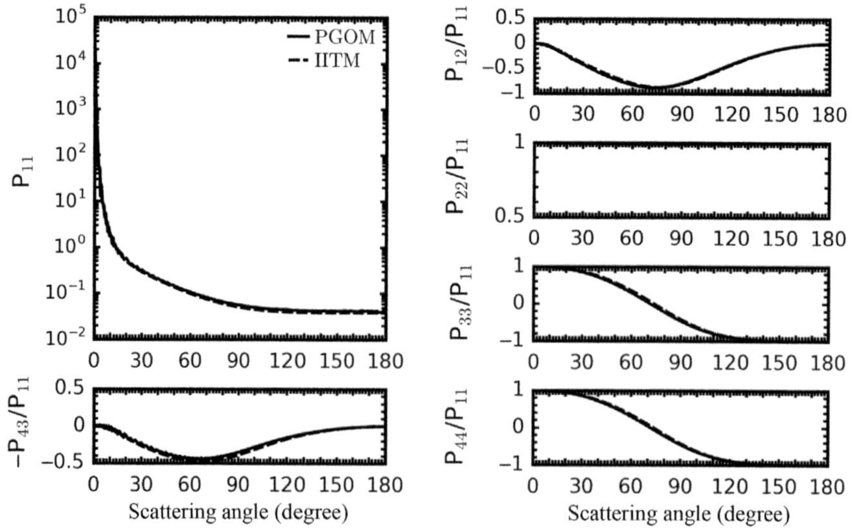


**Fig. 5.21** Scattering phase matrix element comparisons between the PGOM and the IITM. The parameters are the same as for Table 5.26.

**Table 5.27** Scattering and extinction cross section and asymmetry factor comparisons. Other parameters are the same as for Table 5.26

| ($\lambda = 2\pi$) | PGOM | IITM | IITM-PGOM |
|---|---|---|---|
| $C_{sca}$ | 106918.6600 | 106234.8613 | −683.7987 |
| $C_{ext}$ | 192921.9600 | 194184.3110 | 1262.3510 |
| $g$ | 0.9410 | 0.9464 | 0.0055 |

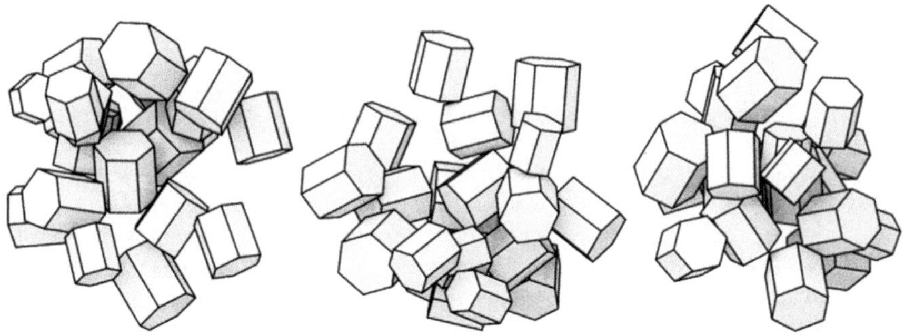

(A) (B) (C)

**Fig. 5.22** Aggregate composed of twenty hexagonal columns; (A), (B), and (C) are the same aggregate but from different viewing directions.

## 5.4.1 Aggregate

An example of aggregate shape is shown in Fig. 5.22. The aggregate has 20 elements that are regular hexagonal prisms with various sizes and aspect ratios. The aggregate with hexagonal prisms is expected to be representative of complex ice crystal adhesions in the atmosphere.

We use IITM and ADDA to compute the orientation-averaged scattering properties of the aggregate shown in Fig. 5.22. The maximum dimension is set to be 30. The vertices of each elements are given in Table 5.28. Two refractive indices $1.308 + i1.43 \times 10^{-8}$ and $1.2762 + i0.4133$ are used in the calculations.

For the case with refractive index $1.308 + i1.43 \times 10^{-8}$, the ADDA calculation has a dpl of 20 and uses 994 orientations to obtain the orientation-averaged result. Comparisons of the six independent scattering phase matrix elements at 19 scattering angles are given in Table 5.29. Fig. 5.23 shows the phase matrix comparisons in all scattering angles. The scattering and extinction cross sections and asymmetry factor comparisons are given in Table 5.30.

When the refractive index is $1.2762 + i0.4133$, the ADDA calculation has dpl set to 20 and uses 4034 orientations to obtain the orientation-averaged result. Comparisons of the six independent scattering phase matrix elements at 19 scattering angles are given in Table 5.31. Fig. 5.24 shows the phase matrix comparisons in all scattering angles. The scattering and extinction cross sections and asymmetry factor comparisons are given in Table 5.32.

**Table 5.28** The vertices in the Cartesian coordinate of the aggregate shown in Fig. 5.22.

| # | | | |
|---|---|---|---|
| 1 | 2.417614, 1.39581, −2.79162 | 0, 2.79162, −2.79162 | −2.417614, 1.39581, −2.79162 |
| | −2.417614, −1.39581, −2.79162 | 0, −2.79162, −2.79162 | 2.417614, −1.39581, −2.79162 |
| | 2.417614, 1.39581, 2.79162 | 0, 2.79162, 2.79162 | −2.417614, 1.39581, 2.79162 |
| | −2.417614, −1.39581, 2.79162 | 0, −2.79162, 2.79162 | 2.417614, −1.39581, 2.79162 |
| 2 | −5.098339, 1.199303, 1.013561 | −3.59261, −0.522932, 3.136895 | −4.800815, −3.240744, 4.083259 |
| | −7.514747, −4.23632, 2.906288 | −9.020475, −2.514085, 0.782953 | −7.812271, 0.203727, −0.16341 |
| | −2.034402, −1.753221, −3.55397 | −0.528674, −3.475456, −1.430636 | −1.736878, −6.193268, −0.484273 |
| | −4.45081, −7.188845, −1.661244 | −5.956538, −5.466609, −3.784578 | −4.748334, −2.748797, −4.730942 |
| 3 | 7.939264, −1.695811, 1.91869 | 7.243595, 0.963908, 2.45015 | 7.968384, 2.98627, 0.654255 |
| | 9.388843, 2.348914, −1.6731 | 10.084512, −0.310804, −2.20456 | 9.359723, −2.333167, −0.408665 |
| | 3.113283, −2.408528, −0.831579 | 2.417614, 0.25119, −0.300119 | 3.142403, 2.273553, −2.096014 |
| | 4.562862, 1.636196, −4.423369 | 5.258531, −1.023522, −4.954829 | 4.533742, −3.045884, −3.158934 |
| 4 | −5.596804, −2.884376, 3.318941 | −5.04486, −0.388848, 2.357007 | −7.084829, 1.292399, 1.671764 |
| | −9.676741, 0.478118, 1.948455 | −10.228685, −2.01741, 2.910389 | −8.188717, −3.698657, 3.595632 |
| | −5.675277, −0.905067, 8.4088 | −5.123333, 1.590461, 7.446866 | −7.163301, 3.271708, 6.761623 |
| | −9.755213, 2.457427, 7.038314 | −10.307157, −0.038101, 8.000249 | −8.267189, −1.719349, 8.685491 |
| 5 | 7.095464, 1.655162, −3.816395 | 5.383496, 1.020393, −5.011705 | 3.386358, 1.760937, −4.536769 |
| | 3.101187, 3.136251, −2.866523 | 4.813155, 3.771021, −1.671213 | 6.810293, 3.030476, −2.146149 |
| | 7.713159, 5.041791, −6.499551 | 6.001191, 4.407021, −7.694861 | 4.004053, 5.147566, −7.219925 |
| | 3.718882, 6.52288, −5.54968 | 5.43085, 7.15765, −4.354369 | 7.427988, 6.417105, −4.829306 |

**Table 5.28** Continued

| # | | | |
|---|---|---|---|
| 6 | 3.427104, −5.517966, 4.169478 | 4.879007, −3.812109, 3.736476 | 4.086406, −1.786615, 3.047547 |
| | 1.841902, −1.46698, 2.79162 | 0.389999, −3.172837, 3.224622 | 1.1826, −5.198331, 3.913551 |
| | 3.125292, −4.15811, 8.514769 | 4.577195, −2.452252, 8.081766 | 3.784594, −0.426759, 7.392837 |
| | 1.54009, −0.107123, 7.136911 | 0.088187, −1.812981, 7.569913 | 0.880788, −3.838475, 8.258842 |
| 7 | −10.189426, −6.669003, −4.6582 | −11.752189, −5.661478, −3.641394 | −12.210907, −6.168541, −1.635479 |
| | −11.106862, −7.68313, −0.646369 | −9.544099, −8.690655, −1.663176 | −9.085381, −8.183592, −3.669091 |
| | −7.425236, −3.761244, −3.291043 | −8.987999, −2.753719, −2.274236 | −9.446717, −3.260782, −0.268321 |
| | −8.342672, −4.775371, 0.720788 | −6.779909, −5.782896, −0.296018 | −6.32119, −5.275833, −2.301933 |
| 8 | 1.097155, 4.091675, −3.45949 | 0.23132, 6.44728, −4.108426 | −1.632149, 6.69702, −5.893019 |
| | −2.629783, 4.591155, −7.028678 | −1.763948, 2.23555, −6.379743 | 0.099521, 1.98581, −4.595149 |
| | −2.503597, 3.79243, 0.258531 | −3.369431, 6.148035, −0.390404 | −5.232901, 6.397775, −2.174998 |
| | −6.230535, 4.29191, −3.310656 | −5.3647, 1.936305, −2.661721 | −3.501231, 1.686565, −0.877128 |
| 9 | 2.096625, −8.527938, 2.745711 | 0.586253, −10.410209, 1.465712 | −1.916295, −9.776758, 0.572093 |
| | −2.908473, −7.261036, 0.958473 | −1.398101, −5.378764, 2.238472 | 1.104448, −6.012216, 3.132092 |
| | 4.204047, −6.96096, −2.045282 | 2.693676, −8.843231, −3.325281 | 0.191127, −8.20978, −4.2189 |
| | −0.80105, −5.694058, −3.83252 | 0.709321, −3.811786, −2.552521 | 3.21187, −4.445238, −1.658901 |
| 10 | −7.007389, 6.734972, 5.607178 | −5.87374, 4.143005, 5.399451 | −4.012045, 3.249183, 7.344131 |
| | −3.284, 4.947328, 9.496538 | −4.41765, 7.539295, 9.704266 | −6.279344, 8.433117, 7.759586 |
| | −11.262203, 4.625313, 8.710775 | −10.128554, 2.033346, 8.503048 | −8.26686, 1.139524, 10.447728 |
| | −7.538815, 2.837669, 12.600135 | −8.672464, 5.429636, 12.807863 | −10.534158, 6.323458, 10.863182 |

*Continued*

**Table 5.28** Continued

| # | | | |
|---|---|---|---|
| 11 | −5.381359, 13.168757, −3.170896<br>−5.952692, 10.057935, −7.213136<br>−5.539848, 9.114029, −0.028062<br>−6.111181, 6.003207, −4.070302 | −7.731784, 12.596588, −4.027609<br>−3.602267, 10.630104, −6.356422<br>−7.890274, 8.54186, −0.884775<br>−3.760756, 6.575376, −3.213588 | −8.017451, 11.041177, −6.048729<br>−3.3166, 12.185516, −4.335302<br>−8.17594, 6.986449, −2.905895<br>−3.475089, 8.130787, −1.192468 |
| 12 | −3.785662, −13.921135, −3.249953<br>−2.161346, −8.85359, −3.639391<br>0.166015, −14.923232, 0.192448<br>1.790331, −9.855687, −0.19699 | −1.995573, −13.232911, −5.104527<br>−3.951436, −9.541814, −1.784817<br>1.956105, −14.235008, −1.662126<br>0.000242, −10.543911, 1.657584 | −1.183415, −10.699139, −5.299247<br>−4.763594, −12.075587, −1.590098<br>2.768263, −11.701236, −1.856845<br>−0.811916, −13.077684, 1.852303 |
| 13 | −10.857787, −7.327569, 1.691308<br>−5.660595, −9.973878, −0.449195<br>−8.060821, −6.233344, 7.129613<br>−2.863629, −8.879653, 4.98911 | −10.39832, −10.376426, 2.068452<br>−6.120061, −6.925021, −0.826339<br>−7.601354, −9.282201, 7.506757<br>−3.323096, −5.830796, 4.611966 | −7.799724, −11.699581, 0.9982<br>−8.718657, −5.601867, 0.243913<br>−5.002759, −10.605356, 6.436506<br>−5.921692, −4.507642, 5.682218 |
| 14 | −1.330102, −2.558145, 14.112331<br>−0.587637, 2.843101, 11.36257<br>−4.989346, −4.367108, 9.571023<br>−4.246881, 1.034137, 6.821262 | −3.23664, −0.255196, 14.731214<br>1.318901, 0.540152, 10.743687<br>−6.895883, −2.06416, 10.189906<br>−2.340343, −1.268811, 6.202379 | −2.865407, 2.445426, 13.356334<br>0.947668, −2.16047, 12.118568<br>−6.524651, 0.636463, 8.815025<br>−2.711575, −3.969434, 7.57726 |
| 15 | 5.369979, −1.619605, −4.40109<br>2.018547, −5.723898, −6.278718<br>9.868865, −4.84481, −5.381328<br>6.517434, −8.949104, −7.258955 | 4.375561, −3.549382, −2.61565<br>3.012964, −3.794122, −8.064158<br>8.874448, −6.774587, −3.595888<br>7.511852, −7.019327, −9.044395 | 2.699846, −5.601528, −3.554464<br>4.68868, −1.741975, −7.125344<br>7.198733, −8.826734, −4.534702<br>9.187567, −4.96718, −8.105582 |

**Table 5.28** Continued

| # | | | |
|---|---|---|---|
| 16 | 5.694204, −8.172377, 7.367236 | 7.686433, −7.85105, 6.514415 | 7.926504, −6.598083, 4.733413 |
| | 6.174347, −5.666443, 3.805232 | 4.182118, −5.987769, 4.658053 | 3.942047, −7.240737, 6.439055 |
| | 6.217347, −4.647925, 9.91727 | 8.209576, −4.326599, 9.064449 | 8.449647, −3.073631, 7.283447 |
| | 6.69749, −2.141991, 6.355266 | 4.705262, −2.463317, 7.208087 | 4.46519, −3.716285, 8.989089 |
| 17 | 0.215815, −7.686915, 9.836102 | −1.893205, −8.473923, 8.612436 | −2.82595, −7.105404, 6.657482 |
| | −1.649675, −4.949876, 5.926193 | 0.459345, −4.162868, 7.149859 | 1.39209, −5.531387, 9.104813 |
| | 3.111999, −10.374432, 6.572946 | 1.002979, −11.161441, 5.34928 | 0.070235, −9.792921, 3.394326 |
| | 1.24651, −7.637394, 2.663038 | 3.355529, −6.850386, 3.886703 | 4.288274, −8.218905, 5.841658 |
| 18 | −10.815121, 2.445366, 5.571715 | −9.839477, 0.578957, 4.007432 | −11.456726, −0.911652, 2.57742 |
| | −14.049617, −0.535853, 2.711691 | −15.025261, 1.330555, 4.275974 | −13.408013, 2.821165, 5.705986 |
| | −10.518238, 5.900562, 1.634349 | −9.542595, 4.034154, 0.070066 | −11.159843, 2.543544, −1.359946 |
| | −13.752735, 2.919343, −1.225675 | −14.728379, 4.785751, 0.338608 | −13.11113, 6.276361, 1.76862 |
| 19 | −1.416031, −0.138788, −10.094143 | −2.813589, 1.197613, −7.789777 | −1.237172, 2.837325, −5.821148 |
| | 1.736803, 3.140635, −6.156886 | 3.134361, 1.804233, −8.461252 | 1.557944, 0.164522, −10.429881 |
| | −2.297061, 4.762162, −13.470745 | −3.694619, 6.098564, −11.166379 | −2.118202, 7.738276, −9.19775 |
| | 0.855773, 8.041586, −9.533488 | 2.253331, 6.705184, −11.837855 | 0.676914, 5.065472, −13.806483 |
| 20 | 6.860756, 3.262692, 11.489416 | 7.913896, 1.122219, 11.479969 | 6.640455, −0.835808, 11.965045 |
| | 4.313874, −0.653362, 12.459568 | 3.260734, 1.487112, 12.469016 | 4.534174, 3.445139, 11.983939 |
| | 5.837697, 2.779792, 6.854393 | 6.890837, 0.639318, 6.844946 | 5.617396, −1.318709, 7.330022 |
| | 3.290815, −1.136262, 7.824545 | 2.237674, 1.004211, 7.833992 | 3.511115, 2.962238, 7.348916 |

The maximum dimension is 30. The first column is the index of the aggregate element. Each element has 12 vertices. Each vertex entry in the table corresponds to $x$-, $y$-, and $z$-coordinates.

**Table 5.29** Scattering phase matrix element comparisons between ADDA and IITM for an aggregate shown in Fig. 5.22.

| $\Theta$(°) | $P_{11}$ | $P_{22}$ | $P_{33}$ | $P_{44}$ | $P_{12}$ | $P_{34}$ |
|---|---|---|---|---|---|---|
| 0.0 | 99.7003 | 99.6886 | 99.6886 | 99.6809 | 0.0000 | 0.0000 |
| | 97.2068 | 97.1925 | 97.1925 | 97.1823 | 0.0000 | 0.0000 |
| | −2.4935 | −2.4960 | −2.4960 | −2.4987 | 0.0000 | 0.0000 |
| 10.0 | 31.3353 | 31.3220 | 31.3194 | 31.3123 | −0.3549 | 0.1996 |
| | 31.6883 | 31.6732 | 31.6715 | 31.6623 | 0.0977 | 0.2861 |
| | 0.3530 | 0.3512 | 0.3521 | 0.3500 | 0.4526 | 0.0865 |
| 20.0 | 2.8494 | 2.8321 | 2.8305 | 2.8264 | −0.0922 | 0.0455 |
| | 2.8098 | 2.7917 | 2.7882 | 2.7826 | −0.0628 | −0.0451 |
| | −0.0396 | −0.0404 | −0.0423 | −0.0437 | 0.0295 | −0.0906 |
| 30.0 | 1.5169 | 1.4959 | 1.4899 | 1.4905 | −0.1085 | 0.0559 |
| | 1.5267 | 1.5058 | 1.5010 | 1.5008 | −0.1010 | 0.0467 |
| | 0.0099 | 0.0099 | 0.0111 | 0.0103 | 0.0075 | −0.0092 |
| 40.0 | 1.3485 | 1.3234 | 1.3077 | 1.3126 | −0.1711 | 0.0908 |
| | 1.3280 | 1.3030 | 1.2855 | 1.2897 | −0.1833 | 0.0836 |
| | −0.0204 | −0.0204 | −0.0221 | −0.0229 | −0.0121 | −0.0072 |
| 50.0 | 0.7253 | 0.6989 | 0.6788 | 0.6872 | −0.1206 | 0.0912 |
| | 0.7318 | 0.7048 | 0.6827 | 0.6900 | −0.1338 | 0.0924 |
| | 0.0065 | 0.0059 | 0.0039 | 0.0028 | −0.0132 | 0.0012 |
| 60.0 | 0.3994 | 0.3727 | 0.3516 | 0.3624 | −0.0754 | 0.0617 |
| | 0.4030 | 0.3758 | 0.3530 | 0.3629 | −0.0844 | 0.0646 |
| | 0.0035 | 0.0031 | 0.0014 | 0.0005 | −0.0091 | 0.0029 |
| 70.0 | 0.2471 | 0.2206 | 0.1991 | 0.2110 | −0.0500 | 0.0338 |
| | 0.2487 | 0.2220 | 0.1994 | 0.2109 | −0.0567 | 0.0356 |
| | 0.0016 | 0.0014 | 0.0003 | 0.0000 | −0.0067 | 0.0018 |
| 80.0 | 0.1495 | 0.1240 | 0.1038 | 0.1157 | −0.0345 | 0.0066 |
| | 0.1508 | 0.1253 | 0.1048 | 0.1165 | −0.0366 | 0.0086 |
| | 0.0014 | 0.0013 | 0.0010 | 0.0008 | −0.0022 | 0.0020 |
| 90.0 | 0.1106 | 0.0864 | 0.0595 | 0.0716 | −0.0338 | −0.0050 |
| | 0.1084 | 0.0844 | 0.0584 | 0.0703 | −0.0343 | −0.0036 |
| | −0.0022 | −0.0019 | −0.0011 | −0.0013 | −0.0005 | 0.0014 |
| 100.0 | 0.0925 | 0.0701 | 0.0321 | 0.0436 | −0.0393 | −0.0130 |
| | 0.0905 | 0.0685 | 0.0311 | 0.0426 | −0.0382 | −0.0125 |
| | −0.0020 | −0.0016 | −0.0009 | −0.0010 | 0.0010 | 0.0005 |
| 110.0 | 0.0755 | 0.0557 | 0.0081 | 0.0183 | −0.0371 | −0.0115 |
| | 0.0752 | 0.0557 | 0.0071 | 0.0173 | −0.0375 | −0.0121 |
| | −0.0003 | 0.0000 | −0.0010 | −0.0010 | −0.0004 | −0.0006 |
| 120.0 | 0.0617 | 0.0445 | −0.0079 | 0.0013 | −0.0261 | −0.0116 |
| | 0.0624 | 0.0453 | −0.0094 | −0.0002 | −0.0273 | −0.0115 |
| | 0.0007 | 0.0009 | −0.0015 | −0.0016 | −0.0012 | 0.0001 |
| 130.0 | 0.0547 | 0.0399 | −0.0199 | −0.0115 | −0.0174 | −0.0130 |
| | 0.0551 | 0.0405 | −0.0216 | −0.0133 | −0.0177 | −0.0128 |
| | 0.0005 | 0.0007 | −0.0017 | −0.0018 | −0.0003 | 0.0002 |
| 140.0 | 0.0537 | 0.0413 | −0.0312 | −0.0240 | −0.0100 | −0.0120 |
| | 0.0545 | 0.0423 | −0.0330 | −0.0259 | −0.0103 | −0.0119 |
| | 0.0008 | 0.0010 | −0.0019 | −0.0019 | −0.0003 | 0.0001 |

**Table 5.29** Continued

| $\Theta(°)$ | $P_{11}$ | $P_{22}$ | $P_{33}$ | $P_{44}$ | $P_{12}$ | $P_{34}$ |
|---|---|---|---|---|---|---|
| 150.0 | 0.0582 | 0.0482 | −0.0425 | −0.0367 | −0.0036 | −0.0093 |
| | 0.0600 | 0.0502 | −0.0449 | −0.0391 | −0.0040 | −0.0090 |
| | 0.0019 | 0.0020 | −0.0024 | −0.0024 | −0.0004 | 0.0003 |
| 160.0 | 0.0660 | 0.0581 | −0.0542 | −0.0493 | 0.0013 | −0.0059 |
| | 0.0693 | 0.0614 | −0.0576 | −0.0526 | 0.0012 | −0.0059 |
| | 0.0033 | 0.0033 | −0.0034 | −0.0033 | −0.0001 | 0.0001 |
| 170.0 | 0.0799 | 0.0705 | −0.0694 | −0.0616 | 0.0029 | −0.0023 |
| | 0.0837 | 0.0743 | −0.0733 | −0.0654 | 0.0029 | −0.0024 |
| | 0.0039 | 0.0038 | −0.0039 | −0.0038 | 0.0000 | 0.0000 |
| 180.0 | 0.0946 | 0.0803 | −0.0803 | −0.0660 | 0.0000 | 0.0000 |
| | 0.0974 | 0.0832 | −0.0832 | −0.0690 | 0.0000 | 0.0000 |
| | 0.0028 | 0.0029 | −0.0029 | −0.0030 | 0.0000 | 0.0000 |

The size parameter in terms of maximum dimension is 30. For each scattering angle, the three lines denote the results from the ADDA, the IITM, and the IITM minus the ADDA. All data are rounded to four decimal points.

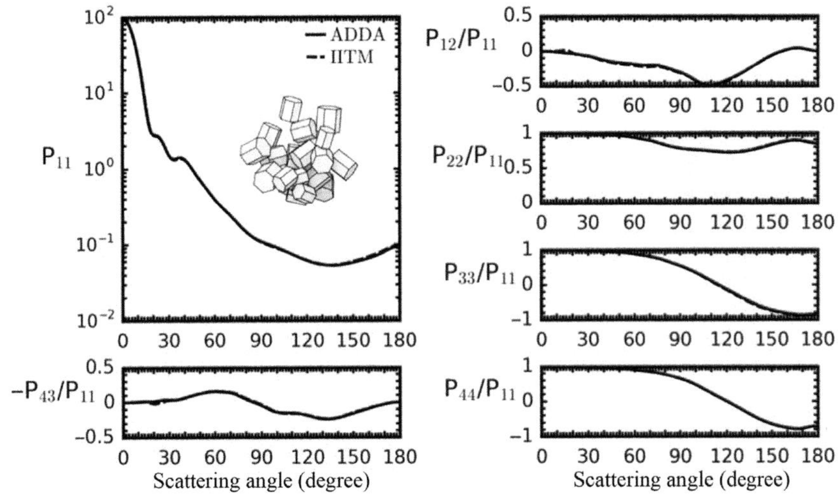


**Fig. 5.23** Scattering phase matrix element comparisons between the ADDA and the IITM. The parameters are the same as for Table 5.29.

**Table 5.30** Scattering and extinction cross section and asymmetry factor comparisons.

| $(\lambda = 2\pi)$ | ADDA | IITM | IITM-ADDA |
|---|---|---|---|
| $C_{sca}$ | 896.0564 | 862.3545 | −33.7019 |
| $C_{ext}$ | 896.0565 | 862.3545 | −33.7020 |
| $g$ | 0.8483 | 0.8478 | −0.0006 |

Other parameters are the same as for Table 5.29.

**Table 5.31** The same as Table 5.29 except that the refractive index is 1.2762 + *i*0.4133.

| $\Theta$(°) | $P_{11}$ | $P_{22}$ | $P_{33}$ | $P_{44}$ | $P_{12}$ | $P_{34}$ |
|---|---|---|---|---|---|---|
| 0.0 | 140.6073 | 140.6006 | 140.6006 | 140.5944 | 0.0000 | 0.0000 |
| | 136.9999 | 136.9927 | 136.9927 | 136.9860 | 0.0000 | 0.0000 |
| | −3.6074 | −3.6079 | −3.6079 | −3.6084 | 0.0000 | 0.0000 |
| 10.0 | 38.9319 | 38.9229 | 38.9114 | 38.9038 | −0.8676 | 0.4166 |
| | 39.3790 | 39.3694 | 39.3507 | 39.3425 | −0.4870 | 1.1027 |
| | 0.4470 | 0.4465 | 0.4394 | 0.4387 | 0.3807 | 0.6861 |
| 20.0 | 1.5536 | 1.5430 | 1.5409 | 1.5338 | −0.0783 | 0.0391 |
| | 1.4682 | 1.4571 | 1.4493 | 1.4416 | −0.1045 | −0.0109 |
| | −0.0854 | −0.0859 | −0.0915 | −0.0922 | −0.0262 | −0.0500 |
| 30.0 | 0.9070 | 0.8946 | 0.8787 | 0.8715 | −0.1396 | 0.0729 |
| | 0.9296 | 0.9176 | 0.8989 | 0.8920 | −0.1451 | 0.0850 |
| | 0.0226 | 0.0230 | 0.0203 | 0.0204 | −0.0055 | 0.0121 |
| 40.0 | 0.8021 | 0.7865 | 0.7479 | 0.7401 | −0.2130 | 0.1096 |
| | 0.8063 | 0.7905 | 0.7528 | 0.7447 | −0.2193 | 0.0956 |
| | 0.0042 | 0.0040 | 0.0049 | 0.0046 | −0.0063 | −0.0140 |
| 50.0 | 0.3335 | 0.3187 | 0.2666 | 0.2600 | −0.1466 | 0.0723 |
| | 0.3410 | 0.3259 | 0.2732 | 0.2667 | −0.1550 | 0.0640 |
| | 0.0076 | 0.0072 | 0.0065 | 0.0067 | −0.0084 | −0.0083 |
| 60.0 | 0.1714 | 0.1590 | 0.0938 | 0.0893 | −0.0980 | 0.0477 |
| | 0.1773 | 0.1650 | 0.0992 | 0.0951 | −0.1041 | 0.0467 |
| | 0.0059 | 0.0060 | 0.0053 | 0.0058 | −0.0060 | −0.0010 |
| 70.0 | 0.0968 | 0.0868 | 0.0263 | 0.0237 | −0.0563 | 0.0201 |
| | 0.0996 | 0.0898 | 0.0286 | 0.0264 | −0.0599 | 0.0211 |
| | 0.0028 | 0.0030 | 0.0023 | 0.0028 | −0.0036 | 0.0010 |
| 80.0 | 0.0593 | 0.0513 | 0.0018 | 0.0000 | −0.0332 | −0.0010 |
| | 0.0597 | 0.0519 | 0.0020 | 0.0006 | −0.0337 | 0.0005 |
| | 0.0004 | 0.0006 | 0.0002 | 0.0006 | −0.0005 | 0.0015 |
| 90.0 | 0.0481 | 0.0417 | −0.0056 | −0.0065 | −0.0290 | −0.0066 |
| | 0.0472 | 0.0411 | −0.0058 | −0.0065 | −0.0284 | −0.0063 |
| | −0.0009 | −0.0007 | −0.0002 | 0.0000 | 0.0007 | 0.0004 |
| 100.0 | 0.0441 | 0.0390 | −0.0136 | −0.0141 | −0.0293 | −0.0049 |
| | 0.0433 | 0.0383 | −0.0133 | −0.0136 | −0.0285 | −0.0056 |
| | −0.0008 | −0.0007 | 0.0003 | 0.0005 | 0.0008 | −0.0008 |
| 110.0 | 0.0415 | 0.0375 | −0.0231 | −0.0231 | −0.0245 | −0.0009 |
| | 0.0412 | 0.0371 | −0.0225 | −0.0224 | −0.0247 | −0.0018 |
| | −0.0003 | −0.0004 | 0.0006 | 0.0008 | −0.0002 | −0.0008 |
| 120.0 | 0.0402 | 0.0370 | −0.0296 | −0.0293 | −0.0175 | 0.0005 |
| | 0.0400 | 0.0367 | −0.0291 | −0.0287 | −0.0179 | 0.0000 |
| | −0.0002 | −0.0003 | 0.0005 | 0.0006 | −0.0004 | −0.0005 |
| 130.0 | 0.0399 | 0.0372 | −0.0337 | −0.0331 | −0.0116 | 0.0008 |
| | 0.0395 | 0.0367 | −0.0331 | −0.0324 | −0.0120 | 0.0004 |
| | −0.0004 | −0.0004 | 0.0005 | 0.0007 | −0.0003 | −0.0004 |
| 140.0 | 0.0396 | 0.0372 | −0.0357 | −0.0349 | −0.0066 | 0.0005 |
| | 0.0393 | 0.0369 | −0.0354 | −0.0344 | −0.0070 | 0.0004 |
| | −0.0002 | −0.0003 | 0.0004 | 0.0005 | −0.0004 | −0.0001 |

**Table 5.31** Continued

| $\Theta(°)$ | $P_{11}$ | $P_{22}$ | $P_{33}$ | $P_{44}$ | $P_{12}$ | $P_{34}$ |
|---|---|---|---|---|---|---|
| 150.0 | 0.0398 | 0.0377 | −0.0371 | −0.0360 | −0.0033 | −0.0001 |
| | 0.0395 | 0.0372 | −0.0366 | −0.0354 | −0.0034 | −0.0001 |
| | −0.0004 | −0.0005 | 0.0005 | 0.0006 | −0.0001 | 0.0000 |
| 160.0 | 0.0406 | 0.0384 | −0.0382 | −0.0367 | −0.0013 | −0.0004 |
| | 0.0403 | 0.0380 | −0.0378 | −0.0362 | −0.0013 | −0.0004 |
| | −0.0003 | −0.0004 | 0.0004 | 0.0005 | 0.0000 | 0.0000 |
| 170.0 | 0.0421 | 0.0398 | −0.0397 | −0.0376 | −0.0003 | −0.0002 |
| | 0.0417 | 0.0392 | −0.0392 | −0.0370 | −0.0003 | −0.0002 |
| | −0.0004 | −0.0005 | 0.0005 | 0.0006 | 0.0000 | 0.0000 |
| 180.0 | 0.0431 | 0.0406 | −0.0406 | −0.0382 | 0.0000 | 0.0000 |
| | 0.0425 | 0.0399 | −0.0399 | −0.0373 | 0.0000 | 0.0000 |
| | −0.0006 | −0.0007 | 0.0007 | 0.0008 | 0.0000 | 0.0000 |

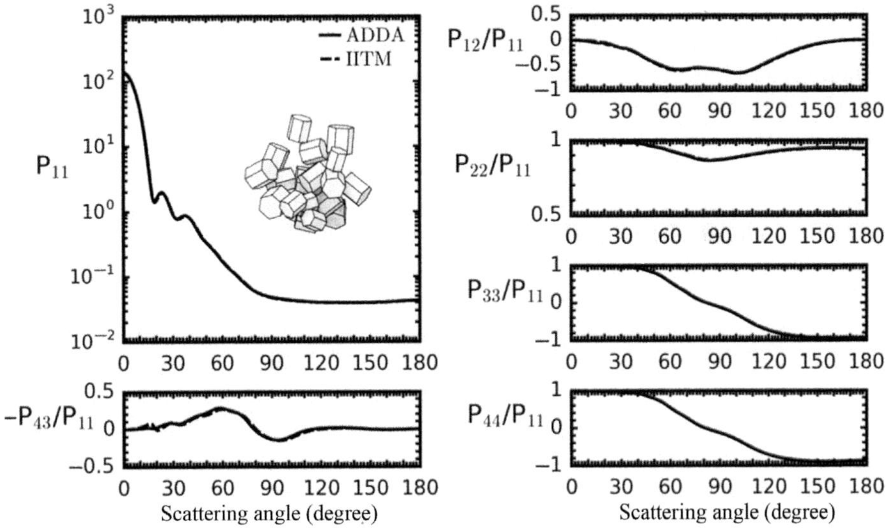


**Fig. 5.24** Scattering phase matrix element comparisons between the ADDA and the IITM. The parameters are the same as for Table 5.31.

**Table 5.32** Scattering and extinction cross section and asymmetry factor comparisons.

| $(\lambda = 2\pi)$ | ADDA | IITM | IITM-ADDA |
|---|---|---|---|
| $C_{sca}$ | 344.0966 | 333.0558 | −11.0408 |
| $C_{ext}$ | 777.9492 | 755.3624 | −22.5868 |
| $g$ | 0.9161 | 0.9157 | −0.0004 |

Other parameters are the same as for Table 5.31.

The earlier comparisons show that the IITM is consistent with the ADDA for an asymmetric aggregate. The absolute differences between the IITM and ADDA results are comparable with their differences for a regular hexagon, as shown in Section 5.3.

### 5.4.2 Hexahedron

Examples of hexahedral particles are shown in Fig. 5.25. The facets of the hexahedra are randomly tilted. The irregular hexahedron was used to model dust aerosol particles.

We use IITM and ADDA to compute the orientation-averaged scattering properties of the hexahedron shown in Fig. 5.25A. The maximum dimension is set to 40. The vertices of each element are given in Table 5.33. The refractive index is $1.53+i0.008$.

In the ADDA calculation, the dpl is 20, and 994 orientations are used to obtain the orientation-averaged result. The comparison of the six independent scattering phase matrix elements at 19 special scattering angles is given in Table 5.34. Fig. 5.26 shows the phase matrix comparisons in all scattering angles. The scattering and extinction cross sections and asymmetry factor comparisons are given in Table 5.35.

The earlier calculation results of the hexahedron show good consistency between IITM and ADDA. For both an aggregate and a hexahedron, the IITM gives smaller scattering and extinction cross sections and asymmetry factors than the ADDA, which may be due to the limited number of orientations in the ADDA calculations to obtain orientation-averaged results.

We also compare the IITM with the PGOM for a large hexahedron, the shape of which is shown in Fig. 5.25B. The maximum dimension is 180, and the refractive index is $1.5+i0.005$. The vertices of the hexahedron are given in Table 5.36.

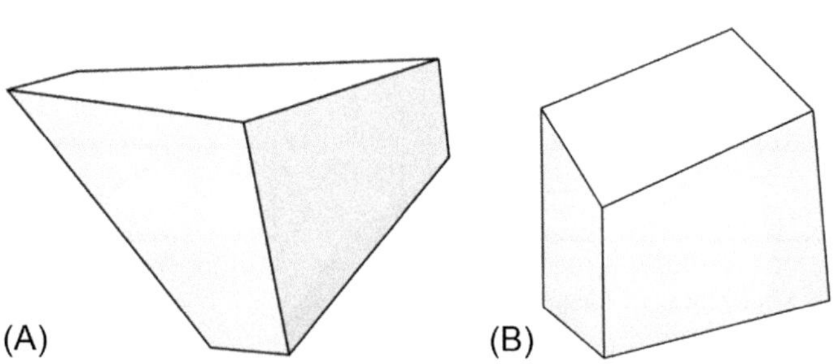


**Fig. 5.25** Two irregular hexahedra.

**Table 5.33** The vertices of the hexahedron shown in Fig. 5.25A in Cartesian coordinates.

| | |
|---|---|
| −13.063069, −10.90927, 8.394027 | 16.410131, −2.608928, 6.257648 |
| 20.13119, 6.238842, 7.5527 | −3.988827, 19.254298, 13.274903 |
| −13.69475, −13.262484, −1.136608 | −6.463066, 10.411761, −10.464675 |
| 1.475217, 5.978498, −12.70568 | 0.941668, −10.491784, −8.146189 |

The maximum dimension is 40. The hexahedron has eight vertices. Each vertex entry in the table corresponds to $x$-, $y$-, and $z$-coordinates.

**Table 5.34** Scattering phase matrix element comparisons between ADDA and IITM for the hexahedron shown in Fig. 5.25A.

| $\Theta$(°) | $P_{11}$ | $P_{22}$ | $P_{33}$ | $P_{44}$ | $P_{12}$ | $P_{34}$ |
|---|---|---|---|---|---|---|
| 0.0 | 127.8125 | 127.7592 | 127.7592 | 127.7150 | 0.0000 | 0.0000 |
| | 124.8422 | 124.7953 | 124.7953 | 124.7579 | 0.0000 | 0.0000 |
| | −2.9703 | −2.9639 | −2.9639 | −2.9571 | 0.0000 | 0.0000 |
| 10.0 | 18.0946 | 18.0346 | 17.9878 | 17.9451 | −0.2665 | 1.1564 |
| | 18.7967 | 18.7425 | 18.6993 | 18.6634 | −0.0568 | 1.1636 |
| | 0.7021 | 0.7078 | 0.7115 | 0.7183 | 0.2097 | 0.0072 |
| 20.0 | 2.9488 | 2.9106 | 2.8785 | 2.8622 | −0.0582 | −0.0374 |
| | 2.8656 | 2.8327 | 2.8005 | 2.7892 | −0.0531 | −0.0501 |
| | −0.0832 | −0.0779 | −0.0781 | −0.0729 | 0.0051 | −0.0127 |
| 30.0 | 1.6445 | 1.6035 | 1.5604 | 1.5515 | −0.0248 | −0.0241 |
| | 1.6264 | 1.5903 | 1.5476 | 1.5427 | −0.0358 | −0.0303 |
| | −0.0181 | −0.0132 | −0.0128 | −0.0087 | −0.0110 | −0.0062 |
| 40.0 | 1.0608 | 1.0148 | 0.9636 | 0.9619 | 0.0631 | −0.0180 |
| | 1.0686 | 1.0263 | 0.9791 | 0.9806 | 0.0582 | −0.0075 |
| | 0.0078 | 0.0115 | 0.0155 | 0.0187 | −0.0049 | 0.0105 |
| 50.0 | 0.7674 | 0.7158 | 0.6692 | 0.6755 | 0.0010 | −0.0575 |
| | 0.7671 | 0.7189 | 0.6721 | 0.6819 | 0.0047 | −0.0616 |
| | −0.0003 | 0.0031 | 0.0029 | 0.0064 | 0.0037 | −0.0041 |
| 60.0 | 0.4935 | 0.4345 | 0.3757 | 0.3921 | 0.0118 | −0.0726 |
| | 0.4956 | 0.4392 | 0.3807 | 0.3993 | 0.0134 | −0.0735 |
| | 0.0021 | 0.0047 | 0.0050 | 0.0073 | 0.0015 | −0.0009 |
| 70.0 | 0.3935 | 0.3232 | 0.2612 | 0.2907 | −0.0069 | −0.0767 |
| | 0.3924 | 0.3241 | 0.2625 | 0.2950 | −0.0053 | −0.0811 |
| | −0.0011 | 0.0008 | 0.0013 | 0.0043 | 0.0016 | −0.0044 |
| 80.0 | 0.3373 | 0.2500 | 0.1801 | 0.2267 | −0.0059 | −0.0831 |
| | 0.3379 | 0.2529 | 0.1823 | 0.2318 | −0.0066 | −0.0863 |
| | 0.0006 | 0.0029 | 0.0022 | 0.0051 | −0.0007 | −0.0032 |
| 90.0 | 0.3039 | 0.1951 | 0.1211 | 0.1906 | −0.0073 | −0.0795 |
| | 0.3049 | 0.1984 | 0.1228 | 0.1945 | −0.0076 | −0.0843 |
| | 0.0010 | 0.0033 | 0.0017 | 0.0038 | −0.0002 | −0.0048 |
| 100.0 | 0.2877 | 0.1585 | 0.0763 | 0.1677 | −0.0126 | −0.0757 |
| | 0.2860 | 0.1591 | 0.0750 | 0.1687 | −0.0119 | −0.0786 |
| | −0.0017 | 0.0006 | −0.0012 | 0.0010 | 0.0007 | −0.0030 |
| 110.0 | 0.2669 | 0.1205 | 0.0346 | 0.1452 | −0.0111 | −0.0631 |
| | 0.2677 | 0.1231 | 0.0337 | 0.1456 | −0.0099 | −0.0679 |
| | 0.0008 | 0.0026 | −0.0010 | 0.0003 | 0.0012 | −0.0048 |
| 120.0 | 0.2466 | 0.0909 | −0.0041 | 0.1177 | −0.0122 | −0.0473 |
| | 0.2473 | 0.0929 | −0.0058 | 0.1178 | −0.0107 | −0.0499 |
| | 0.0008 | 0.0020 | −0.0017 | 0.0001 | 0.0014 | −0.0026 |
| 130.0 | 0.2257 | 0.0749 | −0.0254 | 0.0931 | −0.0091 | −0.0346 |
| | 0.2270 | 0.0775 | −0.0300 | 0.0899 | −0.0086 | −0.0367 |
| | 0.0013 | 0.0026 | −0.0046 | −0.0032 | 0.0004 | −0.0021 |
| 140.0 | 0.2188 | 0.0686 | −0.0451 | 0.0737 | −0.0109 | −0.0203 |
| | 0.2198 | 0.0715 | −0.0473 | 0.0721 | −0.0112 | −0.0239 |
| | 0.0011 | 0.0029 | −0.0023 | −0.0016 | −0.0003 | −0.0037 |

*Continued*

**Table 5.34** Continued

| $\Theta(°)$ | $P_{11}$ | $P_{22}$ | $P_{33}$ | $P_{44}$ | $P_{12}$ | $P_{34}$ |
|---|---|---|---|---|---|---|
| 150.0 | 0.2165 | 0.0601 | −0.0639 | 0.0612 | −0.0042 | −0.0059 |
| | 0.2159 | 0.0632 | −0.0658 | 0.0578 | −0.0041 | −0.0068 |
| | −0.0006 | 0.0030 | −0.0019 | −0.0033 | 0.0001 | −0.0009 |
| 160.0 | 0.2189 | 0.0758 | −0.0647 | 0.0479 | −0.0002 | −0.0042 |
| | 0.2180 | 0.0790 | −0.0666 | 0.0437 | 0.0010 | −0.0020 |
| | −0.0009 | 0.0033 | −0.0019 | −0.0043 | 0.0012 | 0.0022 |
| 170.0 | 0.1885 | 0.0946 | −0.0536 | 0.0058 | 0.0113 | 0.0283 |
| | 0.1893 | 0.0981 | −0.0578 | 0.0013 | 0.0117 | 0.0314 |
| | 0.0008 | 0.0035 | −0.0043 | −0.0044 | 0.0004 | 0.0031 |
| 180.0 | 0.3494 | 0.1391 | −0.1391 | 0.0712 | 0.0000 | 0.0000 |
| | 0.3550 | 0.1477 | −0.1477 | 0.0596 | 0.0000 | 0.0000 |
| | 0.0056 | 0.0086 | −0.0086 | −0.0116 | 0.0000 | 0.0000 |

The size parameter in terms of maximum dimension is 40. The refractive index is $1.53 + i0.008$. For each scattering angle, the three rows denote the results from the ADDA, the IITM, and the IITM minus the ADDA. All data are rounded to four decimal points.

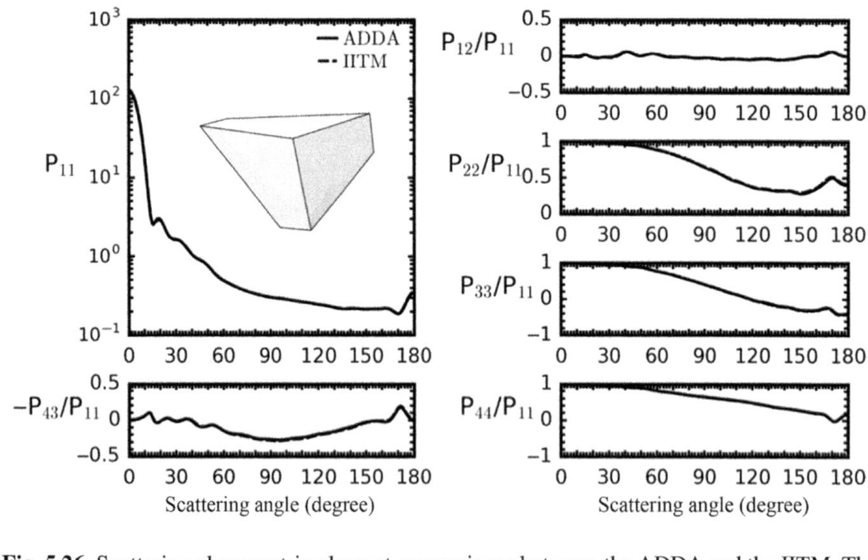


**Fig. 5.26** Scattering phase matrix element comparisons between the ADDA and the IITM. The parameters are the same as for Table 5.34.

**Table 5.35** Scattering and extinction cross section and asymmetry factor comparisons.

| $(\lambda = 2\pi)$ | ADDA | IITM | IITM-ADDA |
|---|---|---|---|
| $C_{sca}$ | 1119.3471 | 1096.9195 | −22.4276 |
| $C_{ext}$ | 1330.2456 | 1301.1434 | −29.1021 |
| $g$ | 0.6893 | 0.6890 | −0.0003 |

Other parameters are the same as for Table 5.34.

**Table 5.36** The vertices in Cartesian coordinates of the hexahedron shown in Fig. 5.25B.

| | |
|---|---|
| −41.154377, −15.070762, 68.240768 | 56.690417, −56.295801, 41.896383 |
| 80.225625, 32.528391, 24.530895 | −0.517173, 67.156493, 46.20264 |
| −80.038884, −36.035598, −20.349684 | −40.704128, 50.540381, −54.953531 |
| 24.035842, 24.372157, −67.049702 | 6.02124, −69.730988, −36.719407 |

The maximum dimension is 180. The hexahedron has eight vertices. Each vertex entry in the table corresponds to *x*-, *y*-, and *z*-coordinates.

**Table 5.37** Scattering phase matrix element comparisons between PGOM and IITM for the hexahedron shown in Fig. 5.25B.

| $\Theta(°)$ | $P_{11}$ | $P_{22}$ | $P_{33}$ | $P_{44}$ | $P_{12}$ | $P_{34}$ |
|---|---|---|---|---|---|---|
| 0.0 | 3335.6932 | 3335.6672 | 3335.6672 | 3335.6453 | 0.0000 | 0.0000 |
| | 3428.2230 | 3428.1570 | 3428.1570 | 3428.0960 | 0.0000 | 0.0000 |
| | 92.5298 | 92.4898 | 92.4898 | 92.4507 | 0.0000 | 0.0000 |
| 10.0 | 7.3346 | 7.3128 | 7.2889 | 7.2823 | −0.0177 | 0.0241 |
| | 7.4728 | 7.4462 | 7.4136 | 7.3993 | −0.0756 | −0.0152 |
| | 0.1381 | 0.1334 | 0.1247 | 0.1170 | −0.0579 | −0.0393 |
| 20.0 | 0.9514 | 0.9395 | 0.9119 | 0.9109 | −0.0868 | −0.0094 |
| | 0.9148 | 0.9020 | 0.8714 | 0.8674 | −0.0786 | −0.0406 |
| | −0.0365 | −0.0375 | −0.0405 | −0.0435 | 0.0081 | −0.0312 |
| 30.0 | 0.4443 | 0.4352 | 0.3966 | 0.3979 | −0.0970 | −0.0295 |
| | 0.4297 | 0.4203 | 0.3804 | 0.3795 | −0.0896 | −0.0501 |
| | −0.0146 | −0.0150 | −0.0162 | −0.0184 | 0.0074 | −0.0206 |
| 40.0 | 0.3140 | 0.3044 | 0.2547 | 0.2577 | −0.0885 | −0.0386 |
| | 0.2984 | 0.2884 | 0.2380 | 0.2389 | −0.0810 | −0.0566 |
| | −0.0156 | −0.0160 | −0.0167 | −0.0188 | 0.0074 | −0.0180 |
| 50.0 | 0.2553 | 0.2435 | 0.1843 | 0.1897 | −0.0757 | −0.0455 |
| | 0.2429 | 0.2306 | 0.1714 | 0.1745 | −0.0696 | −0.0619 |
| | −0.0124 | −0.0129 | −0.0129 | −0.0152 | 0.0061 | −0.0164 |
| 60.0 | 0.2373 | 0.2228 | 0.1531 | 0.1613 | −0.0538 | −0.0561 |
| | 0.2291 | 0.2136 | 0.1447 | 0.1500 | −0.0512 | −0.0688 |
| | −0.0082 | −0.0092 | −0.0085 | −0.0113 | 0.0026 | −0.0127 |
| 70.0 | 0.1908 | 0.1735 | 0.1023 | 0.1136 | −0.0393 | −0.0512 |
| | 0.1893 | 0.1707 | 0.0999 | 0.1084 | −0.0365 | −0.0629 |
| | −0.0016 | −0.0028 | −0.0025 | −0.0052 | 0.0028 | −0.0116 |
| 80.0 | 0.1718 | 0.1513 | 0.0741 | 0.0889 | −0.0334 | −0.0584 |
| | 0.1713 | 0.1499 | 0.0718 | 0.0835 | −0.0326 | −0.0693 |
| | −0.0005 | −0.0014 | −0.0023 | −0.0054 | 0.0009 | −0.0109 |
| 90.0 | 0.1505 | 0.1267 | 0.0517 | 0.0704 | −0.0207 | −0.0498 |
| | 0.1499 | 0.1258 | 0.0517 | 0.0671 | −0.0210 | −0.0568 |
| | −0.0007 | −0.0009 | 0.0000 | −0.0033 | −0.0003 | −0.0070 |
| 100.0 | 0.1352 | 0.1081 | 0.0278 | 0.0504 | −0.0109 | −0.0470 |
| | 0.1328 | 0.1063 | 0.0289 | 0.0476 | −0.0100 | −0.0514 |
| | −0.0024 | −0.0018 | 0.0011 | −0.0028 | 0.0009 | −0.0044 |
| 110.0 | 0.1202 | 0.0899 | 0.0086 | 0.0345 | −0.0064 | −0.0409 |
| | 0.1189 | 0.0897 | 0.0106 | 0.0324 | −0.0052 | −0.0437 |
| | −0.0013 | −0.0002 | 0.0019 | −0.0022 | 0.0011 | −0.0028 |

*Continued*

**Table 5.37** Continued

| $\Theta(°)$ | $P_{11}$ | $P_{22}$ | $P_{33}$ | $P_{44}$ | $P_{12}$ | $P_{34}$ |
|---|---|---|---|---|---|---|
| 120.0 | 0.1051 | 0.0713 | −0.0105 | 0.0189 | −0.0023 | −0.0300 |
| | 0.1046 | 0.0725 | −0.0076 | 0.0173 | −0.0009 | −0.0322 |
| | −0.0005 | 0.0012 | 0.0029 | −0.0016 | 0.0014 | −0.0022 |
| 130.0 | 0.0968 | 0.0581 | −0.0271 | 0.0070 | 0.0017 | −0.0221 |
| | 0.0953 | 0.0585 | −0.0248 | 0.0045 | 0.0025 | −0.0219 |
| | −0.0015 | 0.0004 | 0.0023 | −0.0025 | 0.0009 | 0.0002 |
| 140.0 | 0.0910 | 0.0512 | −0.0371 | −0.0023 | 0.0020 | −0.0146 |
| | 0.0892 | 0.0507 | −0.0341 | −0.0045 | 0.0027 | −0.0142 |
| | −0.0017 | −0.0005 | 0.0031 | −0.0022 | 0.0007 | 0.0004 |
| 150.0 | 0.0886 | 0.0496 | −0.0451 | −0.0119 | −0.0006 | −0.0023 |
| | 0.0842 | 0.0482 | −0.0425 | −0.0160 | 0.0004 | −0.0017 |
| | −0.0044 | −0.0014 | 0.0026 | −0.0041 | 0.0010 | 0.0005 |
| 160.0 | 0.0998 | 0.0576 | −0.0507 | −0.0155 | 0.0001 | 0.0063 |
| | 0.0975 | 0.0571 | −0.0494 | −0.0205 | 0.0008 | 0.0064 |
| | −0.0023 | −0.0005 | 0.0013 | −0.0051 | 0.0007 | 0.0001 |
| 170.0 | 0.1497 | 0.0692 | −0.0667 | 0.0038 | −0.0031 | 0.0065 |
| | 0.1524 | 0.0701 | −0.0681 | −0.0024 | −0.0039 | 0.0049 |
| | 0.0027 | 0.0008 | −0.0014 | −0.0061 | −0.0008 | −0.0016 |
| 180.0 | 0.3653 | 0.1492 | −0.1492 | 0.0621 | 0.0000 | 0.0000 |
| | 0.4417 | 0.1672 | −0.1672 | 0.1073 | 0.0000 | 0.0000 |
| | 0.0764 | 0.0180 | −0.0180 | 0.0452 | 0.0000 | 0.0000 |

The size parameter in terms of maximum dimension is 180. The refractive index is $1.5+i0.005$. For each scattering angle, the three rows denote the results from the PGOM, the IITM, and the IITM minus the PGOM. All data are rounded to four decimal points.

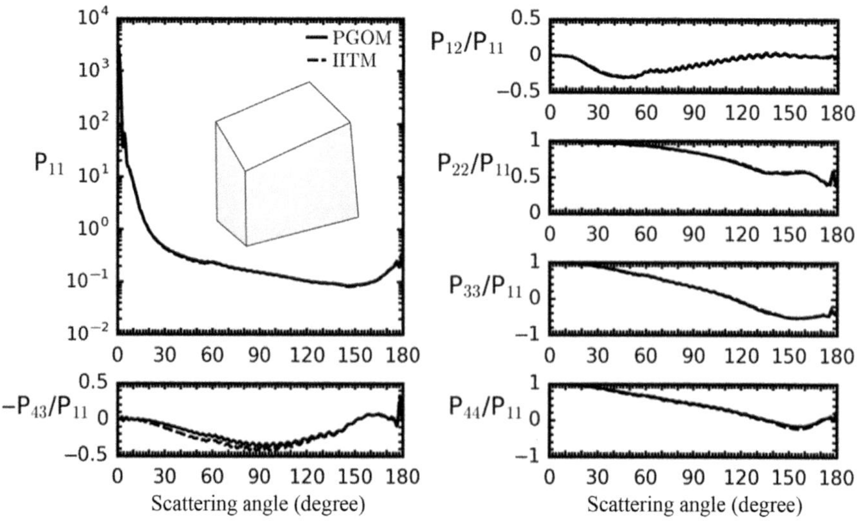


**Fig. 5.27** Scattering phase matrix element comparisons between the PGOM and the IITM. The parameters are the same as for Table 5.37.

**Table 5.38** Scattering and extinction cross section and asymmetry factor comparisons.

| $(\lambda = 2\pi)$ | PGOM | IITM | IITM-PGOM |
|---|---|---|---|
| $C_{sca}$ | 18897.4800 | 18876.5125 | −20.9675 |
| $C_{ext}$ | 27869.0580 | 28277.4773 | 408.4193 |
| $g$ | 0.8609 | 0.8648 | 0.0039 |

Other parameters are the same as for Table 5.37.

The comparisons of the six independent scattering phase matrix elements at 19 scattering angles are given in Table 5.37. Fig. 5.27 shows the phase matrix comparisons at all scattering angles. The scattering and extinction cross sections and asymmetry factor comparisons are given in Table 5.38.

The earlier IITM and PGOM results show good consistency between the two methods. In Fig. 5.27, the difference is only noticeable in $b_2$. Compared with the PGOM, IITM results have a larger extinction cross section but a smaller scattering cross section, suggesting that IITM gives a larger absorption cross section than PGOM for the scattering particle. The absolute difference between the asymmetry factors by the IITM and PGOM is larger than that by the IITM and ADDA, as shown in Tables 5.35 and 5.38.

## 5.5 Inhomogeneous particles

An inhomogeneous particle has more than one refractive index in different volume domains. It is straightforward to apply the IITM to compute the scattering properties of an inhomogeneous particle. Fig. 5.28 shows an inhomogeneous particle with two layers. Both layers are hexahedral shapes.

We compare the IITM with the ADDA for the inhomogeneous particle shown in Fig. 5.28. The outer layer hexahedron shape is the same as Fig. 5.25B, and the inner

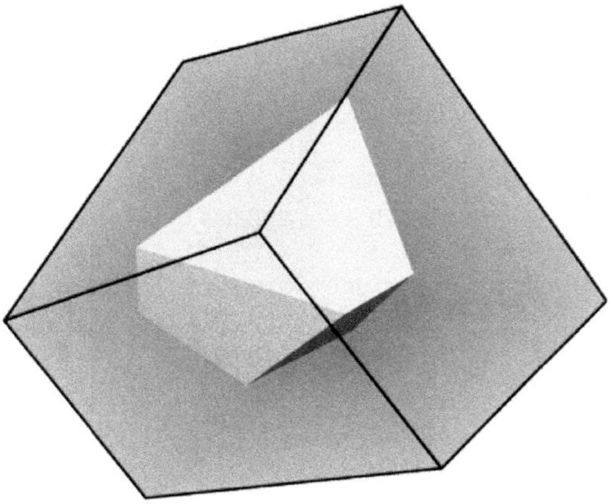

**Fig. 5.28** An inhomogeneous particle with two hexahedral layers.

layer hexahedron shape is the same as Fig. 5.25A. The outer layer shape maximum dimension is 40, and the inner layer shape maximum dimension is 20. The vertices of the two layers can be obtained by scaling the vertices in Tables 5.33 and 5.36 accordingly. The outer and inner layers have refractive index $1.308 + i1.43 \times 10^{-8}$ and $1.53 + i0.008$, respectively.

The comparisons of the six independent scattering phase matrix elements at 19 special scattering angles are given in Table 5.39. Fig. 5.29 shows the phase matrix

**Table 5.39** Scattering phase matrix element comparisons between ADDA and IITM for the inhomogeneous particle shown in Fig. 5.28.

| $\Theta$(°) | $P_{11}$ | $P_{22}$ | $P_{33}$ | $P_{44}$ | $P_{12}$ | $P_{34}$ |
|---|---|---|---|---|---|---|
| 0.0 | 126.2056 | 126.1895 | 126.1895 | 126.1758 | 0.0000 | 0.0000 |
| | 122.8041 | 122.7880 | 122.7880 | 122.7742 | 0.0000 | 0.0000 |
| | −3.4015 | −3.4015 | −3.4015 | −3.4016 | 0.0000 | 0.0000 |
| 10.0 | 16.0731 | 16.0547 | 16.0185 | 16.0078 | −0.3987 | 0.8629 |
| | 16.5026 | 16.4852 | 16.4533 | 16.4433 | −0.2347 | 0.8397 |
| | 0.4295 | 0.4305 | 0.4348 | 0.4355 | 0.1640 | −0.0233 |
| 20.0 | 2.2489 | 2.2237 | 2.2220 | 2.2149 | 0.0062 | 0.0464 |
| | 2.2806 | 2.2576 | 2.2564 | 2.2516 | 0.0172 | 0.0179 |
| | 0.0317 | 0.0339 | 0.0344 | 0.0367 | 0.0110 | −0.0285 |
| 30.0 | 2.5195 | 2.4916 | 2.4755 | 2.4697 | 0.0045 | −0.1055 |
| | 2.5322 | 2.5063 | 2.4871 | 2.4832 | 0.0070 | −0.1474 |
| | 0.0127 | 0.0147 | 0.0115 | 0.0135 | 0.0025 | −0.0418 |
| 40.0 | 1.5400 | 1.5141 | 1.4747 | 1.4749 | 0.1361 | −0.1431 |
| | 1.5143 | 1.4902 | 1.4481 | 1.4503 | 0.1524 | −0.1514 |
| | −0.0257 | −0.0238 | −0.0266 | −0.0247 | 0.0162 | −0.0083 |
| 50.0 | 1.1059 | 1.0820 | 1.0615 | 1.0668 | 0.0951 | −0.0366 |
| | 1.1463 | 1.1235 | 1.1025 | 1.1092 | 0.0896 | −0.0633 |
| | 0.0404 | 0.0416 | 0.0410 | 0.0424 | −0.0055 | −0.0267 |
| 60.0 | 0.6955 | 0.6717 | 0.6399 | 0.6494 | 0.0170 | −0.1003 |
| | 0.6926 | 0.6693 | 0.6317 | 0.6418 | 0.0255 | −0.1208 |
| | −0.0030 | −0.0024 | −0.0082 | −0.0076 | 0.0085 | −0.0206 |
| 70.0 | 0.3956 | 0.3708 | 0.3435 | 0.3561 | 0.0303 | −0.0531 |
| | 0.3968 | 0.3726 | 0.3456 | 0.3586 | 0.0318 | −0.0568 |
| | 0.0011 | 0.0018 | 0.0021 | 0.0025 | 0.0014 | −0.0037 |
| 80.0 | 0.3209 | 0.2923 | 0.2669 | 0.2845 | −0.0178 | −0.0515 |
| | 0.3204 | 0.2925 | 0.2633 | 0.2811 | −0.0191 | −0.0660 |
| | −0.0006 | 0.0001 | −0.0036 | −0.0034 | −0.0012 | −0.0145 |
| 90.0 | 0.2565 | 0.2213 | 0.1922 | 0.2170 | −0.0147 | −0.0565 |
| | 0.2529 | 0.2187 | 0.1875 | 0.2121 | −0.0107 | −0.0619 |
| | −0.0036 | −0.0026 | −0.0046 | −0.0049 | 0.0040 | −0.0054 |
| 100.0 | 0.1992 | 0.1568 | 0.1320 | 0.1651 | −0.0201 | −0.0301 |
| | 0.2005 | 0.1589 | 0.1323 | 0.1650 | −0.0217 | −0.0396 |
| | 0.0014 | 0.0021 | 0.0002 | −0.0001 | −0.0016 | −0.0095 |
| 110.0 | 0.1551 | 0.1072 | 0.0765 | 0.1155 | −0.0222 | −0.0316 |
| | 0.1541 | 0.1068 | 0.0753 | 0.1140 | −0.0179 | −0.0371 |
| | −0.0010 | −0.0004 | −0.0012 | −0.0015 | 0.0043 | −0.0054 |

**Table 5.39** Continued

| $\Theta$(°) | $P_{11}$ | $P_{22}$ | $P_{33}$ | $P_{44}$ | $P_{12}$ | $P_{34}$ |
|---|---|---|---|---|---|---|
| 120.0 | 0.1227 | 0.0691 | 0.0369 | 0.0818 | −0.0223 | −0.0152 |
| | 0.1235 | 0.0703 | 0.0367 | 0.0814 | −0.0225 | −0.0202 |
| | 0.0008 | 0.0013 | −0.0001 | −0.0005 | −0.0003 | −0.0050 |
| 130.0 | 0.1085 | 0.0481 | 0.0083 | 0.0597 | −0.0144 | −0.0152 |
| | 0.1082 | 0.0481 | 0.0078 | 0.0589 | −0.0124 | −0.0176 |
| | −0.0003 | 0.0000 | −0.0005 | −0.0008 | 0.0020 | −0.0024 |
| 140.0 | 0.1113 | 0.0434 | −0.0099 | 0.0478 | −0.0138 | −0.0125 |
| | 0.1122 | 0.0447 | −0.0107 | 0.0466 | −0.0131 | −0.0160 |
| | 0.0010 | 0.0013 | −0.0008 | −0.0012 | 0.0008 | −0.0035 |
| 150.0 | 0.1283 | 0.0436 | −0.0298 | 0.0421 | −0.0107 | −0.0122 |
| | 0.1293 | 0.0456 | −0.0302 | 0.0406 | −0.0092 | −0.0135 |
| | 0.0010 | 0.0020 | −0.0004 | −0.0015 | 0.0015 | −0.0013 |
| 160.0 | 0.1467 | 0.0401 | −0.0606 | 0.0304 | −0.0093 | −0.0071 |
| | 0.1486 | 0.0416 | −0.0622 | 0.0290 | −0.0089 | −0.0082 |
| | 0.0019 | 0.0015 | −0.0016 | −0.0015 | 0.0003 | −0.0010 |
| 170.0 | 0.1364 | 0.0855 | −0.0327 | −0.0015 | 0.0158 | 0.0043 |
| | 0.1368 | 0.0858 | −0.0340 | −0.0040 | 0.0153 | 0.0057 |
| | 0.0004 | 0.0004 | −0.0013 | −0.0025 | −0.0005 | 0.0014 |
| 180.0 | 0.3129 | 0.0970 | −0.0970 | 0.1188 | 0.0000 | 0.0000 |
| | 0.3112 | 0.0973 | −0.0973 | 0.1167 | 0.0000 | 0.0000 |
| | −0.0016 | 0.0002 | −0.0002 | −0.0021 | 0.0000 | 0.0000 |

The size parameters in terms of maximum dimension are 40 and 20 for outer and inner layers, respectively. The outer layer refractive index is $1.308 + i1.43 \times 10^{-8}$, and the inner layer refractive index is $1.53 + i0.008$. For each scattering angle, the three rows denote the results from the ADDA, the IITM, and the IITM minus the ADDA. All data are rounded to four decimal points.

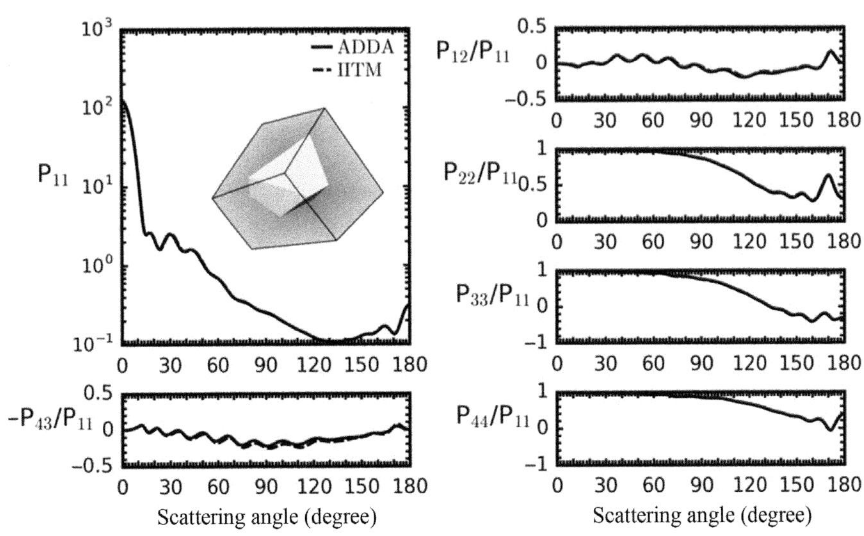


**Fig. 5.29** Scattering phase matrix element comparisons between the ADDA and the IITM. The parameters are the same as for Table 5.39.

**Table 5.40** Scattering and extinction cross section and asymmetry factor comparisons.

| $(\lambda = 2\pi)$ | ADDA | IITM | IITM-ADDA |
|---|---|---|---|
| $C_{sca}$ | 1471.4852 | 1429.8956 | −41.5896 |
| $C_{ext}$ | 1502.4933 | 1459.8579 | −42.6354 |
| $g$ | 0.7412 | 0.7404 | −0.0009 |

Other parameters are the same as for Table 5.39.

comparison for all scattering angles. The scattering and extinction cross sections and asymmetry factor comparisons are given in Table 5.40.

The earlier IITM and ADDA results for the inhomogeneous particle are consistent. The scattering and extinction cross sections and asymmetry factors obtained by the IITM are all smaller than those by the ADDA, which may be due to the limited number of orientations used in the ADDA to obtain the orientation averaged results.

# References

Abhyankar, K.D., Fymat, A.L., 1969. Relations between the elements of the phase matrix for scattering. J. Math. Phys. 10, 1935–1938.

Abramowitz, M., Stegun, I.A., 1965. Handbook of Mathematical Functions: With Formulas, Graphs, and Mathematical Tables, vol. 55. Courier Corporation, New York.

Aden, A.L., Kerker, M., 1951. Scattering of electromagnetic waves from two concentric spheres. J. Appl. Phys. 22 (10), 1242–1246.

Ambarzumian, V.A., 1943. Diffuse reflection of light by a foggy medium. In: Dokl. Akad. Nauk SSSR.vol. 38 (8), pp. 257–261.

Arfken, G.B., Weber, H.J., 1995. Mathematical Methods for Physicists. Academic Press, San Diego.

Atkinson, K.E., 1989. An Introduction to Numerical Analysis. John Wiley & Sons, New York.

Barakat, R., 1981. Bilinear constraints between elements of the $4 \times 4$ Mueller-Jones transfer matrix of polarization theory. Opt. Commun. 38, 159–161.

Barber, P.W., Hill, S.C., 1990. Light Scattering by Particles: Computational Methods. World Scientific Publishing, Singapore.

Bellman, R., Wing, G.M., 1975. An Introduction to Invariant Imbedding. Wiley, New York.

Berenger, J.-P., 1994. A perfectly matched layer for the absorption of electromagnetic waves. J. Comput. Phys. 114, 185–200.

Bi, L., Yang, P., 2013. Modeling of light scattering by biconcave and deformed red blood cells with the invariant imbedding T-matrix method. J. Biomed. Opt. 18 055001.

Bi, L., Yang, P., 2014. Accurate simulation of the optical properties of atmospheric ice crystals with the invariant imbedding T-matrix method. J. Quant. Spectrosc. Radiat. Transf. 138, 17–35.

Bi, L., Yang, P., 2015. Impact of calcification state on the inherent optical properties of *Emiliania huxleyi* coccoliths and coccolithophores. J. Quant. Spectrosc. Radiat. Transf. 155, 10–21.

Bi, L., Yang, P., 2016. Tunneling effects in electromagnetic wave scattering by nonspherical particles: a comparison of the Debye series and physical-geometric optics approximations. J. Quant. Spectrosc. Radiat. Transf. 178, 93–107.

Bi, L., Yang, P., Kattawar, G.W., Mishchenko, M.I., 2013. Efficient implementation of the invariant imbedding T-matrix method and the separation of variables method applied to large nonspherical inhomogeneous particles. J. Quant. Spectrosc. Radiat. Transf. 116, 169–183.

Bi, L., Yang, P., Kattawar, G.W., Mishchenko, M.I., 2015. Optical tunneling by arbitrary macroscopic three-dimensional objects. Phys. Rev. A 92 (1), 013814.

Bi, L., Lin, W., Wang, Z., Tang, X., Zhang, X., Yi, B., 2018a. Optical modeling of sea salt aerosols: the effects of nonsphericity and inhomogeneity. J. Geophys. Res.: Atmos 123, 543–558.

Bi, L., Xu, F., Gouesbet, G., 2018b. Depolarization of nearly spherical particles: the Debye series approach. Phys. Rev. A 98, 053809.

Biedenharn, L.C., Louck, J.D., 1981. Angular Momentum in Quantum Physics Theory and Application. Addison-Wesley Publishing, Reading, MA.

Blatt, J.M., Weisskopf, V.F., 1952. Theoretical Nuclear Physics. John Wiley & Sons, New York.

Bohren, C.F., Huffman, D.R., 1983. Absorption and Scattering of Light by Small Particles. John Wiley & Sons, New York.

Bruning, J.H., Lo, Y.T., 1971a. Multiple scattering of EM waves by spheres. I. Multipole expansion and ray-optical solutions. IEEE Trans. Antennas Propag. 19, 378–390.

Bruning, J.H., Lo, Y.T., 1971b. Multiple scattering of EM waves by spheres. II. Numerical and experimental results. IEEE Trans. Antennas Propag. 19, 391–400.

Chandrasekhar, S., 1960. Radiative Transfer. Dover Publications, New York.

Cockburn, B., Karniadakis, G.E., Shu, C.-W., 2000. Discontinuous Galerkin Methods. Theory, Computation and Applications, Lecture Notes in Computational Science and Engineering, 11. Springer-Verlag, Berlin.

Cruzan, O.R., 1962. Translational addition theorems for spherical vector wave functions. Q. Appl. Math. 20, 33–40.

de Haan, J.F., Bosma, P.B., Hovenier, J.W., 1987. The adding method for multiple scattering calculations of polarized light. Astron. Astrophys. 183, 371–391.

DeVoe, H., 1964. Optical properties of molecular aggregates. I. Classical model of electronic absorption and refraction. J. Chem. Phys. 42, 393.

Doicu, A., Wriedt, T., 2018. The invariant imbedding T matrix approach. In: Wriedt, T., Eremin, Y. (Eds.), The Generalized Multipole Technique for Light Scattering. Springer, Cham, pp. 35–47.

Doicu, A., Wriedt, T., Eremin, Y.A., 2006. Light Scattering by Systems of Particles—Null-Field Method with Discrete Sources: Theory and Programs. Springer-Verlag, Berlin, Heidelberg.

Domke, H., 1974. The expansion of scattering matrices for an isotropic medium in generalized spherical functions. Astrophys. Space Sci. 29 (2), 379–386.

Draine, B.T., Flatau, P.J., 2003. User Guide to the Discrete Dipole Approximation Code DDSCAT.6.0. http://arxiv.org/abs/astro-ph/0300969.

Ergul, O., Gurel, L., 2014. The Multilevel Fast Multipole Algorithm (MLFMA) for Solving Large-Scale Computational Electromagnetics Problems. Wiley, New York.

Farrington, C.C., 1961. Numerical quadrature of discontinuous functions. In: Proceedings of the 1961 16th ACM National Meeting. ACM, pp. 21–401.

Foot, J., 1988. Some observations of the optical properties of clouds. II: Cirrus. Q. J. R. Meteorol. Soc. 114, 145–164. https://doi.org/10.1002/qj.49711447908.

Friedman, B., Russek, J., 1954. Addition theorems for spherical waves. Q. Appl. Math. 12, 13–23.

Fry, E.S., Kattawar, G.W., 1981. Relationships between elements of the stokes matrix. Appl. Opt. 20, 2811–2814.

Fuller, K.A., Kattawar, G.W., 1988a. Consummate solution to the problem of classical electromagnetic scattering by an ensemble of spheres. I: Linear chains. Opt. Lett. 13, 90–92.

Fuller, K.A., Kattawar, G.W., 1988b. Consummate solution to the problem of classical electromagnetic scattering by an ensemble of spheres. II: Clusters of arbitrary configuration. Opt. Lett. 13, 1063–1065.

Fuller, K.A., Mackowski, D.W., 2000. Electromagnetic scattering by compounded spherical particles. In: Mishchenko, M.I., Hovenier, J.W., Travis, L.D. (Eds.), Light Scattering by Nonspherical Particles: Theory, Measurements, and Applications. Academic Press, San Diego, pp. 225–272.

Goody, R.M., Yung, Y.L., 1989. Atmospheric Radiation. Oxford University Press, New York.

Gouesbet, G., Grehan, G., 2011. Generalized Lorenz-Mie Theories. Springer, Berlin, Heidelberg.

Harrington, R.F., 1968. Field Computation by Moment Methods. The Macmillan Company, New York.

Havemann, S., Baran, A.J., 2001. Extension of T-matrix to scattering of electromagnetic plane waves by non-axisymmetric dielectric particles: application to hexagonal ice cylinders. J. Quant. Spectrosc. Radiat. Transf. 70 (2), 139–158.

Hertz, H., 1887. Ueber einen Einfluss des ultravioletten Lichtes auf die electrische Entladung. Ann. Phys. 267, 983–1000.

Hovenac, E.A., Lock, J.A., 1992. Assessing the contributions of surface waves and complex rays to far-field Mie scattering by use of the Debye series. J. Opt. Soc. Am. A. 9, 781–795.

Hovenier, J.W., van der Mee, C.V.M., 1983. Fundamental relationships relevant to the transfer of polarized light in a scattering atmosphere. Astron. Astrophys. 128, 1–16.

Hovenier, J.W., van der Mee, C.V.M., Domke, H., 2004. Transfer of Polarized Light in Planetary Atmospheres: Basic Concepts and Practical Methods. Kluwer Academic Publishers, Dordrecht, Netherlands.

Hu, C., Kattawar, G.W., Parkin, M.E., Herb, P., 1987. Symmetry theorems on the forward and backward scattering Mueller matrices for light scattering from a nonspherical dielectric scatterer. Appl. Opt. 26, 4159–4173.

Iskander, M., Lakhtakia, A., Durney, C., 1983. A new procedure for improving the solution stability and extending the frequency range of the EBCM. IEEE Trans. Antennas Propag. 31 (2), 317–324.

Iskander, M.F., Chen, H.Y., Duong, T.V., 1989. A new sectioning procedure for calculating scattering and absorption by elongated dielectric objects. IEEE Trans. Electromagn. Compat. 31 (2), 157–163.

Jackson, J.D., 1998. Classical Electrodynamics. John Wiley & Sons, New York.

Johnson, B.R., 1988. Invariant imbedding T matrix approach to electromagnetic scattering. Appl. Opt. 27, 4861–4873.

Jones, R.C., 1941. A new calculus for the treatment of optical systems: I. Description and discussion of the calculus. J. Opt. Soc. Am. 31, 488–493.

Jones, R.C., 1947. A new calculus for the treatment of optical systems: V. A more general formulation, and description of another calculus. J. Opt. Soc. Am. 37, 107–110.

Kahnert, M., 2005. Irreducible representations of finite groups in the T-matrix formulation of the electromagnetic scattering problem. JOSA A 22 (6), 1187–1199.

Kahnert, M., 2010. Numerical methods in electromagnetic scattering theory. J. Quant. Spectrosc. Radiat. Transf. 79-80, 775–824.

Kahnert, M., 2013. The T-matrix code Tsym for homogeneous dielectric particles with finite symmetries. J. Quant. Spectrosc. Radiat. Transf. 123, 67–82.

Kahnert, M., Rother, T., 2011. Modeling optical properties of particles with small-scale surface roughness: combination of group theory with a perturbation approach. Opt. Express 19 (12), 11138–11151.

Kahnert, F.M., Stamnes, J.J., Stamnes, K., 2001. Application of the extended boundary condition method to homogeneous particles with point-group symmetries. Appl. Opt. 40 (18), 3110–3123.

Kattawar, G.W., Hood, D.A., 1976. Electromagnetic scattering from a spherical polydispersion of coated spheres. Appl. Opt. 15, 1996–1999.

Kattawar, G.W., Plass, G.N., 1967. Electromagnetic scattering from absorbing spheres. Appl. Opt. 6 (8), 1377–1382.

Kattawar, G.W., Hu, C., Parkin, M.E., Herb, P., 1987. Mueller matrix calculations for dielectric cubes: comparison with experiments. Appl. Opt. 26, 4174–4180.

Khare, V., 1976. Short-Wavelength Scattering of Electromagnetic Waves by a Homogeneous Dielectric Sphere. (Ph.D. thesis) U. Rochester, New York (available from University Microfilms, Ann Arbor, Mich.).

Khlebtsov, N.G., 1992. Orientational averaging of light-scattering observables in the T-matrix approach. Appl. Opt. 31, 5359–5365.

Korn, G.A., Korn, T.M., 2000. Mathematical Handbook for Scientists and Engineers: Definitions, Theorems, and Formulas for Reference and Review. Dover Corporation, New York.

Krumpholz, M., Katehi, L.P.B., 1996. MRTD: new time-domain schemes based on multiresolution analysis. IEEE Trans. Microwave Theory Tech. 44, 555–571.

Kuščer, I., Ribarič, M., 1959. Matrix formalism in the theory of diffusion of light. Opt. Acta: Int. J. Optics 6 (1), 42–51.

Lakhtakia, A., Varadan, V.K., Varadan, V.V., 1984. Scattering by highly aspherical targets: EBCM coupled with reinforced orthogonalizations. Appl. Opt. 23 (20) 3502_1–3504.

Liou, K.-N., 1980. An Introduction to Atmospheric Radiation. Academic Press, New York.

Liu, Q.H., 1997. The PSTD algorithm: a time-domain method requiring only two cells per wavelength. Microw. Opt. Technol. Let. 15, 158–165.

Liu, C., Bi, L., Panetta, R.L., Yang, P., Yurkin, M.A., 2012. Comparison between the pseudospectral time domain method and the discrete dipole approximation for light scattering simulations. Opt. Express 20, 16763–16776.

Liu, J., Yang, P., Muinonen, K., 2015. Dust-aerosol optical modeling with Gaussian spheres: combined invariant-imbedding T-matrix and geometric-optics approach. J. Quant. Spectrosc. Radiat. Transf. 161, 136–144.

Mackowski, D.W., 1994. Calculation of total cross sections of multiple-sphere clusters. JOSA A 11 (11), 2851–2861.

Mackowski, D.W., 2002. Discrete dipole moment method for calculation of the T matrix for nonspherical particles. J. Opt. Soc. Am. A 19, 881–893.

Mackowski, D.W., 2014. A general superposition solution for electromagnetic scattering by multiple spherical domains of optically active media. J. Quant. Spectrosc. Radiat. Transf. 133, 264–270.

Mackowski, D.W., Mishchenko, M.I., 1996. Calculation of the T matrix and the scattering matrix for ensembles of spheres. J. Opt. Soc. Am. A 13, 2266–2278.

Mackowski, D.W., Mishchenko, M.I., 2011. A multiple sphere T-matrix Fortran code for use on parallel computer clusters. J. Quant. Spectrosc. Radiat. Transf. 112, 2182–2192.

Maxwell, J.C., 1861. On physical lines of force. Philos. Mag. 90, 11–23.

Mishchenko, M.I., 1991. Light scattering by randomly oriented axially symmetric particles. J. Opt. Soc. Am. A. 8, 871–882.

Mishchenko, M.I., 1993. Light scattering by size-shape distributions of randomly oriented axially symmetric particles of a size comparable to a wavelength. Appl. Opt. 32, 4652–4666.

Mishchenko, M.I., 2000. Calculation of the amplitude matrix for a nonspherical particle in a fixed orientation. Appl. Opt. 39 (6), 1026–1031.

Mishchenko, M.I., 2002. Vector radiative transfer equation for arbitrarily shaped and arbitrarily oriented particles: a microphysical derivation from statistical electromagnetics. Appl. Opt. 41, 7114–7134.

Mishchenko, M.I., 2014. Electromagnetic Scattering by Particles and Particle Groups. Cambridge University Press, Cambridge.

Mishchenko, M.I., Martin, P.A., 2013. Peter Waterman and T-matrix methods. J. Quant. Spectrosc. Radiat. Transf. 123, 2–7.

Mishchenko, M.I., Travis, L.D., 1994. T-matrix computations of light scattering by large spheroidal particles. Opt. Commun. 109 (1–2), 16–21.

Mishchenko, M.I., Travis, L.D., 1998. Capabilities and limitations of a current FORTRAN implementation of the T-matrix method for randomly oriented, rotationally symmetric scatterers. J. Quant. Spectrosc. Radiat. Transf. 60 (3), 309–324.

Mishchenko, M.I., Yurkin, M.A., 2017. On the concept of random orientation in far-field electromagnetic scattering by nonspherical particles. Opt. Lett. 42, 494–497.

Mishchenko, M.I., Travis, L.D., Mackowski, D.W., 1996. T-matrix computations of light scattering by nonspherical particles: a review. J. Quant. Spectrosc. Radiat. Transf. 55, 535–575.

Mishchenko, M.I., Travis, L.D., Lacis, A.A., 2002. Scattering, Absorption, and Emission of Light by Small Particles. Cambridge University Press, Cambridge.

Mishchenko, M.I., Videen, G., Babenko, V.A., Khlebtsov, N.G., Wriedt, T., 2004. T-matrix theory of electromagnetic scattering by particles and its applications: a comprehensive reference database. J. Quant. Spectrosc. Radiat. Transf. 88, 357–406.

Mishchenko, M.I., Travis, L.D., Lacis, A.A., 2006. Multiple Scattering of Light by Particles: Radiative Transfer and Coherent Backscattering. Cambridge University Press, Cambridge, UK.

Mishchenko, M.I., Videen, G., Babenko, V.A., Khlebtsov, N.G., Wriedt, T., 2007. Comprehensive T-matrix reference database: a 2004-06 update. J. Quant. Spectrosc. Radiat. Transf. 106, 304–324.

Mishchenko, M.I., Videen, G., Khlebtsov, N.G., Wriedt, T., Zakharova, N.T., 2008. Comprehensive *T*-matrix reference database: a 2006-07 update. J. Quant. Spectrosc. Radiat. Transf. 109, 1447–1460.

Mishchenko, M.I., Zakharova, N.T., Videen, G., Khlebtsov, N.G., Wriedt, T., 2010. Comprehensive T-matrix reference database: a 2007-2009 update. J. Quant. Spectrosc. Radiat. Transf. 111, 650–658.

Mishchenko, M.I., Videen, G., Khlebtsov, N.G., Wriedt, T., 2013. Comprehensive *T*-matrix reference database: a 2011-2013 update. J. Quant. Spectrosc. Radiat. Transf. 123, 145–152.

Mishchenko, M.I., Zakharova, N.T., Khlebtsov, N.G., Wriedt, T., Videen, G., 2014. Comprehensive thematic T-matrix reference database: a 2013-2014 update. J. Quant. Spectrosc. Radiat. Transf. 146, 349–354.

Mishchenko, M.I., Zakharova, N.T., Khlebtsov, N.G., Videen, G., Wriedt, T., 2016. Comprehensive thematic T-matrix reference database: a 2014-2015 update. J. Quant. Spectrosc. Radiat. Transf. 178, 276–283.

Mishchenko, M.I., Zakharova, N.T., Khlebtsov, N.G., Videen, G., Wriedt, T., 2017. Comprehensive thematic T-matrix reference database: a 2015-2017 update. J. Quant. Spectrosc. Radiat. Transf. 202, 240–246.

Misner, C.W., Thorne, K.S., Wheeler, J.A., 1973. Gravitation. Freeman & Co., New York.

Mobley, C.D., 1994. Light and Water: Radiative Transfer in Natural Waters. Academic Press, London, England.

Morse, P.M., Feshbach, H., 1953. Methods of Theoretical Physics. McGraw-Hill Book, New York.

Mueller, H., 1948. The foundation of optics (Abstract). J. Opt. Soc. Am. 38, 661.

Mugnai, A., Wiscombe, W.J., 1986. Scattering from nonspherical Chebyshev particles. I: Cross sections, single-scattering albedo, asymmetry factor, and backscattered fraction. Appl. Opt. 25 (7), 1235–1244.

Nieminen, T.A., Rubinszteindunlop, H., Heckenberg, N.R., 2003. Calculation of the T-matrix: general considerations and application of the point-matching method. J. Quant. Spectrosc. Radiat. Transf. 79, 1019–1029.

Nussenzveig, H.M., 1992. Diffraction Effects in Semiclassical Scattering. Cambridge University Press, Cambridge.

Ono, A., 1969. The shape and riming properties of ice crystals in natural clouds. J. Atmos. Sci. 26 (1), 138–147.

Panetta, R.L., Liu, C., Yang, P., 2013. A pseudo-spectral time domain method for light scattering computation. In: Kokhanovsky, A. (Ed.), Light Scattering Reviews 8. Springer Praxis Books. Springer, Berlin, Heidelberg.

Parke III, N.G., 1948. Matrix Optics. (Thesis) MIT.

Parke III, N.G., 1949. Optical algebra. J. Math. Phys. 28, 131.

Perrin, F., 1942. Polarization of light scattered by isotropic opalescent media. J. Chem. Phys. 10, 415–427.

Peterson, B., Ström, S., 1973. T matrix for electromagnetic scattering from an arbitrary number of scatterers and representations of E (3). Phys. Rev. D 8 (10), 3661.

Preisendorfer, R.W., 1965. Radiative transfer on discrete spaces. In: International Series in Pure and Applied Mathematics. Vol. 74. Pergamon Press, Oxford 474 pp.

Press, W.H., Teukolsky, S.A., Vetterling, W.T., Flannery, B.P., 2007. Numerical Recipes 3rd Edition: The Art of Scientific Computing. Cambridge University Press, New York.

Purcell, E.M., Pennypacker, C.R., 1973. Scattering and absorption of light by nonspherical dielectric grains. Astrophys. J. 186, 705–714.

Roth, S.D., 1982. Ray casting for modeling solids. Comput. Graphics Image Process. 18 (2), 109–144.

Rothwell, E.J., Cloud, M.J., 2001. Electromagnetics. CRC Press, Boca Raton.

Saxon, D.S., 1955. Tensor scattering matrix for the electromagnetic field. Phys. Rev. 100, 1771–1775.

Schot, S.H., 1992. Eighty years of Sommerfeld's radiation condition. Hist. Math. 19 (4), 385–401.

Schulz, F.M., Stamnes, K., Stamnes, J.J., 1998. Scattering of electromagnetic waves by spheroidal particles: a novel approach exploiting the T-matrix computed in spheroidal coordinates. Appl. Opt. 37, 7875–7896.

Schulz, F.M., Stamnes, K., Stamnes, J.J., 1999. Point-group symmetries in electromagnetic scattering. JOSA A 16 (4), 853–865.

Siewert, C.E., 1981. On the equation of the transfer relevant to the scattering of polarized light. Astrophys. J. 245, 1080–1086.

Silver, S., 1964. Microwave Antenna Theory and Design. Dover, New York.

Sobolev, V.V., 1975. Light Scattering in Planetary Atmospheres. Pergamon Press, Oxford.

Sommerfeld, A., 1949. Partial Differential Equations in Physics. vol. 1 Academic Press, London, England.

Song, J.M., Lu, C.C., Chew, W.C., 1997. Multilevel fast multipole algorithm for electromagnetic scattering by large complex objects. IEEE Trans. Antennas Propag. 45, 1488–1493.

Stamnes, K., Tsay, S.C., Wiscombe, W., Laszlo, I., 2000. DISORT, A General-Purpose Fortran Program for Discrete-Ordinate-Method Radiative Transfer in Scattering and Emitting Layered Media: Documentation of Methodology. Tech. Rep. Dept. of Physics and Engineering Physics, Stevens Institute of Technology, Hoboken, NJ, p. 112.

Stein, S., 1961. Addition theorems for spherical wave functions. Q. Appl. Math. 19, 15–24.

Stratton, J.A., 1941. Electromagnetic Theory. McGrow-Hill Book Company, Inc., New York/London.

Sun, B., Yang, P., Kattawar, G.W., Zhang, X., 2017. Physical-geometric optics method for large size faceted particles. Opt. Express 25 (20), 24044–24060.

Taflove, A., Hagness, S.C., 2000. Computational Electrodynamics. Artech House, London.

Tai, C.-T., 1994. Dyadic Green Functions in Electromagnetic Theory. IEEE Press, New York.

Toon, O.B., Ackerman, T.P., 1981. Algorithms for the calculation of scattering by stratified spheres. Appl. Opt. 20, 3657–3660.

Tsang, L., Kong, J.A., 1980. Multiple scattering of electromagnetic waves by random distributions of discrete scatterers with coherent potential and quantum mechanical formalism. J. Appl. Phys. 51 (7), 3465–3485.

Tsang, L., Kong, J.A., Shin, R.T., 1985. Theory of Microwave Remote Sensing. Wiley, New York.

Tsang, L., Kong, J.A., Ding, K.-H., 2000. Scattering of Electromagnetic Waves: Theories and Applications. John Wiley & Sons, New York.

van de Hulst, H.C., 1957. Light Scattering by Small Particles. John Wiley & Sons, New York.

van de Hulst, H.C., 1980. Multiple Light Scattering: Tables, Formulas, and Applications. vol. 1 Academic Press, San Diego 317 pp.

Varadan, V.V., Varadan, V.K., 1980. Multiple scattering of electromagnetic waves by randomly distributed and oriented dielectric scatterers. Phys. Rev. D 21 (2), 388.

Varshalovich, D.A., Moskalev, A.N., Khersonskii, V.K., 1988. Quantum Theory of Angular Momentum. World Scientific, Singapore.

Waterman, P.C., 1965. Matrix formulation of electromagnetic scattering. Proc. IEEE 53, 805–812.

Waterman, P.C., 1969. New formulation of acoustic scattering. J. Acoust. Soc. Am. 45, 1417–1429.

Waterman, P.C., 1971. Symmetry, unitarity, and geometry in electromagnetic scattering. Phys. Rev. D 3, 825–839.

Wendisch, M., Yang, P., 2012. Theory of Atmospheric Radiative Transfer: A Comprehensive Introduction. Wiley-VCH, Berlin.

Wigner, E.P., 1993. On the matrices which reduce the Kronecker products of representations of SR groups. In: The Collected Works of Eugene Paul Wigner. Springer-Verlag, Berlin Heidelberg, New York, pp. 608–654.

Wiscombe, W.J., 1980. Improved Mie scattering algorithms. Appl. Opt. 19, 1505–1509.

Wolf, E., 2007. Introduction to the Theory of Coherence and Polarization of Light. Cambridge University Press, Cambridge.

Xu, Y.-L., 1995. Electromagnetic scattering by an aggregate of spheres. Appl. Opt. 34, 4573–4588.

Xu, Y.-L., 1997. Electromagnetic scattering by an aggregate of spheres: far field. Appl. Opt. 36, 9496–9508.

Xu, F., Lock, J.A., 2010. Debye series for light scattering by a coated nonspherical particle. Phys. Rev. A. 81 063812.

Xu, F., Lock, J.A., Gouesbet, G., 2010. Debye series for light scattering by a nonspherical particle. Phys. Rev. A. 81 043824.

Yang, P., Ding, J., Panetta, R.L., Liou, K.N., Kattawar, G.W., Mishchenko, M.I., 2019. On the convergence of numerical computations for both exact and approximate solutions for electromagnetic scattering by nonspherical dielectric particles (invited review). Prog. Electromagn. Res. 164, 27–61.

Yee, S.K., 1966. Numerical solution of initial boundary value problems involving Maxwell's equation in isotropic media. IEEE Trans. Antennas Propag. 14, 302–307.

Yurkin, M.A., Hoekstra, A.G., 2007. The discrete-dipole-approximation: an overview and recent developments. J. Quant. Spectrosc. Radiat. Transf. 106, 558–589.

Yurkin, M.A., Hoekstra, A.G., 2011. The discrete-dipole-approximation code ADDA: capabilities and known limitations. J. Quant. Spectrosc. Radiat. Transf. 112, 2234–2247.

Yurkin, M.A., Hoekstra, A.G., Brock, R.S., Lu, J.Q., 2007. Systematic comparison of the discrete dipole approximation and the finite difference time domain method for large dielectric scatterers. Opt. Express 15, 17902–17911.

Zakharova, N.T., Videen, G., Khlebtsov, N.G., 2012. Comprehensive T-matrix reference database: a 2009-2011 update. J. Quant. Spectrosc. Radiat. Transf. 113, 1844–1852.

# Index

Note: Page numbers followed by *f* indicate figures and *t* indicate tables.

## R



## S

9780128180907